Argentine Precordillera: Sedimentary and Plate Tectonic History of a Laurentian Crustal Fragment in South America

Martin Keller
*Institut für Geologie und Mineralogie
Friedrich-Alexander Universität Erlangen-Nurnberg
Schlossgarten 5
91054 Erlangen
Germany*

341

1999

Copyright © 1999, The Geological Society of America, Inc. (GSA). All rights reserved. GSA grants permission to individual scientists to make unlimited photocopies of one or more items from this volume for noncommercial purposes advancing science or education, including classroom use. Permission is granted to individuals to make photocopies of any item in this volume for other noncommercial, nonprofit purposes provided that the appropriate fee ($0.25 per page) is paid directly to the Copyright Clearance Center, 222 Rosewood Drive, Danvers, MA 01923, USA, phone (978) 750-8400, http://www.copyright.com (include title and ISBN when paying). Written permission is required from GSA for all other forms of capture or reproduction of any item in the volume including, but not limited to, all types of electronic or digital scanning or other digital or manual transformation of articles or any portion thereof, such as abstracts, into computer-readable and/or transmittable form for personal or corporate use, either noncommercial or commercial, for-profit or otherwise. Send permission requests to GSA Copyrights.

Copyright is not claimed on any material prepared wholly by government employees within the scope of their employment.

Published by The Geological Society of America, Inc.
3300 Penrose Place, P.O. Box 9140, Boulder, Colorado 80301

Printed in U.S.A.

GSA Books Science Editor Abhijit Basu

Library of Congress Cataloging-in-Publication Data

Cover: The Rio San Juan valley. Its spectacular landscape is a combination of Tertiary and Quaternary geomorphologic processes superimposed onto the Paleozoic sedimentary succession. Thus the valley provides the most complete geologic transect through the Argentine Precordillera. Above the alluvial fan deposits, a small slice of Ordovician limestone is visible. The dark sediments that follow are part of the Punta Negra Formation, which in turn are overlain by red and white Carboniferous strata. High range in the background is another thrust sheet from the Devonian.

10 9 8 7 6 5 4 3 2 1

Contents

Abstract 1
Introduction 3
  Argentine Precordillera 3
  Scope of the study 4
    Methods 4
Regional setting of the Argentine Precordillera 6
  Neoproterozoic-Paleozoic global paleographic reconstructions and the evolving terrane concept
  for the Argentine Precordillera 8
    Occidentalia terrane 9
    Texas plateau 9
    Precordillera terrane 9
    Cuyania terrane 11
  Historical review and stratigraphy of the Argentine Precordillera 11
    Cambrian 11
    Ordovician 12
    Silurian 12
    Devonian 13
    Post-Devonian 13
  Structural history of the Argentine Precordillera 13
  Subdivision of the Argentine Precordillera 13
    Western basin 14
    Eastern basin 14
Carbonate platform rocks of the Argentine Precordillera 14
  La Laja Formation and equivalents 14
    Lithofacies 15
    Facies associations and depositional environment 16
    Platform configuration 19
    Sedimentary succession of the La Laja Formation 19
    Evolution and sequence stratigraphy of the La Laja Formation 20
  Zonda Formation 23
    Lithofacies 23
    Facies associations and depositional environment 23
    Platform configuration 26
    Sedimentary succession of the Zonda Formation 26
    Evolution and sequence stratigraphy of the Zonda Formation 26
  La Flecha Formation 26
    Lithofacies: Mechanically deposited rocks 28
    Lithofacies: Microbial boundstones and stromatolites 28
    Facies associations and depositional environment 28
    Platform configuration 30
    Sedimentary succession of La Flecha Formation 30

Evolution and sequence stratigraphy of the Zonda and La Flecha Formations 30
La Silla Formation 32
    Lithofacies 32
    Facies associations and depositional environment 34
    Platform configuration 35
Sedimentary succession of La Silla Formation 36
    Talacasto subbasin 36
    San Juan subbasin 36
    Evolution and sequence stratigraphy of the La Silla Formation 36
San Juan Formation 37
    Time frame of deposition 38
    Lithofacies 38
    Facies associations and depositional environment 40
    Platform configuration 44
    Sedimentary succession of the San Juan Formation 44
    Evolution and sequence stratigraphy of the San Juan Formation 46
Rocks overlying the San Juan Formation 49
    Conformable and unconformable successions 49
        Rio Sassito section 49
        Cerro La Chilca section 50
        Las Chacritas section 50
        Las Aguaditas section 50
        Guandacol subbasin 50
        Discussion 50
    Don Braulio section 51
    Gualcamayo Formation 51
        Lithofacies and the sedimentary succession of the Gualcamayo Formation 51
    La Cantera Formation 51
        Sedimentary succession of the La Cantera Formation 51
        Discussion 53
    Don Braulio Formation 53
        Lithofacies 53
        Sedimentary succession of the Don Braulio Formation 54
        Discussion and sequence stratigraphy of the Don Braulio Formation 54
    Rinconada Formation 54
        Lithofacies 55
        Discussion 56
Continental margin sediments and allochthonous Cambrian and Ordovician rocks 56
    La Cruz limestones 56
        Discussion 56
        Lithofacies 59
        Interpretation and depositional environment 60
    Los Sombreros Formation 60
        San Isidro section 60
        Los Sombrebros section 61
        Ojos de Agua section 61
        Los Ratones section 61
        Los Túneles section 61
        Time frame of deposition of the Los Sombreros Formation 71
        Interpretation and depositional environment of the Los Sombreros Formation 73
    Empozada Formation 73
        Interpretation 74
        Discussion 75
Ordovician of Ponon Trehue 76

Contents

Ponon Trehue Formation 76
    Lithofacies 76
    Interpretation, facies associations, and the sedimentary succession
    of the Ponon Trehue Formation 76
    Evolution and sequence stratigraphy of the Ponon Trehue Formation 78
  Lindero Formation 79
    Lithofacies 79
    Interpretation, facies associations, and the sedimentary succession of the Lindero Formation .. 80
Early Paleozoic evolution and paleogeography of the Argentine Precordillera—
general aspects and the carbonate platform 81
  General aspects 81
    Size of the terrane 81
    Biogeographic aspects and faunal provincialism 82
    K-bentonites 84
    Heavy mineral associations 84
  Evolution of the carbonate platform 85
    Early Cambrian 85
    Middle Cambrian 85
    Late Cambrian 85
    Early Ordovician 86
    Sea-level history of the Precordillera 87
    Comparison between the carbonate platform of the Precordillera and the eastern
    margin of Laurentia 91
Early Paleozoic evolution and paleogeography of the Argentine Precordillera—the
Middle Ordovician through Devonian siliciclastic successions 96
  Middle Ordovician 96
  Late Ordovician 97
  Discussion of the Middle and Upper Ordovician evolution of the Argentine Precordillera 98
    Guanadacol event 98
    Silurian 99
  Discussion of the Silurian evolution of the Argentine Precordillera 99
    Devonian 101
    Discussion of the Devonian evolution of the Argentine Precordillera ... 102
Geodynamic evolution of the Argentine Precordillera and its parental terrane 102
  History of the Precordillera 102
    Precordillera as part of Laurentia 102
    Discussion 103
    Argentine Precordillera as a marginal plateau to Laurentia 104
    Continental breakup and the rift-drift transition 106
    Discussion 108
    Approach of the Argentine Precordillera toward Gondwana 110
    Discussion 110
Plate tectonic scenario for the evolution of the Argentine Precordillera 112
  Plate tectonic evolution of the Argentine Precordillera 112
    Precordillera as part of Laurentia 112
    Argentine Precordillera as an independent microplate 112
    Argentine Precordillera in the conservative models 114
    Argentine Precordillera as part of Gondwana: Precordillera in the alternative models 114
    Discussion 115
Epilogue 115
    Remaining questions 115
Acknowledgments 116
Appendix 117
References cited 118

# *Argentine Precordillera: Sedimentary and Plate Tectonic History of a Laurentian Crustal Fragment in South America*

Martin Keller

Institut für Geologie und Mineralogie, Friedrich-Alexander Universität Erlangen-Nürnberg, Schlossgarten 5, 91054 Erlangen, Germany

ABSTRACT

The Argentine Precordillera ("Precordillera") is part of a larger terrane which is exotic to the western margin of Gondwana. This larger terrane is characterized by a Grenvillian-type basement and the presence of Laurentian carbonate-platform rocks of Cambrian–Ordovician age. The original position of the terrane is interpreted to be the Ouachita embayment of southeastern Laurentia.

Terminal Neoproterozoic continental breakup and the formation of the Appalachian-Ouachita margin during the late Precambrian–Early Cambrian established a passive margin to the present east of the Precordillera. This passive margin is the original continental margin of Laurentia, which was located outboard of the Precordillera. Continuous crustal extension during the Cambrian along the Ouachita margin created a marginal plateau that accommodated the Argentine Precordillera carbonate platform. This marginal plateau to Laurentia was separated from mainland Laurentia by a deep graben system, today represented by the offshore Ouachita facies of the southeastern United States. The Cambrian through earliest Middle Ordovician mark the climax of the Laurentian episode because the Argentine Precordillera sediments and fauna are indistinguishible from Laurentia.

In the carbonate platform succession of the Argentine Precordillera, the La Laja Formation marks the transition from rift-related redbeds to carbonate sedimentation. The La Laja Formation was deposited on a platform seaward of the nearshore siliciclastic trap but inboard of a (only partly exposed) carbonate belt which might have marked the platform margin. Several third-order cycles developed; differential rates of sea-level rise were the driving forces for cycle formation. Near the base of the La Laja Formation, a sandy interval is separated from the older rocks by a presumed hiatus. This hiatus might represent the Hawke Bay regression event of the eastern margin of Laurentia.

Upper Cambrian limestones and dolostones of the Zonda and La Flecha Formations record the evolution of vast peritidal flats. The spatial and biostratigraphical relations between both formations are not clear and await further research. At present, it seems that the lower, more calcareous facies of the La Flecha Formation in the Guandacol subbasin is coeval to the dolomite facies of the Zonda Formation in the San Juan subbasin. The last important terrigenous input is noticed in Marjuman sediments of the La Flecha Formation, which marks a major shift in coastal onlap toward the hinterland.

The Tremadocian La Silla Formation reflects a somewhat deeper environment, although open-marine rocks are absent. However, extensive tidal flats are also absent. The rocks are mud-dominated limestones and dolostones with locally abundant oolites.

Small-scale cycles are locally present. The general arrangement of facies is best explained by a tidal-flat island model. The La Silla Formation shows the most uniform facies development of all successions of the carbonate platform in the Argentine Precordillera. The very uniform facies arrangement can be explained either by a rimmed-shelf model or the model of a carbonate ramp.

The San Juan Formation was mainly deposited during the Arenig and consists of open-marine limestones. Two important reef accumulations are present near the base and near the top of the formation. The San Juan Formation represents sediments of a carbonate ramp including inner ramp, mid-ramp, and deep-ramp environments. In the upper part of the formation, a deepening led to the drowning of the formation and the regionally developed carbonate platform. Drowning across the platform was not coeval, the different episodes were caused by the onset of extensional tectonics. Sea-level fluctuations still exerted a major control on facies evolution and the effects of eustasy versus tectonics are difficult to distinguish.

In the carbonate platform succession of the Precordillera, 13 third-order sequences have been recognized. Biostratigraphic control on many of the sequence boundaries is still poor; however, most of them are located in time intervals characterized by eustatic events. Hence it seems that eustasy was the dominant factor in carbonate platform evolution. The effects of the global Lange Ranch eustatic event at the Cambrian-Ordovician boundary are recognized for the first time in the Argentine Precordillera.

From Middle Cambrian time on, a deep-water carbonate environment existed to the west of the platform. Rocks of the deep-water facies are only present as olistoliths in the Ordovician continental-margin successions with the exception of Cerro Pelado, where they overlie an isolated outcrop of platform dolomites of the Cerro Pelado Formation. Within the olistoliths, mass-flow deposits that might have been derived from the platform margin are conspicuously absent. The absence is interpreted as reflecting a very gentle relief between the platform and the basin during the early history of the Precordillera.

Near San Rafael, in the province of Mendoza, Ordovician rocks are present that belong to the same terrane as does the Argentine Precordillera. The Ponon Trehue Formation is composed of limestones and dolostones that show a similar succession of sedimentological events and an identical fauna to the La Silla and San Juan Formations. About 80 m of the Ponon Trehue Formation resemble more than 300 m in the Precordillera and are the expression of a cratonal setting, where onlap of the carbonates did not occur before the Tremadocian.

Sediments of an evolving continental-margin comprise the Los Sombreros Formation and its equivalents. Sedimentation of these deposits started during the Llanvirn; however, the most important events took place during the Llandeilo and early Caradoc with the emplacement of megaolistoliths and megabreccias. The Los Sombreros Formation is interpreted as a continental-rise deposit at the toe of a major fault scarp, locally more than 2100 m high.

The basinal facies shows an overall fining tendency toward the west; there is an overall fining-upward into the early Caradoc that is expressed by the widespread deposition of graptolitic black shales within the *Nemagraptus gracilis* zone. Along the western margin of the Precordillera, pillow basalts were extruded into the shales. These basaltic magmas are taken as evidence of the existence of oceanic crust by some, but certainly mark the culmination of the preceding crustal extension.

In the eastern basin, the Middle and lower Upper Ordovician siliciclastic successions containing abundant conglomerates were deposited in half grabens and grabens attributed to an extensional regime.

Crustal extension dominated the depositional history of the Argentine Precordillera during the Middle and Late Ordovician, and is interpreted to reflect the rifting that led to the final separation of the Precordillera from mainland Laurentia. In this context, the widespread Guandacol tectonic phase and the accompanying unconformity separate synrift siliciclastic from prerift strata. In many sections, the Guandacol event is represented by a hiatus that was caused by erosion of uplifted areas. Locally, horst structures were used for the establishment of an isolated carbonate platform, the Sassito limestones.

Latest Ordovician platform deposits with a chert-pebble conglomerate at the base unconformably cover older rocks. Although the effects of the preceding erosion cannot be determined in detail, erosion and the accompanying hiatus mark this unconformity as the breakup unconformity that separates the synrift rocks from the postrift rocks, which in the Precordillera are represented by siliciclastic platform deposits. From the Late Ordovician into the Silurian, the Precordillera was an independent microplate.

The history of accretion of the Argentine Precordillera to Gondwana is ambigious. A Late Silurian to Early Devonian metamorphic event is observed at the western margin of the Precordillera. Compressional deformation there affected strata as young as Early Devonian. Along the eastern margin of the Precordillera Devonian sedimentary rocks are absent. Deformation there can only be confined to post-Late Silurian and pre-Late Carboniferous time.

Deformation in the west is interpreted to reflect the docking of the Chilenia terrane whereas concomitant structures in the east may be related to the accretion of the Argentine Precordillera to Gondwana. The easterly derived Punta Negra flysch-like sediments record uplift and erosion of an igneous and metamorphic source area, which included the basement of the Precordillera and its parent terrane. Hence the Paleozoic compressional deformation of the Precordillera is the result of complex plate interactions between western Gondwana, the Precordillera, and the Chilenia terrane. The accretion of the Precordillera to Gondwana was terminated prior to the Late Carboniferous, when molasse sediments were deposited, initiating the Gondwanan sedimentary cycle.

INTRODUCTION

Argentine Precordillera

The Precordillera of Western Argentina ("Precordillera") (Figs. 1 and 2) is part of the Andean mountain chain, which extends along the entire Pacific margin of South America from Colombia to the southern tip of Chile. The Precordillera is thus one element of the predominantly Mesozoic-Cenozoic mobile belt, formed along the present western edge of cratonic South America, which in turn represents the western part of Paleozoic Gondwana. The mobile belt, which developed as a response to subduction of the laterally extensive Phoenix plate and its successor (the Nazca plate) beneath South America, incorporates several Paleozoic tectono-stratigraphic terranes, among which are the Precordillera as part of a larger Cuyania terrane (Ramos, 1995) or Occidentalia terrane (Dalla Salda et al., 1992a), and the neighboring terranes of the Puna, Sierra de Famatina, and Sierras Pampeanas (Fig. 2). These terranes were accreted to the western margin of Paleozoic Gondwana during the early Paleozoic Famatinian orogeny (see Appendix 1). The Sierras Pampeanas (Fig. 2) mark the transition zone from the stable craton toward the different mobile belts. Each of the geologic provinces of western and northwestern Argentina had a distinct geologic history during the early Paleozoic. These histories were summarized in time-space diagrams by Bahlburg and Hervé (1997), Pankhurst et al. (1998), and Rapela et al. (1998). The Argentine Precordillera is unique in preserving a Lower Cambrian through lower Middle Ordovician carbonate platform (see Appendix 1). Isolated carbonates of Ordovician age were described from the Sierra Pintada near San Rafael (Wichmann, 1928; Nuñez, 1979) in the province of Mendoza (Fig. 2) and from the province of La Pampa (Limay Mahuida; Baldis et al., 1985). Laterally extensive carbonate platform deposits of that age, however, are absent from any other terrane in western Gondwana (South America, Africa). The presence of carbonate platform rocks in the Precordillera that are similar in age, facies, sedimentology, and paleontology to deposits around the passive margin of Laurentia opened the door for speculations about the origin and provenance of the Precordillera (Borrello, 1971; Bond et al. 1984; Ramos et al. 1986; Keppie, 1991).

On the basis of thermal subsidence curves, Bond et al. (1984) speculated that the Argentine Precordillera until Neoproterozoic time might have been part of Laurentia and that it might be exotic to South America. This idea was put forward by Ramos et al. (1986) and Ramos (1988a, 1988b), who proposed a collisional model for South America that centered around the Precordillera. Various other allochthonous models were developed in subsequent years (e.g., Dalla Salda et al., 1992a, 1992b; Dalziel et al., 1994; Astini et al., 1995, 1996b; Dalziel, 1997); however, there were always objections and models that favored an autochthonous or parautochthonous position of the Precordillera. A right-lateral movement from the area of the province of La Pampa (Limay Mahuida) or south of Buenos Aires

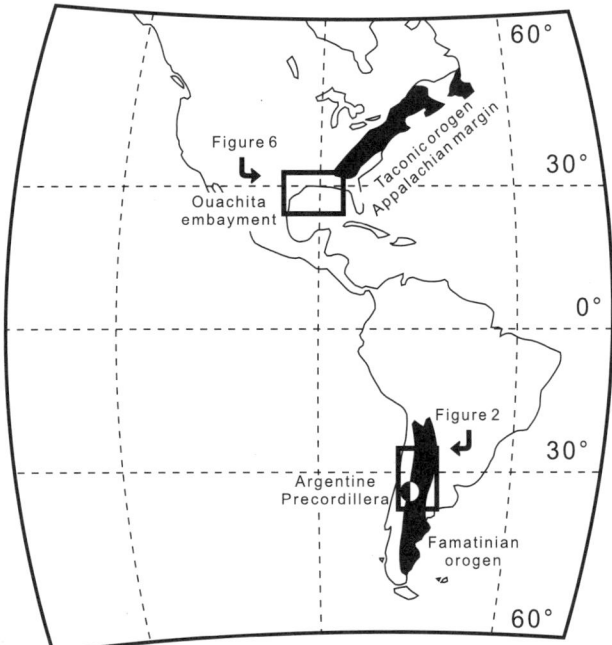

Figure 1. Present-day locations of Precordillera, the Famatinian and Taconic orogens, and Ouachita embayment (modified from Dalla Salda et al., 1992b).

(Sierras de la Ventana and del Tandil, respectively) was discussed by Baldis et al. (1989); however, they concluded that there is neither sufficient sedimentologic nor tectonic evidence for an allochthonous origin of the Precordillera. Aceñolaza and Toselli (1988, 1998) assumed that the Argentine Precordillera originally was located much farther south (present coordinates) along the paleopacific margin of western Gondwana and that it was transferred to its present-day position together with the western Sierras Pampeanas along a system of right-lateral shear zones after Ordovician time. Loske (1992) proposed a model in which the Precordillera carbonate platform originated in a back arc environment related to the Famatinian magmatic arc. Following this model, the Precordillera is supposed to have reached its present-day position during a left-lateral transport with respect to the Famatina system. On the basis of sedimentologic and structural evidences, Gonzalez Bonorino and Gonzalez Bonorino (1991) argued in favor of an entirely autochthonous evolution of the Precordillera.

As a consequence of the fundamentally different interpretations of the evolution of the Precordillera a Penrose Conference was held in San Juan (Argentina) in 1995 and there was general agreement that the Precordillera in its early history was part of Laurentia (discussion in Dalziel et al., 1996). However, the timing and the processes involved in the transfer of the Precordillera from Laurentia to Gondwana are still highly controversial (e.g., Dalla Salda et al., 1992a, 1992b; Astini et al., 1995, 1996a; Dalziel et al., 1994; Dalziel and Dalla Salda, 1996; Keller, 1995, 1996; Keller and Dickerson, 1996; Dalziel, 1997).

Scope of the Study

In order to discuss the hypothesis of an allochthonous origin of the Argentine Precordillera and its presumed Laurentian provenance, this paper describes, documents, and interprets the evolution of the carbonate platform succession developed during Cambrian through early Middle Ordovician time, when more than 2000 m of carbonate sediments accumulated. Although these carbonate rocks have been studied since the nineteenth century (Kayser, 1867; Stelzner, 1873), there is still no integrated synthesis to interpret the entire succession. Many publications deal with one or two sections (e.g., Espisua, 1968; Beresi and Bordonaro, 1984; Beresi et al., 1987; M. Keller et al., 1989), or rarely with one formation on a regional scale (Beresi, 1986a; Cañas, 1995b). In addition to the carbonate platform rocks, the Middle and Upper Ordovician continental margin facies and their transition into basinal deposits in the Precordillera have been studied. The interpretation of the geotectonic environment in which the sediments of the Precordillera were deposited forms the basis for the discussion about allochthony versus autochthony of the Precordillera and its parental crustal fragment. Subsequently, a plate tectonic scenario is presented to explain the probable provenance of this parental terrane, the Cuyania (Ramos, 1995) or Occidentalia terrane (Dalla Salda et al., 1992a), and the processes involved in the transfer.

Methods. This study is based on 68 sections with Cambrian and Ordovician strata measured at 33 localities (Fig. 3) in the Argentine Precordillera and the area of San Rafael (Fig. 2), southeast of Mendoza. A measured section was restricted to one formation, although there are many localities where the exposed succession contains two or more formations. Consequently, two or more sections are counted in these localities. The measured sections total more than 14000 m of strata, which were analyzed bed by bed. About 1300 thin sections have been studied for petrography, facies, and microfacies. Diagenesis was studied under a luminiscope using about 10% of the samples.

An important part of field work was dedicated to biostratigraphic sampling, especially for conodonts. These samples were treated by O. Lehnert (Erlangen) and many of the results are published (e.g., Lehnert, 1993, 1995a, 1995b, 1995c; Lehnert and Keller 1993a, 1993b; Lehnert et al., 1997, 1998). Sections in the Guandacol area have not been studied to avoid overlap with the doctoral thesis of F. Cañas (Córdoba), who worked on the Upper Cambrian–Lower Ordovician succession. As Fernando completed part of his thesis with us in Erlangen, there was no need to repeat his efforts in the field. The results are published (Cañas 1995a, 1995b) and I believe adequately cited here.

The data obtained in the field and from thin-section analysis were the base for constructing a stratigraphic panel (Fig. 4) and sedimentary columns, which, together with the new biostratigraphic data from the Ordovician rocks, served for the correlation of sections and for the interpretation of the geotectonic history of the Precordillera.

The data were also used for the sequence-stratigraphic anal-

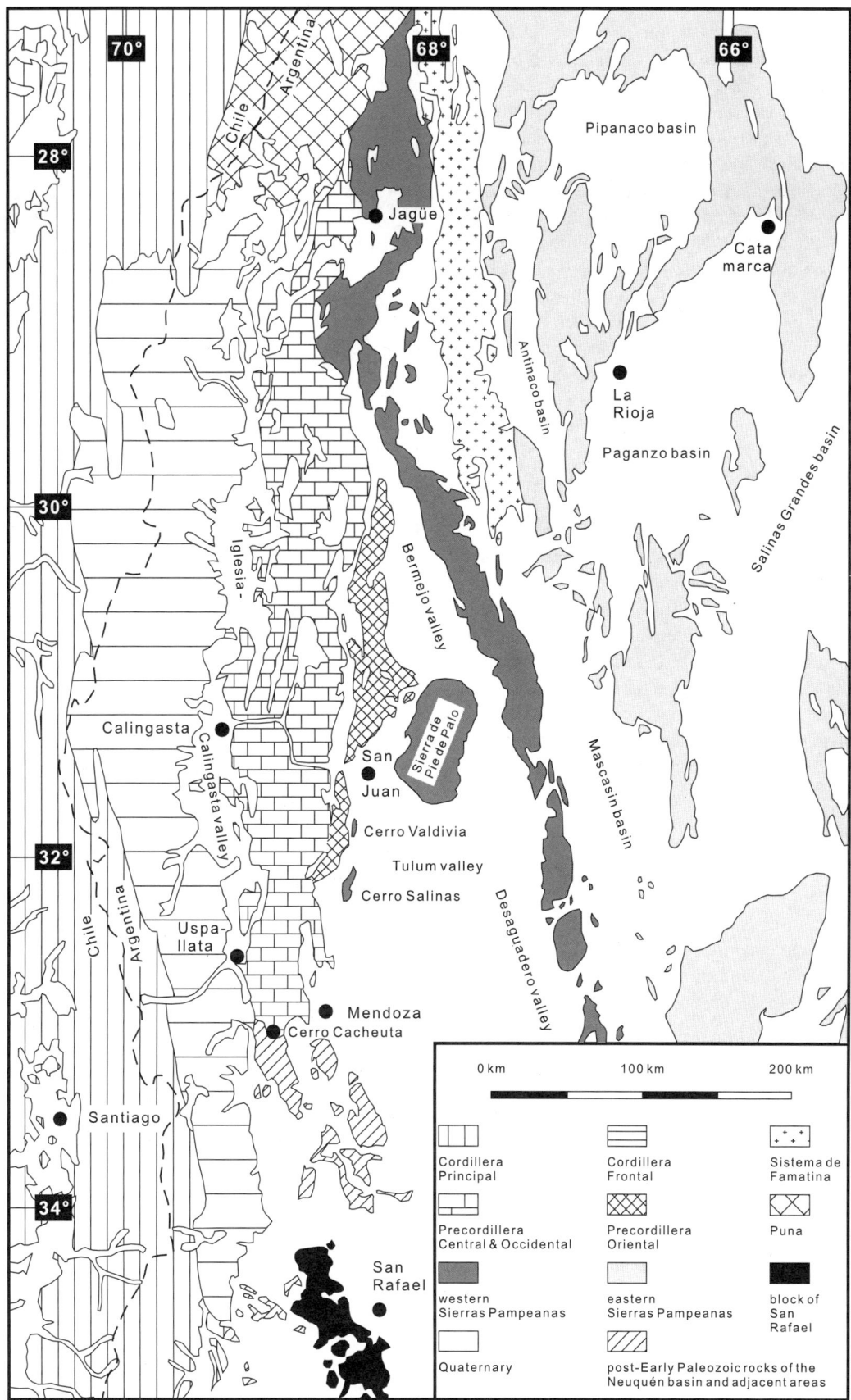

Figure 2. Geologic provinces and morphostructural units of northwestern Argentina and adjacent areas.

ysis of the carbonate platform succession. However, this analysis is hampered by the following. Sequence stratigraphy is based on the recognition of depositional sequences and their bounding surfaces (Vail et al., 1977; Van Wagoner et al., 1988; Posamentier and Vail, 1988; Sarg, 1988). Although many boundaries have been identified in the field, their character (onlap, toplap, downlap) could not be evaluated in the field. Sequence stratigraphy in carbonates (Sarg, 1988; Handford and Loucks, 1993) is based on sea-level changes in relation to the platform-bank margin. In the Precordillera, the position of the platform margin for most of the history of the carbonate platform is unknown; there is an idea about its location only during the Arenig. However, as stated by James and Mountjoy (1983, p. 189), "... the nature of the shelf-slope break is commonly interpreted rather than observed, and synthesized on the basis of information from surrounding facies". Similarly, the cratonward and basinward migration of coastal onlap is almost impossible to determine, because the relation of the platform to the basement is unknown. With all these restrictions of the applicability of sequence stratigraphy in mind, I am still convinced that the concept of sequence stratigraphy can be a valuable tool in interpretations of basin configuration and the geodynamic history of the Precordillera.

For the following discussion, third-order sequences and sequence boundaries in the carbonate platform succession have been labeled from 1 to 13. Because the nature of a sequence boundary (type 1 vs. type 2) determines the nature of the overlying systems tract and the overlying sequence, a sequence boundary plus the overlying strata form an entity. Sequence boundary 1 is not exposed because it underlies the lowermost exposed strata of the La Laja Formation.

Regional Setting of the Argentine Precordillera

The Precordillera forms the eastern slope of the Andes in western Argentina. It trends north-south between 28°45′S and 33°15′S over more than 400 km and is separated from the main Andes (Cordillera Frontal) by the longitudinal valley of Iglesias-Calingasta-Uspallata (Fig. 2). Toward the south, the Precordillera is in fault contact with Mesozoic and Cenozoic rocks of the main Andes.

To the east, the Precordillera is separated from the western Sierras Pampeanas by the Bermejo-Tulum Valley (Fig. 2), which is filled by several thousand meters of Tertiary and Quaternary sediments. This valley trends obliquely (~ northwest–southeast) to the main structures of the Precordillera, following a presumed right-lateral shear zone (Aceñolaza and Toselli, 1988). Consequently, the northern end of the Precordillera overlaps in latitude with the Sierra de Famatina and, in plane view, forms a wedge between the Cordillera Frontal and the Sierra de Famatina (Fig. 2).

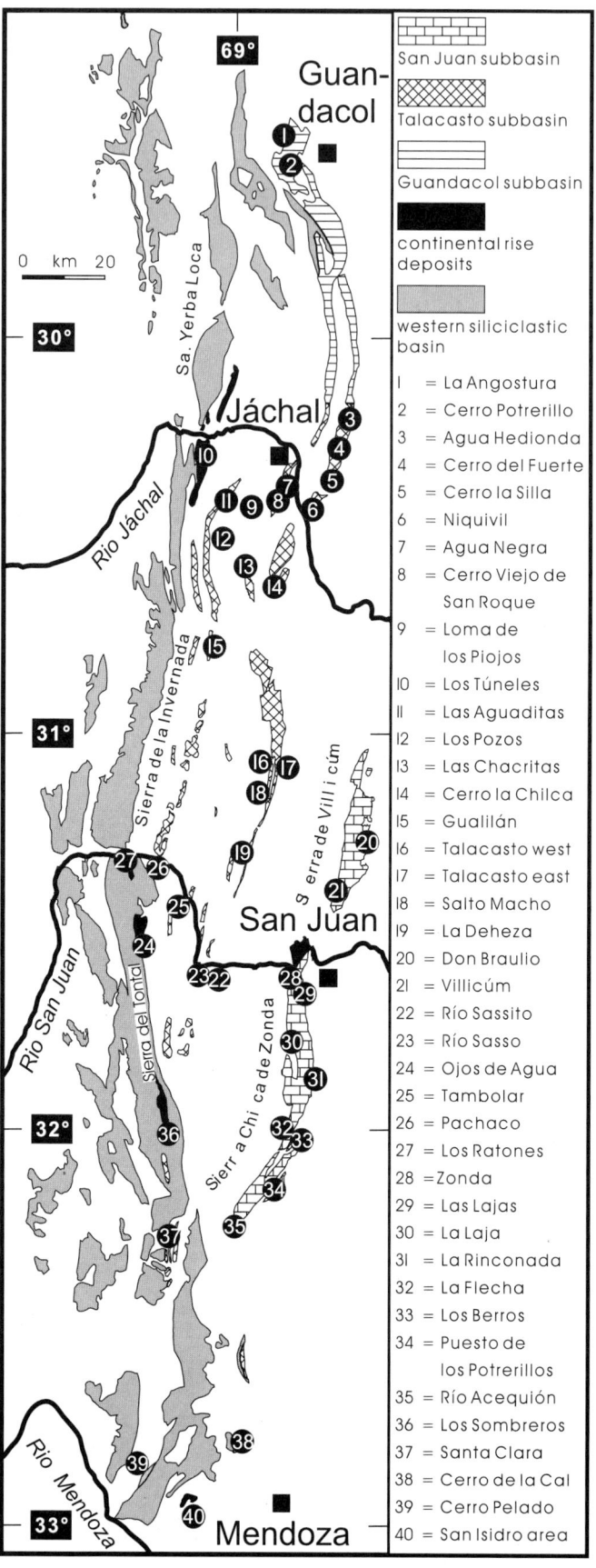

Figure 3. Distribution of Cambrian-Ordovician sedimentary rocks in Precordillera and localities mentioned in text. Also shown is subdivision of Precordillera into basins and subbasins as applied in this paper.

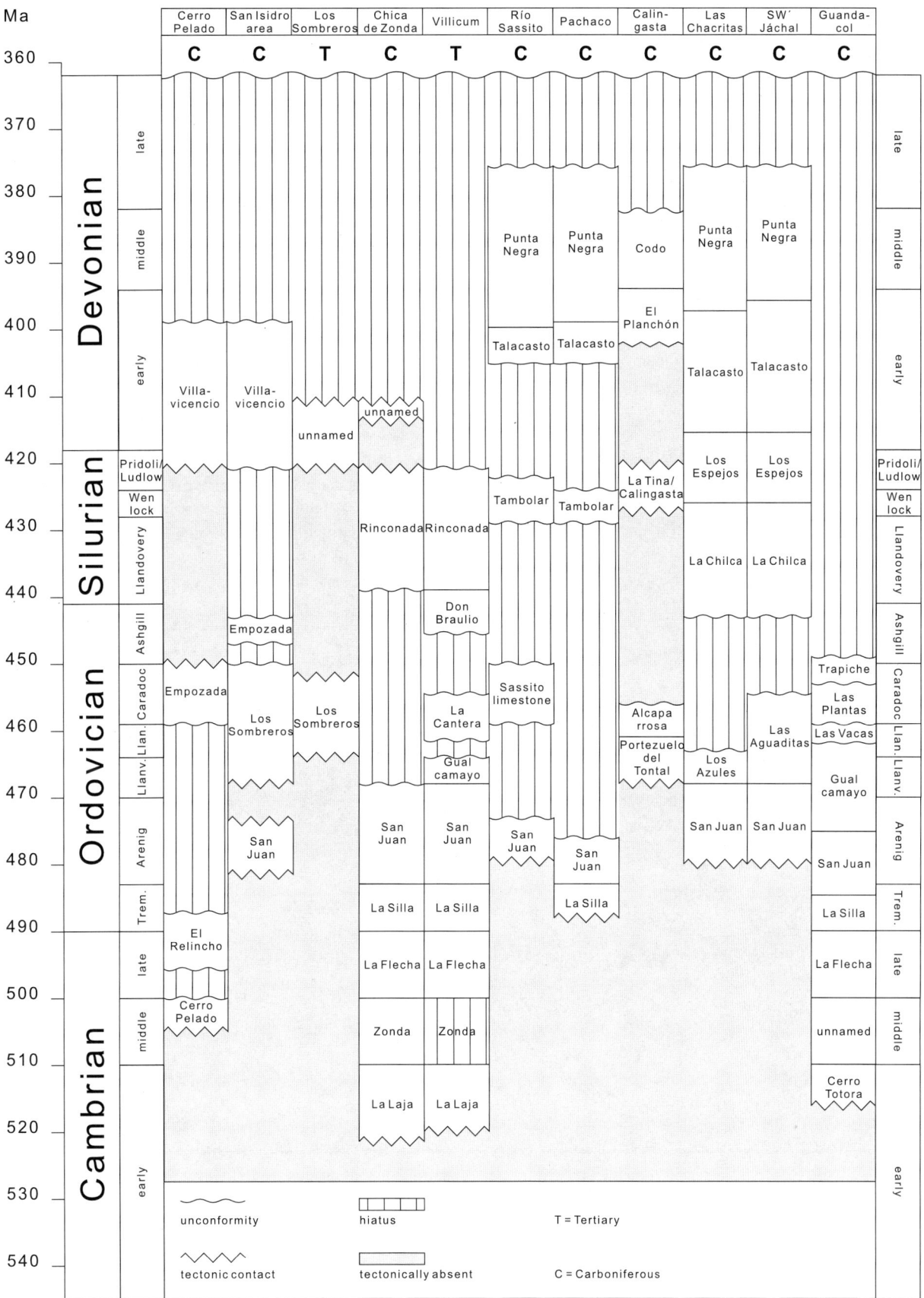

Figure 4. Stratigraphic chart for Cambrian through Devonian rocks of Precordillera. Absolute ages are from Bowring and Erwin (1998) for Cambrian and Early Ordovician; Ordovician, Silurian, and Devonian ages are from Tucker et al. (1990, 1998) and Fordham (1992).

The term Precordillera is strictly an orographic and morphostructural term; however, because of its unique rock suite, the term is almost exclusively used as that of a geologic province. This has led to much confusion in the recent discussion about the allochthony of the Precordillera and the existence of a Precordillera terrane.

Historically, a distinction was made between the "Precordillera of San Juan" or "San Juan faunal province" and the "Precordillera of Mendoza" or "Mendoza faunal province" (e.g., Borrello, 1971). Although based on political boundaries, this distinction also implied lithostratigraphic differences between both areas, especially for the Cambrian-Ordovician succession. Cambrian-Ordovician strata of the San Juan Precordillera record the evolution of a carbonate platform, whereas in the Precordillera of Mendoza the corresponding continental-margin sediments are preserved.

Neoproterozoic-Paleozoic Global Paleogeographic Reconstructions and the Evolving Terrane Concept for the Argentine Precordillera

Conventional plate tectonic models for the early Paleozoic show Laurentia as an isolated continent, its Appalachian-Caledonian margin facing the Iapetus ocean, and bounded on its opposite (conjugate) margin by Baltica and northwestern Gondwana (Scotese and McKerrow, 1990; McKerrow et al., 1991, 1992; Fortey and Cocks, 1992; Torsvik et al., 1995, 1996; Van der Pluijm et al., 1995; MacNiocaill et al., 1997). Although in some of these models (e.g., MacNiocaill et al., 1997) the Precordillera and its accretion to Gondwana are discussed, the Precordillera is not regarded as an Ordovician pinpoint because a later accretion is inferred. A detailed discussion of these models, which in this paper is referred to as "conventional plate tectonic models," was given by Dalziel (1997).

Recently proposed alternative models for Neoproterozoic and early Paleozoic plate tectonic assemblages center around speculations that various supercontinents may have existed during mid-Neoproterozoic, terminal Proterozoic, and Ordovician time, respectively. These hypothesis were developed from the realization that East Antarctica and South America might have been juxtaposed with the Pacific margin of Laurentia (SWEAT hypothesis of Moores, 1991; see also Dalziel, 1991; Hoffman, 1991) during Neoproterozoic time. Laurentia is supposed to have formed the nucleus of a supercontinent (Fig. 5) termed Rodinia by McMenamin and McMenamin (1990). Following these models, breakup of this supercontinent essentially left eastern Gondwana on the one side and on the other side western Gondwana,

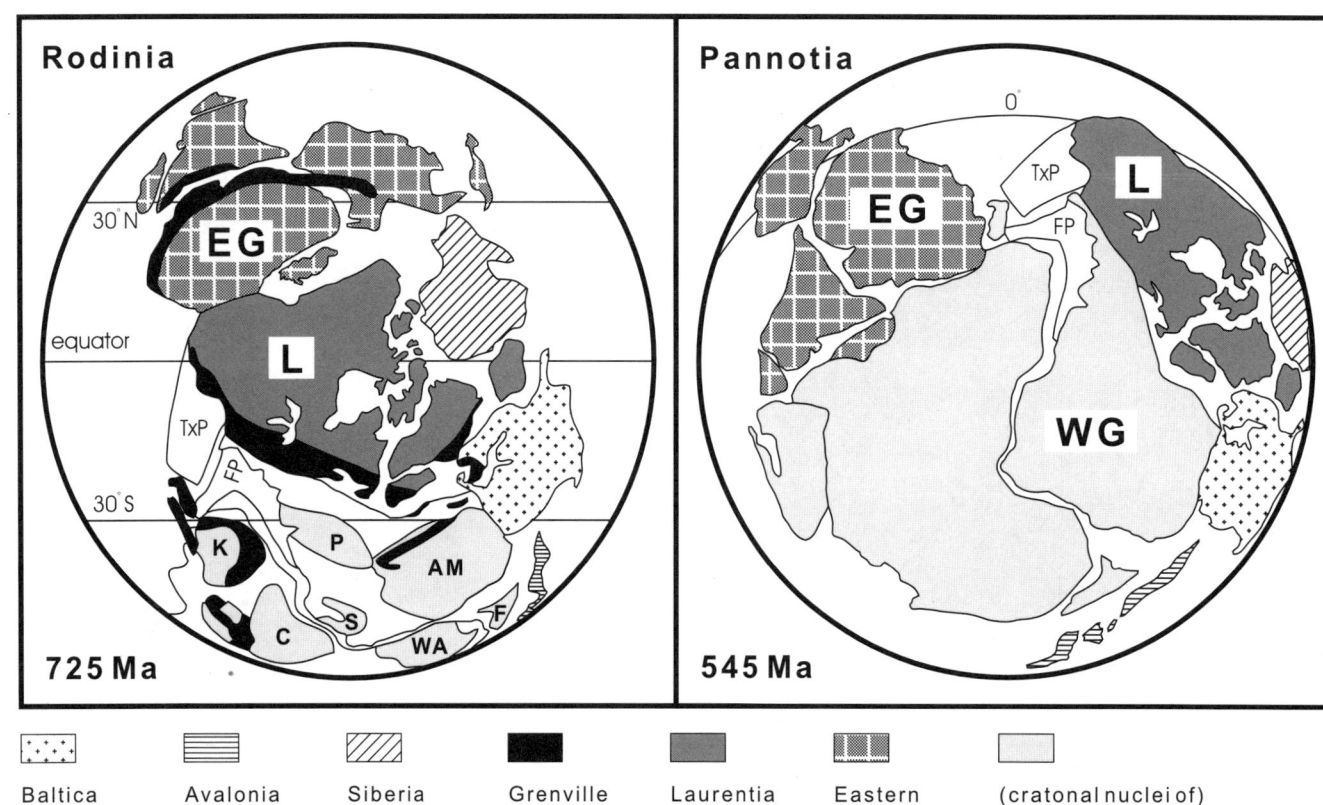

Figure 5. Two hypothetical Neoproterozoic supercontinents of Rodinia and its successor Pannotia in latitudinally controlled reconstructions. TxP is Texas plateau of Dalziel (1997), which includes Precordillera. FP = Falkland Malvinas Plateau, AM = Amazonas craton, C = Congo craton, EG = eastern Gondwana, F = Florida, K = Kalahari craton, L = Laurentia, P = Rio de la Plata craton, S = Sao Francisco craton, WA = West African craton, WG = western Gondwana. Both reconstructions are from Dalziel (1997).

still connected to eastern Laurentia. The Brazilian–Pan-African and related orogenies of latest Neoproterozoic and earliest Cambrian time reflect the final amalgamation of Gondwana. Dalziel (1992), Powell (1995), and Powell et al. (1995) pointed out that, if western Gondwana and Laurentia were still connected, the assembly of Gondwana might have involved the formation of another short-lived terminal Neoproterozoic supercontinent (Fig. 5), which they called Pannotia (see also Dalziel, 1997, Fig. 12). In this scenario, severence of Laurentia from western Gondwana led to the formation of the proto-Appalachian margin of Laurentia. After this separation, Laurentia is supposed to have rotated clockwise around the proto-Andean margin of South America and to have interacted with this margin at different times (Dalziel, 1991) until it reached its position within Pangea.

More or less coeval to these evolving alternative paleogeographic reconstructions was the realization that the western margin of Gondwana might consist of a complex mosaic of different, often poorly defined terranes. Early on, the distinct deformational and stratigraphic history of the Argentine Precordillera led Ramos et al. (1986) and Ramos (1988b) to the assumption that the Precordillera is not indigenous to South America and that it might represent a Paleozoic (exotic) terrane. The first workers who recognized the close similarities between the eastern margin of Paleozoic Laurentia and the Precordillera were Bond et al. (1984). On the basis of thermal subsidence curves and faunal comparisons, they speculated on the possibility that Laurentia and southern South America were combined in a Neoproterozoic supercontinent. Continental breakup of this hypothetical Neoproterozoic supercontinent along the Appalachian margin of Laurentia was responsible for the separation of the Precordillera from Laurentia in their model.

On the basis of the alternative paleogeographic reconstructions and within their general framework, Dalla Salda et al. (1992a) promoted the hypothesis that during the Ordovician, a Laurentia-Gondwana collision may have formed the Famatinian orogen of western South America. In this model, a large sliver of Laurentian continental crust stayed attached to western Gondwana after the renewed separation of both continents during the Ordovician; i.e., the Occidentalia terrane.

Occidentalia terrane. In its original definition, this terrane included all basement fragments from Peru to Patagonia that are thought to have been accreted to this part of South America: the Arequipa massif, the Arequipa-Antofalla terrane, Chilenia (see Appendix 1), and the North Patagonian massif, among others. In west-central Argentina, the Occidentalia terrane includes the western part of the Sierras Pampeanas (Sierra de Pie de Palo, Cerro Valdivia, and Cerro Salinas; Fig. 2) and the Argentine Precordillera (Dalla Salda et al., 1993). This model is based on a main deformational event during the Early Ordovician and peak magmatic activity between 480 and 460 Ma (data from Willner and Miller, 1982; Rapela et al., 1989), which, according to Dalla Salda et al. (1992a, 1992b, 1993), may reflect the collisional event that formed a short-lived Ordovician supercontinent during the Middle Ordovician. On the basis of a roughly similar geotectonic evolution of both the Famatinian mobile belt and the Taconic Appalachians, this model was modified subsequently (Dalla Salda et al., 1992b) in proposing that both orogens might have formed a single mountain chain as the result of the Laurentia-Gondwana collision during the Middle Ordovician. In this successor model, the Arequipa-Antofalla terrane was excluded from Occidentalia. Instead, its southern tip was used as a piercing point, where the Taconic orogen might have been separated from the Famatinian system during Middle to Late Ordovician rifting. Later (Dalziel, 1993; Dalziel et al., 1994) the Precordillera was taken as a tectonic tracer in the reconstruction of the presumed Ordovician supercontinent.

On the basis of several lines of evidence, Dalla Salda et al. (1992b) pointed out that the Argentine Precordillera might have been derived from the Ouachita embayment of the southeastern Laurentian craton (Fig. 6). A major conclusion of the Occidentalia terrane model (Dalla Salda et al., 1992b, 1993) is the assumption that the Precordillera was separated from Laurentia during a late Middle Ordovician to Late Ordovician rifting event, as documented by the extrusion of abundant pillow lavas into the western basin during the Caradocian.

Texas plateau. A modification of the Occidentalia terrane model and the presumed Ordovician continent-continent collision is the model of the Texas Plateau (Dalziel, 1997). The Texas Plateau, mainly composed of the Argentine Precordillera and its parental terrane, was attached to Laurentia outboard of the Ouachita embayment until the Late Ordovician. The plateau would have had a dimension of 1500 km by 800–1000 km, its length being almost the same as that of the Cuyania terrane (Ramos, 1995). The plateau had attenuated crust, and between Laurentia and the plateau, deep rift-related basins may have existed (Dalziel, 1997), floored by oceanic crust. Middle Ordovician collision of this plateau was then responsible for the Oclóyic orogeny (see Appendix 1) of the Famatinian orogenic cycle. Subsequent rifting left the plateau attached to western Gondwana. In this model, the Precordillera is a tectonic tracer fixing the relative position between Laurentia and western Gondwana during the Ordovician.

Precordillera terrane. Astini et al. (1995) presented a model based on detailed sedimentologic studies and the interpretation of faunas, in which the Precordillera is regarded as a terrane that rifted and drifted away from Laurentia during Early Cambrian time, and collided with cratonic Gondwana during Middle Ordovician time. The critical points in this model are as follows. The Precordillera was an integral part of Laurentia through Early Cambrian time and was the southern continuation of the Appalachian margin. Passive-margin sedimentation on the microplate (i.e., the formation of the carbonate platform) continued into the Arenig. On the basis of paleontologic arguments, Astini et al. (1995) claimed that by late Arenigian time the Precordillera was separated from Laurentia by an ocean 2500 – 3000 km wide and that only a narrow ocean existed east of the Precordillera toward the Sierra de Famatina. During the Llanvirn, this ocean was finally closed by east-directed subduction and the Precordillera

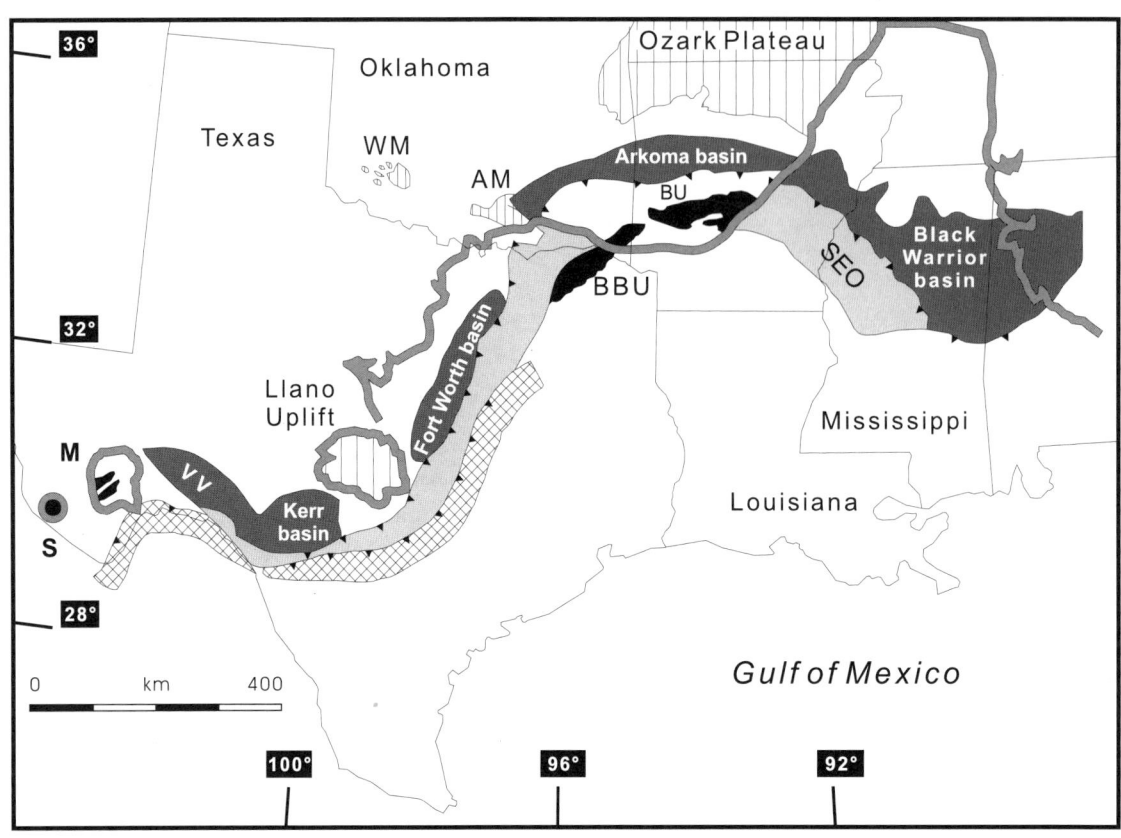

Figure 6. Ouachita margin (embayment) of the southeastern United States, most likely area of provenance of Precordillera, and its tectono-sedimentary components (modified from Viele, 1989).

accreted to the Sierra de Famatina. This subduction is thought to be responsible for the Famatinian magmatic activity and the coeval strong deformation. The subsequent Ordovician sedimentary history, according to Astini et al. (1995), reflects the formation of a foreland basin and a peripheral bulge in the Precordillera. From Caradocian time on, west-directed subduction to the west of the Precordillera is assumed, related to the approach of Chilenia, which in turn is thought to have produced another forebulge in Silurian time. An update of this model was given by Astini et al. (1996b).

In his analysis of the evolving Appalachian margin following latest Neoproterozoic rifting, Thomas (1991) postulated the existence of a major continental fragment that was removed during the opening of the Ouachita embayment. At that time, this continental fragment still awaited identification, although Miser (1921) and King (1937) noted that an enigmatic landmass ("Llanoria") had to have existed outboard of the Marathon-Solitario portion of the Ouachita embayment (Fig. 6). Realizing that the Argentine Precordillera might have been this fragment, and combining their data and their views from either side of the southern Iapetus, Thomas and Astini (1996) claimed that their "Precordillera terrane" was separated from Laurentia during the

Cambrian, then traveled independently across the intervening ocean basin to be accreted to western Gondwana during the Middle Ordovician.

Cuyania terrane. It was demonstrated that the basement rocks of the westernmost section of the Sierras Pampeanas were deformed and metamorphosed during the Grenvillian orogeny. In the Sierra de Pie de Palo (Fig. 2), a Grenvillian age (see Appendix 1) was demonstrated by Varela and Dalla Salda (1993) with 1027 ± 59 Ma rocks. McDonough et al. (1993) reported radiometric ages between 1092 and 938 Ma, and Varela et al. (1996) recorded an age of 1030 ± 30 Ma from the Sierra de Umango. Similarly, radiometric dating of xenoliths (Abbruzzi et al., 1993; Kay et al., 1996) within Miocene lava domes of the Argentine Precordillera showed the presence of Grenvillian rocks underneath the Precordillera with ages of 1188 ± 122 Ma, 1099 ± 3 Ma, and 1102 ± 6 Ma. In addition, these rocks show a whole-rock Pb isotopic signature almost identical to that of the Grenvillian basement along the eastern margin of North America (Kay et al., 1996), especially to the Llano uplift (Fig. 6), but distinctly different from any other South American basement. Basement with similar petrological characteristics seems to be present in the San Rafael area (Fig. 2; Criado Roque and Ibañez 1979; Ramos, 1995, personal commun.), where it forms the foundation for a thin succession of carbonate platform rocks (Wichmann, 1928).

The limestones of San Rafael, which unconformably overlie this basement, have been dated (Bordonaro et al., 1996; Lehnert et al., 1998) as Early Ordovician, and it was demonstrated that the fauna is identical to that of the coeval Precordilleran carbonates. Bordonaro et al. (1996) concluded that these limestones represent a part of the same platform as the carbonates of the Precordillera. The similarities between the basement rocks beneath the Precordillera, in the Sierra de Pie de Palo, and south of San Rafael, together with the similarities between the carbonates of the Precordillera and San Rafael, led Ramos (1995) to the model of the Cuyania terrane. This terrane incorporates all crustal fragments along the western margin of South America that contain Grenvillian-type basement (see Appendix 1) and/or remnants of a Cambrian-Ordovician carbonate platform with Laurentian faunas. In consequence, the basement rocks of the western Sierras Pampeanas together with the basement and the Ordovician rocks of the Sierra Pintada near San Rafael are part of the Cuyania terrane, of which the Precordillera forms by far the largest fragment (Fig. 2).

Looking at the definitions of the Cuyania terrane (Ramos, 1995) and the Precordillera terrane (Astini et al., 1995; Thomas and Astini, 1996), it becomes evident that the latter definition is an areally more restricted one and that the Precordillera terrane constitutes one component of the larger Cuyania terrane. Consequently, the term Cuyania terrane will be used here.

Historical Review and Stratigraphy of the Argentine Precordillera

The Precordillera consists of a thick succession of mainly Paleozoic strata of varying thicknesses. Cambrian and Early Ordovician sedimentation was dominated by carbonate deposition, whereas younger strata are almost entirely composed of siliciclastic rocks. A regional stratigraphic and geologic framework has been known since the late nineteenth century (e.g., Stelzner, 1873; Kayser, 1876) and the early twentieth century (e.g., Bodenbender, 1902; Stappenbeck, 1910; Keidel, 1921). A major paleontological contribution is Kobayashi's (1937) description of fossils of the San Juan limestones. Heim's publications (1948, 1952) are the basis for modern structural interpretation of the Precordillera. More recent interpretations of the regional geology, the geotectonic setting, and the stratigraphy are those of Baldis et al. (1982, 1984a), Cuerda et al. (1985a), Ramos et al. (1986), Cingolani et al. (1987, 1989), Ramos (1988a, 1988b), Astini (1991), Aceñolaza (1992), Loske (1992), Bordonaro (1992), Astini et al. (1995, 1996b), and Keller et al. (1998).

The stratigraphic panel given in Figure 4 is a compilation of a multitude of individual data sets and papers. As paleontologic and sedimentologic research in the pre-Carboniferous rocks continues, there will certainly be revisions in the future.

Cambrian. Cambrian rocks are present in the Sierras de Villicum and Zonda in the San Juan area and from Jáchal toward Guandacol in the northern part of the Precordillera (Figs. 3 and 4). In addition, there is an isolated occurrence of carbonate platform rocks at Cerro Pelado (west of Mendoza; Fig. 3), which was first described by Varela (1973). The presence of Cambrian rocks in the Sierra del Tontal is well established (e.g., Cuerda et al., 1985b; Bordonaro and Baldis, 1987; Banchig and Bordonaro, 1990). All of these rocks, which are of Middle and Late Cambrian age (Bordonaro and Banchig, 1996), are of allochthonous origin and are embedded in Ordovician basinal deposits as olistoliths. Similarly, the Cambrian of San Isidro (Borrello, 1971; Bordonaro, 1985; Pina et al., 1986) was shown to have been resedimented into Ordovician deposits (Bordonaro, 1992; Bordonaro et al., 1993).

In the Precordillera, the Cambrian was a major period of carbonate deposition (e.g., Baldis and Bordonaro, 1981a, 1985). However, a siliciclastic-evaporite succession (Cerro Totora Formation; Fig. 4) of possibly middle to late Early Cambrian age has been described from the Guandacol area (Fig. 3; Astini and Vaccari, 1996). These deposits together with siltstone intercalations in the Middle Cambrian La Laja Formation (Bordonaro, 1980; Bercowski et al., 1990) constitute the only noncarbonate rocks of Cambrian age in the Precordillera.

Although in the need of serious modernization and revision, many of the elements of the Cambrian fauna were described by Rusconi in a series of papers between 1948 and 1962, mainly published in the Revista del Museo de Historia Natural de Mendoza (Volumes 2–14) and various issues of the Boletín Paleontológico de Buenos Aires (Volumes 24–32). More recent contributions to the Cambrian fauna include Bordonaro (1980, 1986, 1989, 1990a, 1990b), Bordonaro and Banchig (1990, 1995, 1996), Borrello (1962, 1971), Poulsen (1960), Shergold et al. (1995), Tortello and Bordonaro (1997), Vaccari (1994), and Vaccari and Bordonaro (1993).

Ordovician. Ordovician strata are the most widespread deposits in the Argentine Precordillera and record an eventful history (platform drowning, rifting, evolution of the western basin, glaciation), especially during Middle and Late Ordovician time (Fig. 4). Carbonate deposition persisted well into the Middle Ordovician; however, the laterally extensive carbonate platform was drowned during the earliest Llanvirn.

Tectonic movements during Middle and Late Ordovician time were attributed to the Guandacol and the Villicum orogenic phases (Baldis et al., 1982; Furque and Cuerda, 1982) and were deduced mainly from the presence of conglomerate units within the Ordovician succession. Tectonic events are also responsible for the differing depositional histories in the former carbonate platform area and in the area to the west (eastern and western tectofacies of Astini, 1992). The contact between both areas is marked by a transition zone in which fine-grained sediments host giant olistoliths and mass-flow deposits (Los Sombreros Formation; Cuerda et al., 1983).

In the former platform area near Guandacol (Fig. 3), conglomerates of the Las Vacas Formation (Fig. 4) overlie black shales, which in this area stopped carbonate deposition at the base of the late Arenig (Hünicken and Sarmiento, 1982, 1985; Ortega et al., 1985). Clastic sedimentation continued well into the Late Ordovician (Las Plantas and Trapiche Formations). South of Jáchal, carbonate sedimentation locally continued into the *Nemagraptus gracilis* zone (Los Azules and Las Aguaditas Formations: Baldis et al., 1982; Cuerda and Furque, 1986; Keller et al., 1993b). In the Sierra de Villicum (Fig. 3), a thick succession of Middle and Upper Ordovician conglomerates, sandstones, and diamictites (see Appendix 1) is preserved (La Cantera and Don Braulio Formations; Fig. 4). Similar deposits are rare in many other areas of the former platform area, because at least two different erosional events affected the pre-Silurian strata. The Ashgillian diamictites of the Don Braulio Formation are interpreted to be of glaciomarine origin (Peralta and Carter, 1990a) related to the Upper Ordovician Gondwana glaciation.

Autochthonous sedimentary rocks older than Middle Ordovician are not present in the western basin. Most units yield Llanvirnian or Llandeilian graptolites (e.g., Cingolani et al., 1987, 1989; Cuerda et al., 1987; Fig. 4). Caradocian sediments include the Alcaparossa Formation, which forms the matrix for abundant basic dikes and pillow basalts (Haller and Ramos, 1984). One important problem in the interpretation of the sedimentary succession of the western clastic basin is the lack of stratigraphic contacts between the formations. Thus, only a few units have been studied in detail (Spalletti et al., 1989).

The difficulties of correlation within the western basin are exemplified by units like the Yerba Loca and Sierra de la Invernada Formations, which crop out in the sierras of those names (Fig. 4). They are of similar lithology and contain a variety of mass-flow deposits. Graptolite ages obtained from various sections vary between early Llanvirn and late Caradoc time (Brussa, 1994, Fig. 22; Albanesi et al., 1995), which at least in places is due to determination of fossils contained in olistoliths.

At San Isidro, near Mendoza (Fig. 3), a structurally complex succession of lower Paleozoic rocks is present. The Empozada Formation of Middle and Late Ordovician age is sedimentologically similar to the Los Sombreros Formation (Gallardo et al., 1988; Bordonaro et al., 1993). However, its relations to the adjacent sediments of the Villavicencio Group are not well established, mainly because the Villavicencio Group contains Ordovician and Devonian rocks and possibly some Silurian sedimentary rocks (Cuerda et al., 1985a, 1987).

There is a wealth of paleontologic and biostratigraphic information on the Ordovician strata of the Argentine Precordillera that cannot be listed in its entirety. Hence only some relevant citations for each group are given as follow: (1) brachiopods: Benedetto (1986, 1987, 1990), Benedetto and Herrera (1986), Herrera and Benedetto (1989, 1991), Levy and Nullo (1974); (2) conodonts: Hünicken (1971, 1985), Lehnert (1990, 1993, 1994, 1995a), Lehnert et al. (1997, 1998), Sarmiento (1986, 1990), Sarmiento and Garcia Lopez (1993), Sarmiento and Rábano (1992); Sarmiento et al. (1986), Serpagli (1974); (3) graptolites: Alfaro (1988), Blasco and Ramos (1976), Brussa (1994, 1995), Cuerda et al. (1983, 1985a, 1986, 1987), Cuerda and Furque (1986), Peralta (1986, 1993), Varela et al. (1982); (4) trilobites: Baldis and Blasco (1975), Baldis and Pöthe de Baldis (1995), Harrington and Leanza (1957), Vaccari (1994, 1995); (5) miscellaneous: Aceñolaza et al. (1976), Albanesi et al. (1995), Carrera (1994), Hünicken and Ortega (1987), Hünicken et al. (1990), Keller and Flügel (1996), Kobayashi (1937), Lehnert and Keller (1993b), Ortega (1987).

Silurian. Silurian deposits are mainly restricted to the former platform area. Two contrasting successions are present. At the eastern boundary of the Sierras Chica de Zonda and Villicum (Fig. 3), the Rinconada and the Mogotes Negros Formations (Fig. 4) are thick siliciclastic units, in which olistoliths are common (Amos, 1954; Cuerda, 1981). In the Rinconada section, huge olistoliths of the San Juan Formation are present, whereas in the Don Braulio section olistoliths of conglomerates are most abundant. These deposits are part of the eastern facies of Cuerda (1985). The other succession, which is much more widespread, consists of varying amounts of sandstone, siltstone, and shale. According to their sedimentological and ichnological inventory (Aceñolaza and Peralta, 1985a, 1985b; Peralta and Carter, 1990b), these rocks were deposited on a stable platform. The basal Silurian is characterized by light colored sandstones (La Chilca Formation), that unconformably overlie Ordovician strata. Lateral facies changes in the Silurian deposits are documented by different stratigraphic names (Fig. 4), i.e., the Tambolar Formation of Heim (1952) in the Río San Juan valley versus La Chilca and Los Espejos Formations (Cuerda, 1969) farther north. Both successions form the central facies, according to Peralta (1986).

Phyllites near Calingasta (Harrington, 1957) have a doubtful Ludlovian age (Xicoy, 1963, cited in Peralta, 1990). These deposits, together with the La Tina Formation (Quartino et al., 1971) represent the western facies (Peralta, 1986) characterized by predominantly fine-grained sediments.

Graptolites and brachiopods are the most important biostratigraphic tools for the Silurian. The former were described by Cuerda (1965, 1969, 1971, 1981, 1985) and Peralta (1984, 1986, 1990), the latter by Baldis (1964), Benedetto (1995), Benedetto et al. (1986, 1992), and Herrera (1985, 1993). In addition, isolated occurrences of conodonts have been described (Hünicken, 1975; Hünicken and Sarmiento, 1988).

Devonian. In the central facies area of the Silurian strata, sedimentation continued into the Devonian without a major depositional break (Fig. 4); however, the deposits become successively coarser grained. The lower unit (Talacasto Formation) still maintains a platform character (Padula et al., 1967; Baldis, 1975), whereas the overlying Punta Negra Formation is turbidite dominated (Gonzalez Bonorino, 1975a), and the entire succession has a flysch-like appearance (Gonzalez Bonorino, 1975a, 1975b). A deep-water depositional environment is also indicated by the ichnofaunas (Peralta, 1986; Peralta and Aceñolaza, 1989). Parts of the Villavicencio Group at San Isidro (Fig. 3), the Canota Formation of Cuerda et al. (1988a), are lithologically similar to the Punta Negra Formation; however, there remain many uncertainties with respect to age and depositional environment of the former (discussion in Kury, 1993).

In the western basin, there are several units that are assigned a Devonian age; however, in many cases there is no agreement about lateral extent, age, and names of the units (Quartino et al., 1971; Sessarego, 1983; Selles Martinez, 1986).

Major paleontologic contributions to the knowledge of the Devonian include those of Baldis (1967, 1975), Baldis and Longobuco (1977), Herrera (1993), Kerlleñevich (1967), and Padula et al. (1967).

Post-Devonian. A large hiatus, in places with an angular unconformity, separates Carboniferous strata from older rocks. This hiatus comprises the majority of the Late Devonian, the Early Carboniferous, and most of the Namurian strata. Carboniferous strata are known not only from the Precordillera but from all surrounding areas. Hence, these deposits are the Paleozoic overstep sequence, indicating that by no later than mid-Carboniferous time accretion of the terranes to Gondwana was completed.

Permian-Triassic magmatic rocks of the Choiyoi Group and Triassic sedimentary rocks are restricted to the western and eastern borders of the Precordillera. Their distribution and interpretation were discussed by Baraldo et al. (1990). During Tertiary and Quaternary time, thick continental successions were deposited in longitudinal valleys (Fig. 2) and along the eastern border of the Precordillera, where they represent the filling of the Andean foredeep (Beer and Jordan, 1989; Beer, 1990). Andean magmatic rocks are present within the Precordillera (Leveratto, 1968, 1976; Bercowski and Figueroa, 1987) and, although not widely distributed, bear important clues to the history of the Precordillera (Abbruzzi et al., 1993).

Structural History of the Argentine Precordillera

Today, the Precordillera represents a thin-skinned, high-level thrust-and-fold belt (Heim, 1952; Baldis and Chebli, 1969; Von Gosen, 1992) in which the entire sedimentary column of Paleozoic and Tertiary rocks is affected by deformation. This Andean crustal shortening (see Appendix 1) was triggered by flat and shallow subduction of a part of the Nazca plate beneath South America (Jordan et al., 1983). Crustal shortening as a result of thrusting and resulting east-directed imbrications is estimated to be on the order of 50% and more (Allmendinger et al., 1990; Von Gosen, 1992). The timing of the various deformation events and their bearing on sedimentary patterns in the foreland basin(s) was documented by Beer and Jordan (1989, and references therein).

There is no record of a thermal overprint, that might have accompanied Andean deformation. However, pre-Carboniferous sedimentary rocks show an increasingly higher thermal alteration from east to west (Keller et al. 1993c), where rocks with a greenschist facies overprint are widespread. Metamorphism and the corresponding deformation are of Late Silurian–Devonian age (Paleozoic [see Appendix 1] deformation; Von Gosen, 1992, 1997; Buggisch et al., 1994b). In addition, near Uspallata at the southwestern end of the Precordillera (Fig. 2), pre-Permian-Triassic compression is observed (Von Gosen, 1995).

Subdivision of the Argentine Precordillera

The early Paleozoic record of the Precordillera displays a complex interplay of facies deposited in various basins and subbasins. Many names have been used to characterize the different sedimentary successions. Astini (1992) subdivided the Ordovician deposits into an eastern and a western tectofacies, and only the central and western facies are still recognized from the Devonian (Peralta and Baldis, 1990).

Considering the entire Phanerozoic history, the recognition of areas with distinct sedimentary successions accompanied by unique structural styles led to a morphostructural subdivision of the Argentine Precordillera (Baldis and Chebli, 1969; Ortiz and Zambrano, 1981) into eastern, central, and western parts (see also Baldis et al., 1982). As pointed out by Von Gosen (1992), some of the structural arguments for such a subdivision are doubtful. The outlines of these morphostructural provinces partly coincide with the limits of the sedimentary basins and subbasins of the Precordillera.

In the pre-Carboniferous sedimentary history of the Precordillera, it can be demonstrated that there are a number of areas that throughout the early Paleozoic had their own distinct tectono-sedimentary history. These areas are not tectonostratigraphic terranes in the sense of Howell et al. (1985), which would facilitate their discussion as discrete areas. However, in a purely descriptive sense they are sedimentary basins and/or subbasins (Bally and Snelson, 1980; Allen and Allen, 1990) and this terminology will be applied in this paper.

For the purpose of this paper the early Paleozoic includes all Cambrian through Devonian strata below the Upper Carboniferous overstep succession; two main basins are recognized, an eastern basin, characterized by the presence of a thick Cambrian-Ordovician carbonate platform succession, and a western basin filled with Ordovician through Devonian siliciclastic deposits.

Western basin. The western basin (Fig. 3) extends from Cerro Cacheuta (Fig. 2), which forms the southernmost exposed part of the Precordillera southwest of Mendoza, to the northernmost outcrops of the Precordillera northwest of Jagüe (Fig. 2). To the west, it is separated from the Cordillera Frontal by the longitudinal valley of Uspallata-Calingasta-Iglesias (Fig. 2). The eastern boundary of the western basin is at the boundary of the carbonate platform and is marked by a fault along which the Los Sombreros Formation was deposited.

The only uncertainty with respect to the boundary between the western and the eastern basin concerns the area of Jagüe. There, carbonate rocks similar to those of the San Juan Formation are present (Aceñolaza, 1969); however, nothing is known about a possible presence of continental-margin facies or about the nature of the siliciclastic Ordovician rocks there.

Eastern basin. The eastern basin (Fig. 3) comprises the morphostructural provinces of the Precordillera Oriental (Ortiz and Zambrano, 1981) and the Precordillera Central (Baldis and Chebli, 1969; Baldis et al., 1982). To the east and north, the eastern basin is bounded by the longitudinal valleys that separate it from the western Sierras Pampeanas (Fig. 2). Its western limit coincides with the westernmost outcrops of carbonate platform rocks. To the south, the eastern basin disappears beneath the northern Mendoza piedmont.

Three subbasins are recognized within the eastern basin: the San Juan, the Talacasto, and the Guandacol subbasins.

San Juan subbasin. The pre-Carboniferous rocks of the Precordillera Oriental, which are exposed only within the easternmost thrust sheets around the city of San Juan, constitute the San Juan subbasin (Fig. 3). Today, these pre-Carboniferous rocks are exposed in the Sierra de Villicum, Sierra Chica de Zonda, and Cerro Pedernal de los Berros. The San Juan subbasin is characterized by the presence of isolated outcrops of Lower Cambrian limestones, a continuous succession of Middle Cambrian through lower Llanvirnian carbonate platform rocks, and Middle and Upper Ordovician siliciclastic deposits with several stratigraphic gaps of varying magnitude. The Silurian is represented by thick sedimentary melange deposits with olistoliths and olistostromes, whereas Devonian rocks are conspicuously absent. From a tectonic point of view, there are two main features in the San Juan subbasin: (1) the main thrusts are located in the lower–Middle Cambrian limestones, and (2) the entire succession was affected by a pre-Carboniferous deformation, in places demarcated by Carboniferous deposits overlaying almost vertical Cambrian dolomites. The San Juan subbasin includes the Villicum graben of Loske (1992) and the eastern facies of Cuerda (1985) and Peralta (1986). The Ordovician rocks of the San Juan subbasin are part of the eastern tectofacies of Astini (1992).

Talacasto subbasin. The Talacasto subbasin (Fig. 3) is laterally much more extensive than the San Juan subbasin and has representatives from the Río Jáchal in the north to Cerro de la Cal near Mendoza in the south (Fig. 4). The Talacasto subbasin has a sedimentary record that starts with Upper Cambrian through lower Llanvirnian carbonate platform rocks. The drowning of the platform and/or the transition into the overlying black shales, where preserved, is everywhere dated as early Llanvirn. Locally (Las Chacritas and Las Aguaditas sections; Figs. 3 and 4) carbonate production continued, but a dramatic change to deep-water environments is evident. In many places, different episodes of pre-Silurian erosion caused stratigraphic gaps between deeply eroded carbonate platform rocks, Middle and Upper Ordovician siliciclastic deposits, and the overlying Silurian succession. A chert-pebble conglomerate at the base of the overlying succession marks the unconformity. The Silurian shows the return to deposition of (siliciclastic) platform sediments, but during the Devonian, a shift toward flysch-like deposition is observed. In most sections, Carboniferous strata overlie the Devonian with a paraconformable contact; however, in the Agua Hedionda section (Fig. 3), they overlie the San Juan Formation (see Buggisch et al., 1994a).

There are two main tectonic features in the Talacasto subbasin: (1) the main thrusts are located at varying levels within the Upper Cambrian and Lower Ordovician limestones, and (2) the succession was only moderately affected by pre-Carboniferous tectonics.

The northern limit of the Talacasto subbasin is arbitrarily drawn along an almost east-west–trending transect connecting the Los Tuneles section (which belongs to the western basin) with the Agua Hedionda section. To the west and southwest, it is bounded by continental margin deposits, which, during the Ordovician, outlined the transition from the platform to the basin. The Talacasto subbasin comprises most of the Precordillera Central sensu Baldis and Chebli (1969). The Ordovician rocks constitute a part of the eastern tectofacies of Astini (1992), whereas the Silurian rocks represent the facies central of Peralta (1986).

Guandacol subbasin. The Guandacol subbasin (Fig. 3) is a continuation of the Talacasto subbasin; however, there are marked differences in stratigraphy. It has an almost complete record of Cambrian and Lower Ordovician carbonate platform rocks, and some redbeds and evaporites at the base of the succession. However, during the Arenig, the carbonate platform was drowned. The most characteristic feature of the Guandacol subbasin is a very thick succession of Middle and Upper Ordovician siliciclastic deposits, which are overlain by Carboniferous rocks. Silurian and Devonian rocks are absent in this subbasin (Fig. 4).

The Guandacol subbasin is part of the Precordillera Central and its Ordovician rocks are part of Astini's (1992) eastern tectofacies.

CARBONATE PLATFORM ROCKS OF THE ARGENTINE PRECORDILLERA

La Laja Formation and Equivalents

The La Laja Formation (Borrello, 1962; Fig. 4) is the oldest unit of the carbonate platform succession. Outcrops of the La

Laja Formation are restricted to the San Juan subbasin (Fig. 3), where they occur in a single belt parallel to paleodepositional strike. In addition to the six sections measured in the La Laja Formation, several outcrops have been visited, especially those exposing the El Estero Member. In many places, major thrust planes developed in the lowermost exposed beds of the La Laja Formation during Andean orogeny (see Appendix 1). Deformation caused intense folding and faulting of the thinly bedded rocks, and hence prevented a detailed measurement and interpretation of this part of the section.

The La Laja Formation is a mixed carbonate-siliciclastic system, subdivided into four members (Bordonaro, 1986; Fig. 7). The oldest rocks within the La Laja Formation are those of the El Estero Member, which previously was believed to crop out only at isolated places along a major thrust. These rocks were dated as Early Cambrian by Borrello (1962, 1971). Our field work, however, has shown that there are several sections in which the lowermost beds exposed have characteristics of the El Estero Member. No biostratigraphic data are available from these newly discovered horizons. Hence the attribution of these rocks to the El Estero Member is entirely based on sedimentology. The Soldano, Rivadavia, and Juan Pobre Members form the remainder of the formation and were assigned an early Middle Cambrian through latest Middle Cambrian age (Baldis and Bordonaro, 1981a; Bordonaro, 1986). A reevaluation of the trilobites, however, seems to indicate that the trilobites of the Soldano Member belong to the *Glossopleura* zone (A. R. Palmer, 1997, personal commun.; Fig. 7). Trilobites of the *Plagiura-Polliela* and the *Albertella* zones are apparently absent. The youngest trilobites of the Juan Pobre Member belong to the *Bolaspidella* zone; consequently, the three members were deposited during the late Delamaran and early Marjuman of the North American subdivision of the Cambrian (Palmer, 1998).

Coeval, but as yet unnamed, sedimentary rocks in the Guandacol subbasin (Cañas 1988; Figs. 4 and 7) consist of peritidal carbonates with well-developed cyclicity. About 35 m of this succession in stratigraphic contact with dated Lower Cambrian rocks have been described. Within the carbonate succession, two intercalations of fine-grained siliciclastic strata, mainly siltstones, are observed. The entire succession apparently is much thicker (Astini and Vaccari, 1996) than the fault-bounded 35 m described by Cañas (1988).

The Cerro Pelado Formation, formerly assigned to the Late Cambrian (Heredia, 1990), has recently shown to be of late Middle Cambrian age (Holmer et al., 1999). Hence it is coeval with the Juan Pobre Member. Lithologically, it is similar to the rocks exposed near Guandacol with well-developed cyclicity and abundant thrombolites.

Lithofacies. Among the siliciclastic sediments there are variegated shales and siltstones with a clayey or marly matrix. Marlstones are present only in a few places. Some of the shales and siltstones are bioturbated; horizontal trace fossils are the most common. In others, trilobite fragments, eocrinoids, and hyolithids are found (6 in Fig. 8). Ripple-drift cross-lamination and linsen and flaser bedding are also observed. Glauconite is locally abundant in these rocks.

Fine- to coarse-grained quartz sandstones and sandstones containing bioclasts and wave and current ripples are also present. In places, impure sandstones containing feldspar and dolomite clasts have a calcareous matrix; in others carbonate cement is observed. Locally, these impure sandstones laterally grade into sandy limestones. Sorting varies from moderately sorted in the sandstones to well sorted in the quartz sandstones.

Among the carbonates, thick-bedded dark gray to black lime mudstones are abundant. They are rich in organic matter and some of them are moderately to highly biologically mottled. The burrows are easily recognized because they have a filling of brownish dolomicrosparite. Trilobites and hyolithid fragments are the dominant faunal elements (1 in Fig. 8).

Fossiliferous wackestones show well-preserved trilobites, brachiopods, eocrinoids, *Hyolithes*, and sponge spicules. In other wackestones, the fossils are broken or abraded. Bioturbation is locally present, but is not pervasive and may be one cause for the destruction of the fossils.

Oolitic and oncolitic wackestones are also observed in the La Laja Formation. Some of them contain scattered fragments of trilobites and other organisms such as brachiopods or hyolithids (3 in Fig. 8).

Fossiliferous packstones (5 in Fig. 8) are composed of trilobite hash and additional eocrinoids, brachiopods, and *Hyolithes*. Locally, hyolithid packstones are developed (2 in Fig. 8). In some samples, *Girvanella* clasts have been observed. Other components are ooids, oncoids, intraclasts, and detrital quartz. In addition, some of the rocks contain glauconite. The rocks are gray to dark gray and thin to medium bedded.

Oolitic or oncolitic packstones (4 in Fig. 8) in places are cross-bedded or fill erosional channels. They commonly show additional bioclasts. Oncolitic packstones are composed of oncoids with a diameter of only a few millimeters to 2 cm. The main component of the oncoids is *Girvanella*, which is commonly well preserved. In most of the oncoids there is no visible nucleus; however, trilobites or detrital quartz served as a primary substrate for algae. Some oncoids are also found to be completely micritized.

Two varieties of intraclast-bearing packstones have been observed. In many of them the intraclasts consist of broken burrow or worm tubes with micritic and calcified walls. In the others, the intraclasts are white to light gray dolomitic mud chips. The intraclast packstones are present as laterally continuous beds with scours and grading, and as lenses or channel fills to 1 m across. Bed thickness varies from thin to thick.

Pure grainstones are rare in the La Laja Formation and most are oolitic. The oolite grainstones occur as thick-bedded units that often have well-developed herringbone cross-stratification. In many samples, the ooids are replaced almost entirely by large calcite crystals; however, the outer layers are preserved and in a few examples show the typical concentrical arrangement of the individual layers. Oolites containing lithoclasts show a variety of

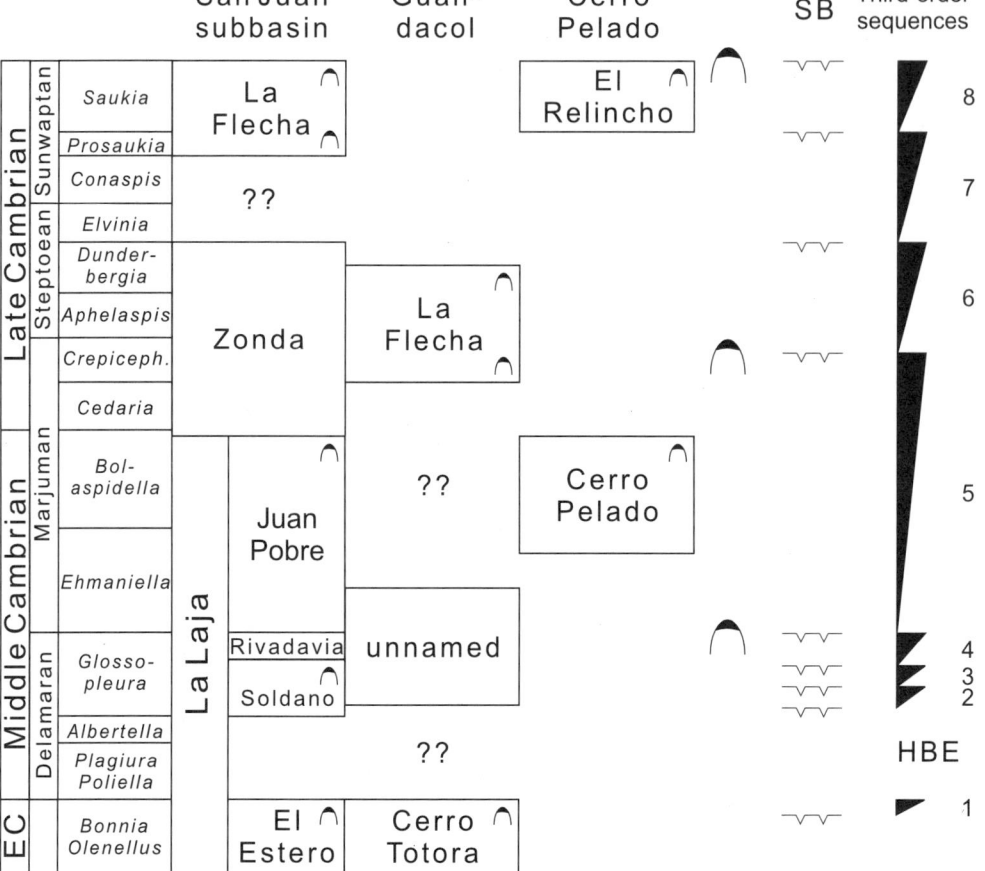

Figure 7. Stratigraphy of Cambrian carbonate platform deposits in Precordillera. Stages are those of Laurentia following Palmer (1998). EC = Early Cambrian, SB = sequence boundary, HBE = Hawke Bay regression event. Trilobite symbols mark fossil control. See text for details.

microfacies within the clasts, mainly from underlying rock units (intraclasts), but detrital quartz and dolomite chips (extraclasts) have also been observed.

Bioclastic grainstones are composed of either eocrinoids or trilobite hash with minor amounts of other bioclasts. Components of mixed grainstones are bioclasts, mainly trilobites and eocrinoids, intraclasts, ooids, and oncoids. Many of the grainstones are cross-bedded or channelled.

Peloidal, strongly biologically mottled grainstones are characterized by low fossil content, mainly trilobites. The dark gray to black color of the lime mud is contrasted by the yellow to orange-brown filling of the burrows. This filling is mainly a dolomitic microspar, but silt has also been observed. Where bioturbation is absent, a very thin internal stratification and ripple-drift cross-lamination are preserved. Peloids are commonly so densely packed that the rock appears to be a lime mudstone.

Facies associations and depositional environment. Lime mudstones and whole-fossil wackestones are characteristic of the lime mudstone–wackestone–intraclast packstone association. The sediments were deposited under marine conditions below fair-weather wave base, where thick successions of carbonate mud with well-preserved fossils accumulated. As pointed out by Osleger and Montañez (1996), these deeper or quieter water conditions do not necessarily imply truly deep water. Instead, those sediments may also form in a deep shelf lagoon that is protected from the open ocean by a well-developed shelf-margin rim (Fig. 9). The effects of storms are represented by intraclast packstones. The intraclasts are mainly of the reworked-burrow variety. Most of these tempestites are medial to distal deposits (Aigner, 1982, 1985), because amalgamation is absent, scouring and other erosional features are scarce, and the beds are mainly thin intercalations (5–40 cm) within the lime mudstones and wackestones.

A variation of this assemblage of rocks is the wackestone-oncolite association, present only in the La Laja section. There, beds of cross-bedded oncolites as thick as 50 cm are present within lime mudstones and wackestones. In places, erosional features are found at the base of the oncolites. The oncolites are also interpreted as tempestites; their source area is not exposed, but may have been a nearby oncolite shoal system.

In the peloidal grainstone-mudstone association most of the lime mudstones are strongly biologically mottled. In some of the grainstones, sedimentary structures (lamination, ripples) are visible. Similar rocks have been described from both ancient (Pratt and James, 1986; Osleger and Montañez, 1996) and Holocene (Garrett, 1977) environments. Pratt and James (1986) interpreted them as lower intertidal deposits; however, distinct indicators of an intertidal setting (mudcracks, birdseye structures; Shinn, 1983b; Goldhammer et al., 1993) are absent. A shallow subtidal,

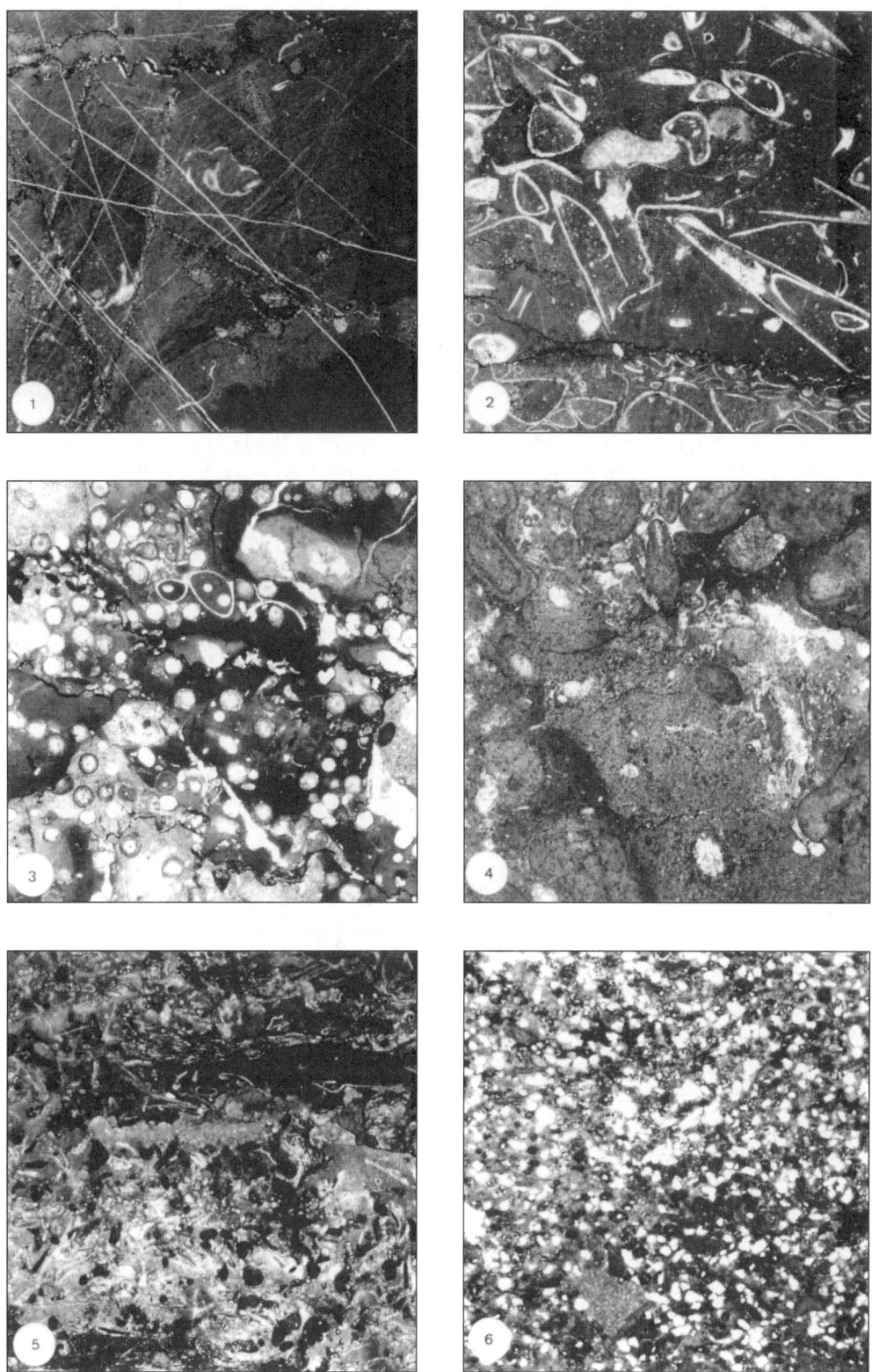

Figure 8. Sedimentary rocks of La Laja Formation. 1: Dark trilobite mudstone. Mudstone-wackestone-intraclast packstone association. 2: Packstone composed of *Hyolithes* sp. with few trilobite fragments. Mudstone-wackestone-intraclast packstone association. Magnification is 4.5. **3:** Oolitic wackestone-packstone with *Hyolithes* sp. Only outer layers of ooids are preserved. Sediment is of tempestitic origin. Mudstone-wackestone-intraclast packstone association. Magnification is 4.5. 4: Oncolitic packstone of wackestone-oncolite association. Magnification is 4.5. 5: Packstone-grainstone composed of trilobites, hyolithids, and phosphatic brachiopods. In addition, silt-sized quartz and ore particles are present. Packstone-grainstone association. Magnification is 4.5. 6: Siltstone with few bioclasts. This is typical sediment of the siltstone intervals. Magnification is 4.5.

18   M. Keller

somewhat restricted depositional setting is deduced from the intense bioturbation, the lack of an open-marine fauna, and the lack of high-energy sedimentary structures. In addition, the lamination points to episodic rather than continuous sedimentation. Hence this association may have formed landward of large sand shoals that were able to protect the platform interior from water exchange and to dampen the effects of waves and tidal currents.

The oolite-oncolite association represents winnowed sand bars, barriers, and subtidal shelf sands. It shows abundant cross-bedding or is present as channel fills. This association was deposited in the shallow subtidal to lowermost intertidal area, where wave and current activity removes most of the lime mud. Tidal influence is clearly visible in the oolites with herringbone cross-stratification. In modern environments, such shoals commonly form near the platform margins in the highest energy zone (e.g., Hine, 1977; Harris, 1979). Unlike the ooids, however, the

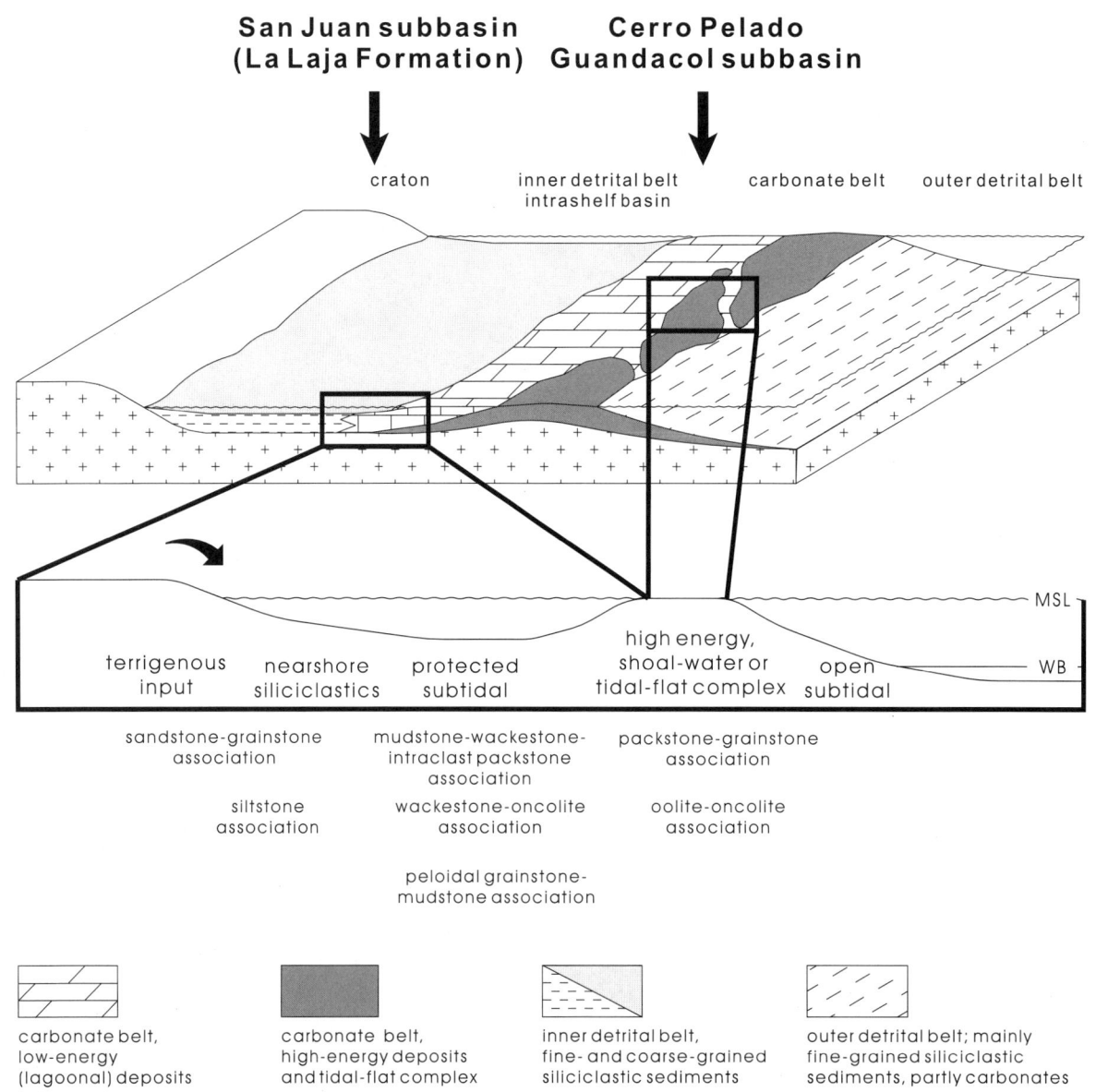

Figure 9. Depositional environments and paleoenvironmental interpretation of La Laja Formation and of coeval rocks in Guandacol subbasin and at Cerro Pelado. Facies belts according to Palmer (1971a, 1971b). MSL = mean sea level, WB = wave base.

primary site of oncoid formation was not the site of later deposition. From the Jurassic to the Holocene, oncoids normally form in slightly hypersaline intertidal ponds or tidal channels (Flügel, 1982), whereas in older sediments oncoids seem to have formed in muddy environments with reduced rates of sedimentation (Peryt, 1981; Bowman, 1983). The latter interpretation implies that the accumulation of oncolite shoals like those in the La Laja Formation was the result of high-energy events like major storms.

An environment similar to the oolite-oncolite association is represented by the packstone-grainstone association, which is best explained as submarine sand belts. These formed in response to the frequent reworking of shallow-marine habitats by storm and wave activity. The presence of mud in the sediments indicates that no constant winnowing took place. Destruction of nearby shoals led to mixing of ooids and oncoids with skeletal debris (Tucker and Wright, 1990) to give a variety of packstones and grainstones. The diverse fauna points to normal marine conditions in a shallow-platform environment and, consequently, deposition of the packstones and grainstones took place on the offshore side of the shoal systems (Fig. 9).

The siltstone association, which is present as discrete intervals within the La Laja Formation, is recognized by its brown to yellow color. Siltstones are the dominant lithotype, but there are also marly siltstones, marlstones, and thin beds of eocrinoidal grainstone. In the La Flecha section, oolitic grainstones are present in discrete beds and as channel fills. Linsen and flaser structures, ripples, and laminations are observed in the rocks. The combination of sedimentary structures and the absence of indicators of intermittent exposure point to a tidally influenced (lower intertidal? to) shallow subtidal depositional environment (Reineck and Singh, 1980).

The sandstone-grainstone association is restricted to the El Estero Member of the La Laja Formation. It is composed of a variety of thin-bedded siliciclastic sandstones, siltstones, and lime grainstones. Shales are uncommon. The grainstones are almost exclusively trilobite coquinas with large (>10 cm) olenellid fragments. The association was deposited under marine high-energy conditions and the intermittent influx of terrigenous detritus. Abundant wave ripples and the absence of intertidal features point to a shallow subtidal setting. The quartz sands are commonly coarse grained and indicate that the source area was not very distant.

Shales, wackestones, and packstones, together with few siliciclastic siltstones and sandstones, are part of the shale-packstone association. The depositional environment was the shelf, but was probably slightly deeper or more protected than during formation of the sandstone-grainstone association.

There is a succession of white quartzitic sandstones, several meters thick, associated with black shales. This association has been observed only in three localities, and everywhere the rocks are tectonically strongly deformed. Hence an environmental discussion or their interpretation in the context of the sedimentary succession of the La Laja Formation can only be speculation. However, the section in the northern part of the Sierra Chica de Zonda still reveals some information. About 15 m of black shales are overlain by quartzitic sandstones (Pereyra, 1986) that exhibit a faint cross-bedding. The important information included in this succession is that from its position within the section and within the resolution of biostratigraphy, it is coeval with the major sand interval at the top of the El Estero Member.

Platform configuration. The facies associations of the La Laja Formation can be grouped into three depositional systems (Fig. 9): a carbonate shoal complex, a subtidal, open marine environment that includes both back-barrier lagoonal deposits and open platform limestones, and a shallow subtidal sand- and silt-dominated shelf. Near Guandacol (Cañas, 1988) and at Cerro Pelado (Heredia, 1990), remnants of a peritidal carbonate system are preserved.

Around the Cambrian-Ordovician Laurentian margins, the lateral relationships of the siliciclastic and carbonate lithosomes were the basis for the concept of an inner detrital belt, a carbonate belt, and an outer detrital belt (Palmer, 1971a, 1971b; Fig. 9). Coarse terrigenous material is trapped in nearshore areas (inner detrital belt), the carbonate factory is located close to the platform margin (carbonate belt), and the outer detrital belt is composed of shales, deep-water carbonates, and mass-flow deposits.

Distinct features of intertidal and supratidal environments (e.g., Goldhammer et al., 1993) are absent in the La Laja Formation (Bercowski et al., 1990). The presence of nearby tidal flats, however, is demonstrated by the small dolomitic mudchips in some of the intraclast conglomerates. These peritidal facies are exposed near Guandacol and at Cerro Pelado. These two outcrop areas indicate that a continuous carbonate belt might have been located farther west. Sand and silt within the La Laja Formation have been interpreted as being shore derived, and there is little doubt that the siltstone intervals represent tongues of the inner detrital belt. Consequently, the La Laja Formation was deposited in a position where both lithosomes interfingered; landward of the carbonate factory, but seaward of the main siliciclastic trap.

Sedimentary succession of the La Laja Formation. The La Laja Formation consists of several large-scale sedimentary successions which have two main components, a siliciclastic interval and a calcareous interval (Fig. 10; see oversized plate at the back of the volume). The siliciclastic intervals can be used as marker horizons and were previously used to define the four members of the La Laja Formation (Bordonaro, 1980). The calcareous intervals consist of one or more shallowing-upward successions, each of them with a thickness between 10 m and more than 100 m. These successions commonly begin with sediments of the mudstone-wackestone-intraclast grainstone association, which represents the deepest environments present in the La Laja Formation. Many of them end with rocks of the oolite-oncolite association.

The El Estero Member consists of sediments of the oolite-oncolite association which upsection pass into sandstones, calcareous sandstones, and lime grainstones (Fig. 10). Locally, white quartz arenites and black shales are developed near the top of the member.

The Soldano Member contains several calcareous shallowing-upward successions and is characterized by a prominent siltstone interval that marks the top of the member. This siltstone interval is present everywhere in the Sierra Chica de Zonda, but is tectonically absent in the section measured in the Sierra de Villicum (Fig. 10). In the La Laja sections, an additional siltstone interval is present within the Soldano Member, which has not been found elsewhere.

The Rivadavia Member is a succession of limestones that consists of two and locally of three (Zonda and La Laja sections; Fig. 10) calcareous shallowing-upward successions. These successions characteristically contain a large amount of dark mudstones and peloidal grainstones, so the member is easily recognized in the field by its black color.

The base of the Juan Pobre Member coincides with the base of the uppermost siltstone interval of the La Laja Formation. Upsection, lime mudstones and wackestones grade into packstones and grainstones. The top of the member consists of oolite shoals in which well-developed herringbone cross-stratification is present.

Evolution and sequence stratigraphy of the La Laja Formation. Considerations about sequence stratigraphy are hampered by the few outcrops, which do not allow a three-dimensional view of the rocks of the La Laja Formation or the reconstruction of the cross-platform architecture. In addition, it is not yet possible to correlate outcrops of the La Laja Formation with those near Guandacol and at Cerro Pelado that might have been close to the platform margin.

The calcareous shallowing-upward successions originated on surfaces across which there are indications of an abrupt deepening (Fig. 11). These surfaces are located either on top of the preceding calcareous shallowing-upward succession, or on top of a siltstone interval. The internal architecture of the shallowing-upward successions demonstrates a dominantly aggradational pattern. Small-scale (1–5 m) cycles within the successions are conspicuously missing.

In the siltstone intervals, abundant glauconite indicates that they are stratigraphically condensed successions characterized by starved sedimentation. Trilobites and bioclastic grainstones point to a marine depositional environment. The grain size (sand to silt) of the siliciclastic intervals leaves little doubt that they represent tongues of shoreward-derived terrigenous detritus (Fig. 9). The intervals have sharp upper boundaries. Most of them also have sharp lower boundaries, but more gradational transitions from the underlying oolites have also been observed. The lower bounding surface of the siltstone intervals does not mark a major environmental shift.

Provenance of the silt in the siltstone units. Terrigenous material was provided by a barren craton and either intermittently or continuously delivered to the shallow shelf. The origin of abundant silt in Middle Cambrian strata of the southern Great Basin was discussed in detail by Osleger and Montañez (1996), who concluded that much of the silt is of wind-blown origin. It was later reworked and redeposited in intertidal and subtidal areas. Modern analogues include the Persian Gulf, where wind-blown silt is transported onto the tidal flats (Kukal and Saadal-

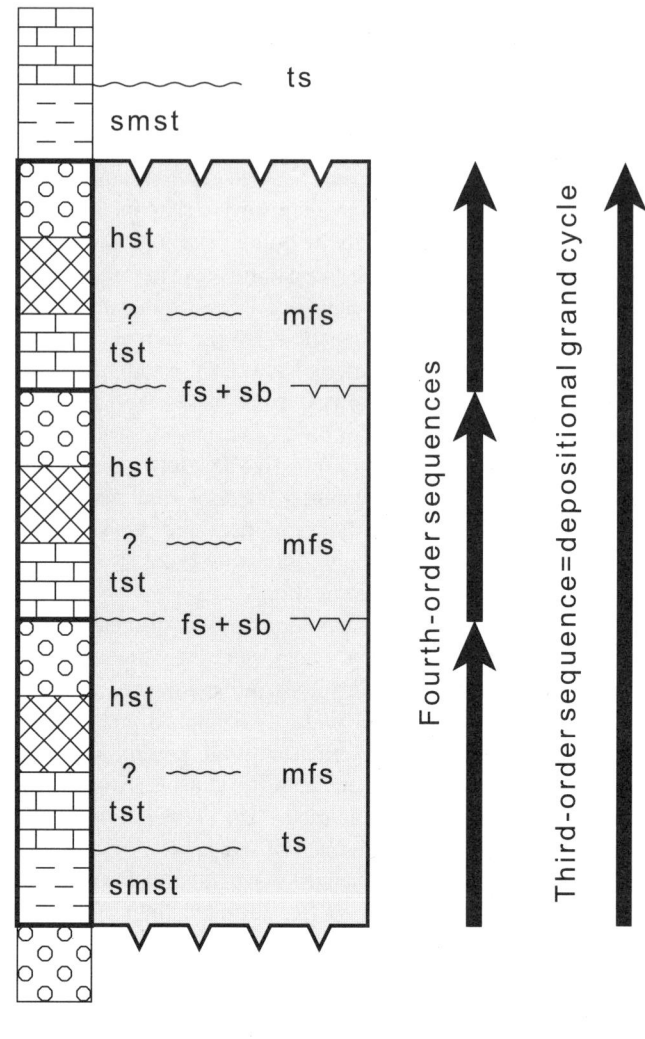

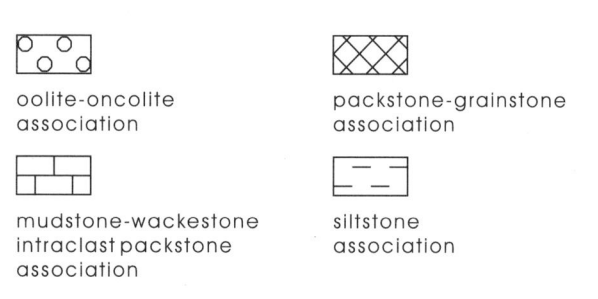

Figure 11. Grand cycle and sequence stratigraphic framework of cycles within La Laja Formation. In this sketch, sequence boundaries are all type 2 boundaries. In contrast to their North American counterparts, siliciclastic interval represents shallowest facies.

lah, 1973). The absence of siliciclastic detritus in most of the carbonate rocks of the La Laja Formation, however, indicates that this detritus was effectively trapped in nearshore areas during periods of relative sea-level rise. When the rate of sea-level rise slowed or came to a stillstand, the relative amount of siliciclastic material on the shelf increased. Accommodation space was filled and the fine-grained detritus was redistributed across the shelf by waves and tides.

Relative sea-level history. Starting arbitrarily at the sharp upper surface of a siltstone association, the surface between the siltstone interval and the overlying limestones marks a rapid rise of relative sea level. This relative rise is documented by the transition from wave-induced structures in the siltstones to muddy, calcareous sediments with intercalated tempestites of a deeper water environment. The rise led to an instantaneous shut-off of siliciclastic supply. Subsequently, relative sea level continued to rise and accommodation was slowly filled with carbonate sediment. Within each calcareous shallowing-upward succession, there is a change from mud-dominated textures near the base toward grain-supported textures near the top. This aggradation toward sea level and the accompanying change in sediment character is interpreted to reflect a steady state in the rate of relative sea-level rise. The widespread presence of oolites on top of the shallowing-upward successions indicates the change from aggradational to progradational sedimentation patterns (Sarg, 1988; Chow and James, 1989).

Many of the shallowing-upward successions are bounded at the top by a flooding surface, above which a new, entirely calcareous, shallowing-upward succession was deposited (Fig. 9). However, there are shallowing-upward successions which at their top show a switch to shallow-water siliciclastic deposition with sedimentation of the siltstone association. In places, a gradational transition between both associations is present; however, sharp contacts between the calcareous shallowing-upward succession and the overlying siltstone interval have been observed. In these cases, the bounding surface marks a very subtle basinward shift of facies.

With the assumption of continuous subsidence, the change to siliciclastic deposition indicates that relative sea-level rise had come to a stop or started to fall slowly, thus increasing the availability of siliciclastic detritus. Sea-level fall, however, was never extensive enough to expose that part of the platform on which the La Laja Formation was deposited.

In this scenario, the siltstone associations are sandwiched between the top of a shallowing-upward succession below and the base of a subsequent shallowing-upward succession above (Fig. 11). A major deepening event separates the siltstone association from the overlying carbonates. From their sedimentologic characteristics, the siliciclastic intervals represent the shallowest facies present in the La Laja Formation. Consequently, the siltstone intervals cannot be the basal part of the overlying shallowing-upward succession. Neither can they be regarded as the upper part of a shallowing-upward succession, because they are often separated from the underlying carbonates by a very sharp surface.

Sequence-stratigraphic interpretation. The basal surface of the calcareous shallowing-upward successions is a major flooding surface and, because it separates the base of a new sequence-stratigraphic entity from the underlying strata (Van Wagoner et al., 1988), it is regarded as a transgressive surface (Fig. 11). The transgressive surface is present either on top of a shallowing-upward succession or on top of a siltstone interval.

Where the transgressive surface separates two calcareous shallowing-upward successions without intervening siliciclastic strata, this surface separates an older highstand systems tract from the subsequent transgressive systems tract. In this case, in the absence of a lowstand systems tract (in the case of a type 1 sequence) or a shelf-margin wedge (in the case of a type 2 sequence), the transgressive surface and the sequence boundary are represented by the same surface (Fig. 11). As pointed out earlier, there is no way to work out of cross-platform architecture, to recognize basinward and or landward shifts of facies, or to trace surfaces perpendicular to depositional strike. Consequently, the only hints to the nature of the surfaces are the absence of subaerial exposure and the lack of siliciclastic input. Hence, with all necessary precaution these surfaces can be regarded as type 2 sequence boundaries. Consequently, each of the calcareous shallowing-upward successions can be regarded as an individual type 2 sequence. The internal architecture (shallowing upward, absence of small-scale cycles) is in agreement with a catch-up system (Kendall and Schlager, 1981; Sarg, 1988), which characterizes the early highstand. In the sequences of the La Laja Formation, the transgressive systems tract seems to be relatively thin, and the maximum flooding surface is a cryptic feature. It might coincide with the change from predominantly muddy textures to grain-supported textures in the carbonates. The highstand deposits are relatively thick. One characteristic feature of catch-up systems is that shoal deposition is restricted to the upper part of the sequence and that the shoals form during the late highstand. In most sequences within the La Laja Formation, sediments of the oolite-oncolite association were deposited during this interval.

There are sequences in which a siliciclastic interval separates older, calcareous highstand deposits from the subsequent transgressive systems tract (Fig. 11). If the interpretation is correct that the underlying oolites represent late highstand deposits related to sea-level stillstand or relative sea-level fall (assuming that subsidence continued), then the siltstone units represent a time of relative sea-level lowstand. During this lowstand, the source area for terrigenous detritus increased. Most of the siliciclastic detritus seems to have been trapped in nearshore areas (not exposed in the La Laja Formation) and filled accommodation there. As carbonate production had essentially ceased during the late highstand, some of the terrigenous detritus might have been washed across the platform during the latest highstand. This is true especially for those units where there is a gradational contact between oolites and siltstones. However, the majority of the siltstone intervals does not belong to the preceding highstand systems tract.

Conventional wisdom dictates that a sequence boundary is

present at the top of the highstand deposits; it also dictates that a transgressive surface is present at the base of the subsequent transgressive systems tract. If the sequence boundary and the transgressive surface are represented by two different surfaces, either a lowstand systems tract (in the case of a type 1 sequence) or a shelf-margin wedge (in the case of a type 2 sequence) must exist between the surfaces.

Position and nature of the sequence boundary between siltstone units and underlying carbonates. The top of the siltstone interval is consistently a transgressive surface; however, the position and the nature of the corresponding sequence boundary are more difficult to assess. In successions in which there is a sharp surface between the oolites and the siltstones, this contact is the most likely candidate for the sequence boundary. However, where there is a gradational contact, the siltstone intervals in their lower part may include calcareous sediments of the youngest portion of the previous late highstand (oolites).

The deposition of the siltstone units testifies to an almost complete shut-off of the carbonate factory. As pointed out by Kendall and Schlager (1981), most carbonate factories do not immediately return to production after the onset of an ensuing relative rise in sea level. In pure carbonate systems, this lag phase is often characterized by condensed intervals (in the sense of sediment starvation) or hardgrounds. In the La Laja Formation, however, siliciclastic strata are present in nearshore areas, and it is here assumed that these silts and sands were redistributed and washed across the carbonates during the transition from relative sea-level fall to rise. A possible mechanism for this redistribution is the increasing effectiveness of waves and tidal currents during the incipient sea-level rise. Although not representing a carbonate system, the siltstone intervals are interpreted to be condensed successions as shown by the presence of abundant glauconite. Hence, they not only represent the lowstand deposits but also the lag phase of the carbonate system. Consequently, the sequence boundary must be located somewhere *within* (the lower part) the siltstone interval. It is likely that in these cases the sequence boundary is not a well-defined single surface, but rather a sequence boundary zone or interval (see also Osleger and Montañez, 1996). Even where there is a discrete surface between the carbonates and the overlying siltstones, there is almost no shift of depositional environment. On the basis of the presence of sequence boundary intervals or surfaces that record only a very minor (if any) basinward shift of facies and the absence of indicators of subaerial exposure, the sequence boundaries are interpreted as type 2 sequence boundaries. Consequently, the siltstone units are interpreted to be shelf-margin wedges rather than lowstand systems tracts.

The fundamental difference between the complete sequences containing a shelf-margin systems tract and those without such a systems tract is the presence of terrigenous detritus. In sequences that do *not* contain a siliciclastic interval, carbonates built up toward sea level, filling accommodation. Once the system was near sea level and had changed from aggradation to progradation during the highstand, the productive area was greatly diminished. This indicates that the rate of relative sea-level rise had slowed. A renewed acceleration of relative sea-level rise led to flooding of the deposits and, after a short lag time, to the return of carbonate production. Consequently, these sequences might be explained by a simple autocyclic model in which sedimentation is governed by availability of carbonate sediment as a function of the productive area.

In contrast, the siliciclastic input combined with the total shut-off of the carbonate factory implies that there was an additional factor operating in formation of the siltstone intervals. This factor was probably eustatic sea-level variations. During the Cambrian, a time of barren cratons, the slowing in the rate of coastal onlap (sea-level rise) or its stillstand must have had two effects. It helped the carbonate factory to build more rapidly toward sea level and to fill accommodation, and it increased the availability of siliciclastic material. This terrigenous material was washed from the craton into the depositional environment. Sands and silts were trapped in the inner detrital belt. Once this sediment trap was filled, the terrigenous material could prograde across the entire platform, especially at the point relative sea-level fall changed to sea-level rise. At that point, before the recovery of the carbonate platform, increasing water depth may have facilitated sediment transport by increasing the effectiveness of wave and tidal current activity. Hence the sequences containing a siliciclastic interval are of allocyclic origin.

The evidence cited here indicates that the differences in internal architecture of the sequences in the La Laja Formation and the causal mechanisms are best explained by differential rates of sea-level rise, an explanation also offered by Palmer and Halley (1979), Chow and James (1987), and Mount et al. (1991), among others. In addition, the sequences within the La Laja Formation seem to reflect the laws of reciprocal sedimentation, i.e., siliciclastic input during sea-level lowstands and carbonate production during rising sea level and sea-level highstand.

One sequence boundary, however, does not fit the model described here (sequence boundary no. 2; Fig. 7). The top of the El Estero Member shows a succession of sandy limestones, sandstones, and locally white quartz arenites and black shales. In comparison with the underlying carbonates, these rocks mark a basinward shift in facies that is characteristic of a type 1 sequence boundary. An important hiatus is present between strata with trilobites of the *Olenellus* zone in the El Estero Member and the *Glossopleura* zone in the Soldano Member (Figs. 7 and 10). As discussed later, this sequence boundary and the accompanying hiatus probably represent the Hawke Bay regression event in the Appalachians (Palmer and James, 1980).

In the La Laja Formation, the sandstone-grainstone association on top of the El Estero Member represents a lowstand wedge of a lowstand systems tract, which seems to have been deposited on an erosional surface. This is indicated by the lower boundary of the sandstone-grainstone association, which is in contact with the siltstone association in the La Laja section, with oolites in the La Flecha section, and with packstones and grainstones in the Villicum section (Fig. 10). This sandstone-grainstone association belongs to the basal sequence within the Soldano Member. The upper boundary of the lowstand wedge coincides with the marked flooding and the onset of carbonate production at the base of the Soldano Member. Consequently, the contact between the El Estero Member and the Soldano Member is marked by the transgressive surface of the type 1 sequence boundary discussed here.

In the La Laja Formation, four successions are present (sequences 2–5; sequence 5 continues into the Zonda Formation), comprising the sediments between two allocyclical events represented by the siliciclastic intervals. Given an approximate duration of the Middle Cambrian of 10 m.y., each of these sequences between two shelf-margin systems tracts spans about 2 to 3 m.y., which is on the order of the third-order cycles of Vail et al. (1977). The individual calcareous shallowing-upward successions that compose the third-order sequences are regarded as fourth-order sequences (Fig. 11).

Zonda Formation

The Zonda Formation is the second unit within the carbonate platform succession and consists almost entirely of dolomite. Four sections have been measured (Figs. 3 and 12) of which only one (Villicum section; Fig. 12) is complete. No biostratigraphic data are available from the Zonda Formation, hence regional considerations and correlations are difficult. These will be discussed within the context of the La Flecha Formation. Both formations show a diversity of microbial boundstones. There are numerous approaches to the description of these rocks (e.g., Logan et al., 1964; Aitken, 1967; Monty, 1967; Burne and Moore, 1987). In this paper the terminologies of Logan et al. (1964) and Aitken (1967) are applied.

Lithofacies. In the Zonda Formation, dolomitization is pervasive and accounts for the replacement of most primary structures in most of the rocks. Equigranular and nonequigranular dolosparites often do not give any information about the original sediment. However, the rocks of the Zonda Formation reveal aspects of primary depositional textures and fabrics. The most important features are microbial dolostones and stromatolites. Among the stromatolites, laterally linked hemispheroids (LLH stromatolites) prevail. Stacked hemispheroids (SH stromatolites) are also present, as well as mud mounds with fenestral fabrics. In places, small mound-like structures are observed. Due to complete silicification, no internal structures are preserved within the mounds. The only clue to their microbial-mound origin is the onlap of surrounding sediment onto the flanks of the buildups.

Microbial laminites (cryptalgal laminites of Aitken, 1967) are abundant; many consist of alternations of thin micrite laminae and layers of peloids. Desiccation cracks (5 in Fig. 13) and fenestral fabrics (3 in Fig. 13) are frequently observed in the laminites. Silt-sized quartz is present in some laminites, where it was trapped by the microbial mats.

Dolomudstones, dolomitic peloidal grainstones, and wackestones are white, dark gray, or dark brown. In places, scattered trilobite fragments are present in these sediments. Bioturbation is present but rarely pervasive. In some of the mudstones desiccation cracks are developed.

Oolitic grainstones are almost always replaced by dolosparite. However, cross-bedding is preserved in several beds. In thin sections, relict structures are visible, which under luminescence are revealed to be ooids. Some of these rocks contain small, dark micritic intraclasts or oncoids with trilobite fragments or ooids as a nucleus.

Intraclast conglomerates (6 in Fig. 13) and flat-pebble conglomerates are present in thin layers. Intraclasts consist of reworked (peloidal) mudstones and laminites. Both grainstones and packstones are developed. In some of these rocks, trilobites or ooids are observed. Flat-pebble conglomerates are monomict; the tabular clasts are moderately rounded along their edges. The grainstones and conglomerates commonly fill scour structures or small channels. In addition, some of the grainstones onlap stromatolite mounds.

Few monomict breccias with angular clasts are present in the Cerro La Silla section (Fig. 12). They are cemented by chert or dolospar. In the upper part of the La Flecha section, which has not been measured, several breccias are present. These breccias are yellow, their clast content is monomict to oligomict and reflects the lithologies of the overlying rocks. In several horizons, chert clasts are present.

Facies associations and depositional environment. Bioturbated mudstones, wackestones, and peloidal grainstones were deposited in shallow subtidal areas. The scarceness of the fauna and of intraclast layers together with the few fossils indicate that this wackestone association formed in a tranquil environment with probably moderately restricted conditions. In modern environments on the Bahamas, these sediments are formed in back-barrier areas within the shallow subtidal and lower intertidal areas (Garrett, 1977), where the muds are often strongly biologically mottled by a low-diversity fauna.

Interbedded oolites and laminites (oolite-laminite association) form a thick succession in the Cerro La Silla section. The oolites are mostly cross-bedded; in the laminites, mudcracks and brecciation are abundant. These rocks record the evolution of an oolite shoal complex and the adjacent tidal flats. The association is very similar to rocks described from Cambrian strata of the Appalachians (Chow and James, 1987), where similar rocks form unpredictable meter-scale assemblages. A comparable setting is invoked for the Zonda Formation (Fig. 14). Chow and James (1987, p. 422) stated that these successions do not form ordinary shallowing-upward cycles. Instead, the facies succession reflects changing hydraulic and topographic conditions on the platform, which lead to migrating facies.

The mudstone-boundstone association is composed of birdseye mudstones, microbial boundstones, and flat-pebble conglomerates. In addition, thin oolite beds may be present. The dominance of low-relief stromatolites and microbial laminites together with the birdseye structures are clear indications of a higher intertidal to lower supratidal environment (Fig. 14; Shinn,

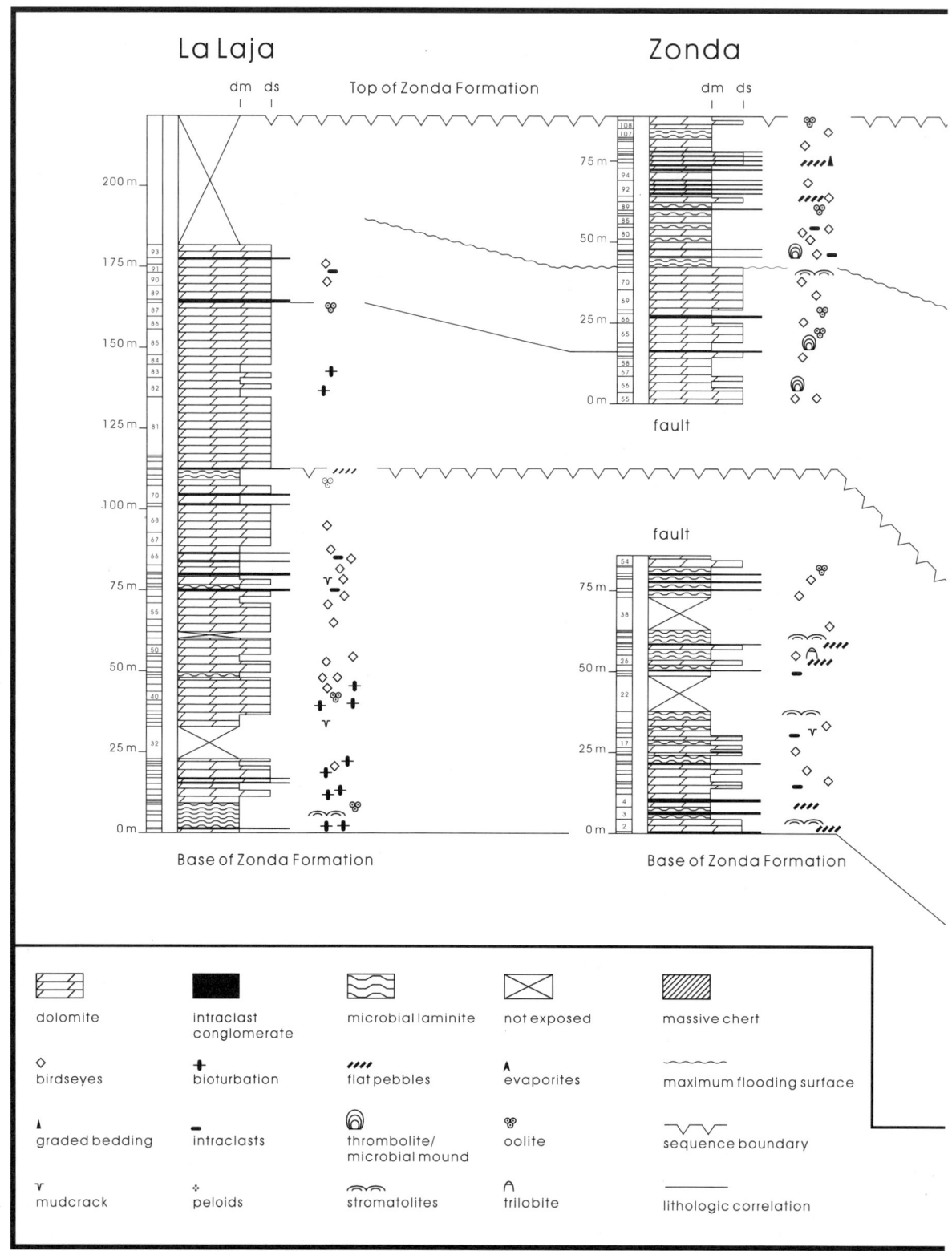

Figure 12. Stratigraphic columns of Zonda Formation, lithologic correlations, and sequence stratigraphic interpretation.

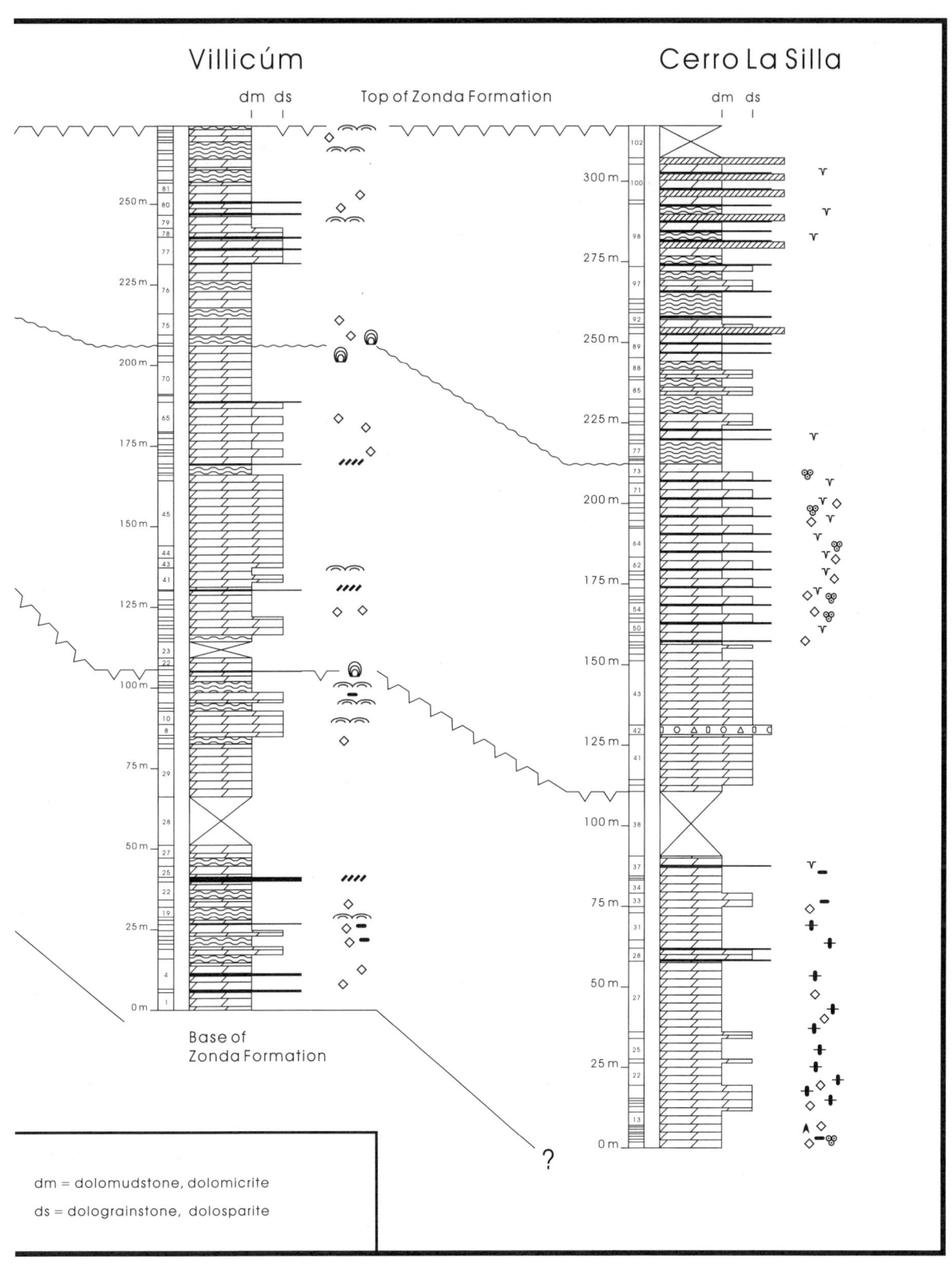

dm = dolomudstone, dolomicrite
ds = dolograinstone, dolosparite

1983b). Long and repeated periods of exposure in the supratidal are also indicated by abundant mudcracks and breccias. The supratidal area also provides lithified sediment that is reworked during major storms. Thin laterally extended layers of intraclasts may represent supratidal storm layers, whereas imbricated clasts in laterally restricted bodies are interpreted as channel fills (Shinn, 1983a). The thin oolite beds also represent storm layers. Silt-sized quartz, trapped within microbial mats, is common in modern tidal flats of the Persian Gulf. Winds are the major transport agent (Kukal and Saadallah, 1973).

Platform configuration. Regional platform configuration is difficult to determine because the outcrops of the Zonda Formation are limited. In addition, the Zonda and the La Flecha Formations seem to interfinger laterally. Consequently, the platform configuration during the time of deposition of the Zonda Formation is discussed in connection with the La Flecha Formation.

Sedimentary succession of the Zonda Formation. The Zonda Formation is composed of three units. In the Sierras de Villicum and Chica de Zonda (Fig. 2), the basal succession consists mainly of the mudstone-boundstone association above the massive oolites on top of the La Laja Formation. This interval is characterized by light gray and white colors. The thickness of the lower unit is on the order of 110–120 m. At Cerro La Silla, the same interval is dominated by dolomudstones with bioturbation or birdseye structures (wackestone association).

The second unit is composed of thick-bedded dark gray dolosparites and abundant mudstones (wackestone association) which in the southern sections only rarely exhibit sedimentary structures. The dark color is caused by a relatively high content of organic matter. Upsection, birdseye mudstones become more abundant. Microbial mounds are restricted to this interval. At Cerro La Silla, a similar succession is present, which in the upper part shows a pseudo-cyclic development (oolite-laminite association). The thickness of the second unit varies between 100 m in the Villicum section and ~130 m in the Cerro La Silla section.

The third, upper succession of the Zonda Formation shows abundant microbial laminites and intraclast conglomerates (Fig. 12). In addition, dolomudstones and coarse dolosparites are present (mudstone-boundstone association). Black chert is present in nodules and, in the Cerro La Silla section, as massive (to 2 m thick) beds (Fig. 12). Desiccation features are frequently observed in this part of the Zonda Formation which again is mainly composed of light gray and white rocks. The upper succession is 65–100 m thick.

In the southern sections, the basal unit is clearly a continuation of the uppermost cycle in the La Laja Formation. In contrast to the underlying cycles within the La Laja Formation, shallowing continues into the intertidal and supratidal environment. No small-scale cycles are visible; instead, the facies associations and their vertical arrangement support an island tidal flat model (Pratt and James, 1986) for the deposition of the basal unit of the Zonda Formation. In this model, sediment distribution and vertical arrangement of facies are governed by changing topography and hydraulics on a steadily subsiding platform. The small-scale sedimentary successions are the result of vertical accretion and lateral migration of peritidal flats. One important characteristic of this model is that cycles are difficult to correlate from one outcrop to another. An actualistic counterpart to this model is provided by the Bahama platform (Schlager and Enos, 1996), where sedimentation takes place in different environments that laterally interfinger across short distances.

In all sections, the second unit indicates a sudden deepening and the onset of dominantly subtidal sedimentation. No dramatic changes within this unit are recognized in the southern sections, but in the Cerro La Silla section, the presence of the oolite-laminite association in the upper part indicates the return to a peritidal environment.

With the beginning of unit 3, the entire platform records the transition toward intertidal and supratidal environments with the formation of thick successions of microbial laminites. Consequently, units 2 and 3 together form a major shallowing-upward succession, which from its thickness and the time involved qualifies it as a third-order cycle (sequence 6 of the carbonate platform succession; Fig. 7).

Evolution and sequence stratigraphy of the Zonda Formation. Because the Zonda Formation in the San Juan subbasin is partly coeval to the lower part of the La Flecha Formation in the Guandacol subbasin, evolution and sequence stratigraphy of all Upper Cambrian rocks is discussed after the description of the La Flecha Formation.

La Flecha Formation

The La Flecha Formation is a highly cyclical succession of predominantly dolomitic rocks exposed from the Guandacol subbasin in the north to the southern tip of the Sierra Chica de Zonda (San Juan subbasin) in the south (Fig. 3). Following the redefinition of the La Flecha Formation by Keller et al. (1994), the Upper Cambrian rocks that are between the Zonda Formation and the La Silla Formation (Keller et al., 1994) are included. In this sense,

Figure 13. Aspects of Zonda and La Flecha Formations. 1: "Not-so shallowing-upward cycle." Basal intraclast grainstones have mudstone cap. When lithified this mudstone served as settleground for SH-stromatolite (stacked hemispheroid). This SH-stromatolite in turn formed base for thrombolite. Both microbial structures are embedded in oolitic grainstone. La Flecha Formation, La Flecha section. 2: Fenestral mud mound growing on sharp surface of thinly bedded dolomitic mudstones. Note onlap of peloidal grainstone on mound. Structure is ~20 cm across. La Flecha Formation, Cerro La Silla section. 3: Thin section of intraclast-peloidal grainstone which grades into dolomitic mudstone. Boundary between both microfacies is characterized by abundant fenestral fabrics. Oolite-laminite association of Zonda Formation. Magnification is 4.5. 4: Large thrombolite mound is cut by erosional channel. This channel with overhanging walls is filled with oolitic grainstone. La Flecha Formation, La Flecha section. 5: Microbial laminite with thin layers of peloidal grainstone. Oolite-laminite association of Zonda Formation. Magnification is 4.5. 6: Graded intraclast grainstone of channel fill in Zonda Formation. Clasts are composed of different microfacies. Magnification is 4.5.

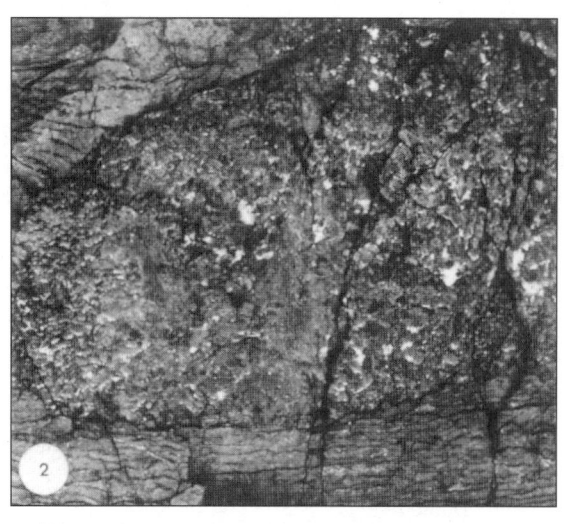
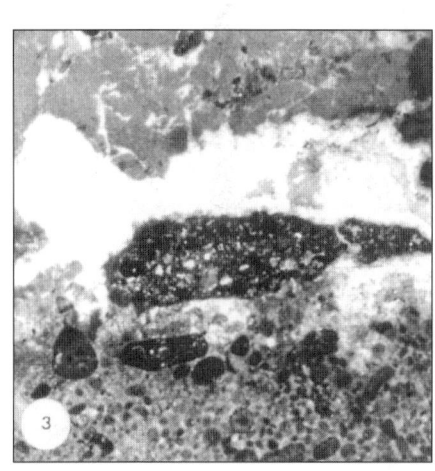

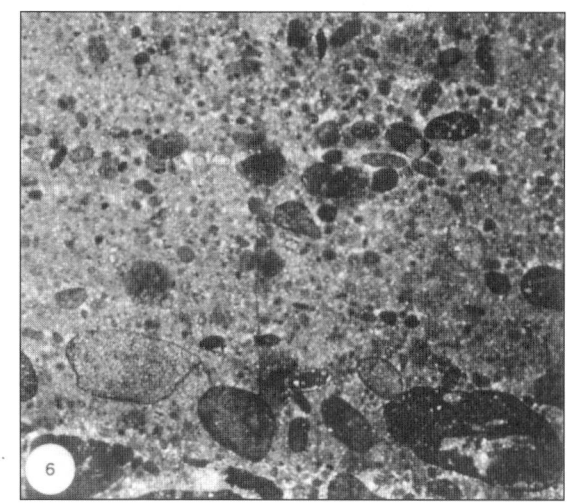

most of the sediments formerly included within the San Roque Formation and the Los Sapitos Formation (Baldis et al., 1981a; Hünicken and Pensa, 1985) are discussed here.

Biostratigraphic control on the Upper Cambrian succession of the Argentine Precordillera is still poor. In the San Juan subbasin, deposition of the Zonda Formation started during the early Marjuman (*Bolaspidella* zone, Bordonaro, 1990a). The next datum is given by *Plethopeltis cf. P. saratogensis* found at the base of the overlying La Flecha Formation in its type section (discussion in Keller et al., 1994), attributed to the (late) Sunwaptian by Vaccari (1994).

On the basis of their lithologic succession, all sections in the Guandacol subbasin are attributed to the La Flecha Formation. These sections are neither complete toward the base or toward the top. In the La Angostura section, Marjuman fossils have been found near the base of the exposed succession (Vaccari, 1994), whereas the top is of Steptoean age (*Dunderbergia*? zone). Hitherto, neither fossils attributed to the latest Steptoean (*Elvinia* zone) nor to the early Sunwaptian have been described from the Zonda or La Flecha Formations. Hence, it is likely that the rocks in the type section of the La Flecha Formation do not have a coeval representative in the Guandacol subbasin, as all rocks there are older and probably correspond to the Zonda Formation.

Lithofacies: Mechanically deposited rocks. Among the mechanically deposited carbonates, nonfossiliferous lime mudstones are widespread and many of them show mudcracks and/or pseudomorphs after evaporitic minerals.

Light gray fossil-bearing mudstones and wackestones are rare in the sections south of the Río Jáchal but are more abundant near Guandacol (Cañas, 1995a). Only trilobites and rare hyolithids have been observed in the rocks. Some of these sediments are bioturbated. Few trilobite packstones were deposited in the La Flecha section, where they form coquinas of broken individuals.

Oolitic grainstones are very abundant and individual beds locally are 1 m thick. Scouring at the base and cross-bedding are present in many intervals. Within intraclast grainstones, clasts represent the entire spectrum of rocks present in the La Flecha Formation. With increasing amounts of ooids there is a transition to oolitic-lithoclast grainstones and lithoclast-bearing oolites. Rudstones are developed where intraclasts were mainly derived from the erosion of thrombolites or stromatolites.

Flat-pebble breccias are mainly monomict and the clasts are barely rounded. Commonly, the clasts are tabular and in places still show fitting fabrics. Oligomict breccias consist of irregularly shaped clasts of different lithologies. All of these described facies are affected locally by brecciation. The clasts are cemented by chert, chalcedony, or calcite.

A sediment with unusual features shows a muddy matrix with oval spheres that consist of a few layers several millimeters thick. The layers show an alternation between brown and white colors. The individual spheres, which have been compared with caliche pisoids (M. Keller et al., 1989), in places coalesce to form enigmatic structures.

Lithofacies: Microbial boundstones and stromatolites. Microbial laminites consist of alternations of lime mud, layers of peloidal grainstone, organic tissue, and small, partly rounded, reworked dolomite crystals. The laminae are subparallel to undulating and there is a transition to low-relief LLH stromatolites. Similar laminae are observed in SH stromatolites and in LLH types. SH morphotypes (1 in Fig. 13) are much less abundant than are LLH morphotypes; however, in the Río Acequión section they form spectacular structures almost 1 m high. SS stromatolites, rounded to elongated microbial structures, are present only locally. Their size varies considerably between a few centimeters and almost 1 m. The latter correspond to Osagia-type stromatolites described from the La Flecha Formation by Baldis et al. (1981b).

Thrombolites (Aitken, 1967) are abundant (4 in Fig. 13; see also Armella 1989a, 1989b). These clotted, nonlaminated structures form individual mounds, in places 2 m high, or they form laterally coalesced biostromes.

Other structures of probable organic origin are small mud mounds (2 in Fig. 13) that are characterized by abundant fenestral fabrics (birdseyes and stromatactis). The mounds are as wide as 2 m and as high as 1 m and some show overhanging walls.

The peritidal nature of the La Flecha Formation implies the presence of a variety of sedimentary and early diagenetic features (Fig. 15) discussed in detail by M. Keller et al. (1989) and Cañas (1995a). Mudcracks, tepee structures, birdseye structures, stromatactis, sheet cracks, and pseudomorphs after evaporites have been described.

Facies associations and depositional environment. The preceding description gives a rather static subdivision of rocks of the La Flecha Formation; however, there is a complex interplay between mechanically deposited carbonates and the microbial boundstones.

The mudstone-wackestone association represents shallow subtidal environments (Fig. 14). The scarceness of the fauna indicates restricted conditions, probably hypersaline, unfavorable for a flourishing fauna. However, burrowing organisms adapted to these conditions were locally abundant. Trilobite packstones represent sporadic high-energy events, probably storms, during which the sediment and the fossils were stirred up, reworked, and redeposited.

The intraclast grainstone-oolite-thrombolite association records constantly agitated water and repeated erosional events. This is demonstrated by channels cutting into individual thrombolite mounds (4 in Fig. 13) or eroding them laterally. Intraclast grainstones and rudstones, commonly containing ooids, and oolitic grainstones fill the channels or onlap the mounds (Fig. 15).

This facies association was deposited in lower intertidal and shallow subtidal areas. Thrombolites were able to grow vertically; however, there are no signs that their growth was terminated by subaerial exposure. Oolites, which in modern-day environments form near mean sea level, onlap and cover the mounds. There are also places where oolites onlap the mounds and the thrombolites, in turn, laterally expand across the onlapping sediment. This shows that the depositional environments of both oolites and thrombolites were closely connected. Early lithification of the mounds is proven by the presence of steep channels within the mounds and the formation of overhanging walls.

Another association is composed of stromatolites with varying amounts of mudstones and grainstones (stromatolite association). Growth forms and environment of the stromatolites within the La Flecha Formation have repeatedly been interpreted in comparison with their modern-day counterparts (Armella, 1989a; M. Keller et al., 1989; Cañas, 1995a). Following these interpretations, the stromatolite association represents shallow subtidal, intertidal, and probably lowermost supratidal environments (Fig. 14). Columnar stromatolites (as a counterpart of SH morphotypes) grow in shallow subtidal areas with strong currents (1 in Fig. 13). This was demonstrated in Shark Bay by Burne and James (1986) and in the Bahamas by Dill et al. (1986). LLH morphotypes are much more common in low-energy intertidal environments (Logan et al., 1964; Hoffman, 1976). In the La Flecha Formation, mudcracks and evaporite crystals in the accompanying sediment as well as their upward grading into microbial laminites are evidence of an intertidal setting. Similar arguments for an intertidal origin were used by Pratt and James (1982, 1986) for Ordovician rocks of North America.

Local erosion of stromatolites and deposition of oolitic and intraclast grainstones indicate intermittent high-energy events, most probably storms or exceptional tides. Their effects are well known in modern environments from the classic studies of Ball (1967), Ball et al. (1967), Shinn et al. (1969), and Hardie (1977).

The microbial laminite-breccia association shows ample evidence of prolonged periods of desiccation and evaporation (mudcracks, tepee structures, pseudomorphs). Microbial laminites as products of supratidal areas have long been known from modern (Hardie, 1977) and ancient environments (Aitken, 1967; Pratt and James, 1986). In supratidal environments, carbonate mud is deposited almost exclusively after major storms, forming thick layers that are later cracked by desiccation. In addition, these lime muds form the matrix in which evaporites crystallize. Cracking of indurated layers produces rip-up clasts, which in the La

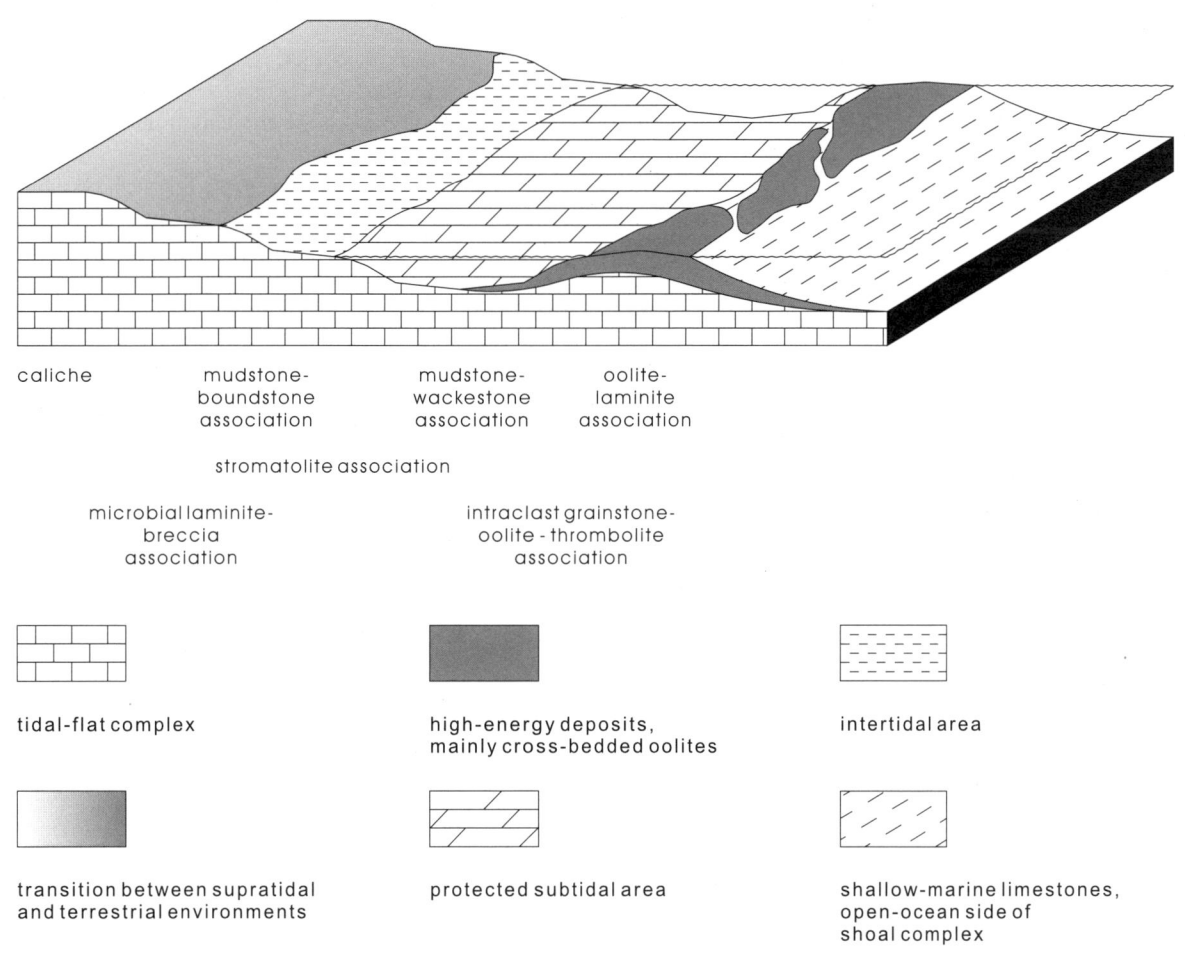

Figure 14. Depositional environments and facies associations of Zonda and La Flecha Formations.

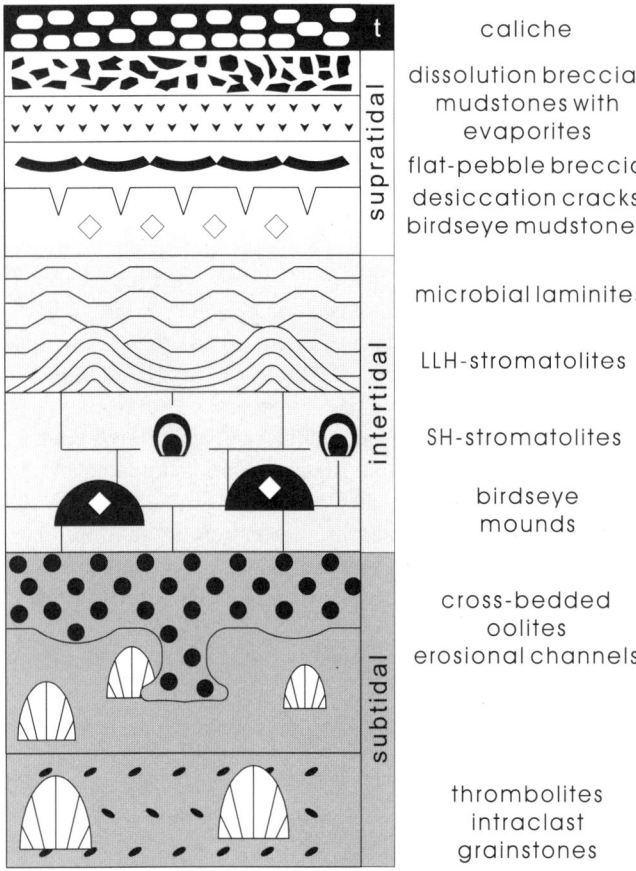

Figure 15. Elements of small-scale shallowing-upward cycles in La Flecha Formation (modified from M.Keller et al., 1989). LLH = laterally linked hemispheroids, SH = stacked hemispheroids, t = terrestial.

Flecha Formation are present as flat-pebble breccias. Dissolution of laterally extensive layers of evaporites (gypsum mush) is most likely the source for oligomict breccias.

The dolomitic mudstones with the unusual pisoid-like spheres in the La Flecha Formation are most likely calcretes (M. Keller et al., 1989). There is no direct clue to the origin of these sediments, because the spheres lack the characteristic radial-concentric fabric of pisoids; however, the spheres probably represent recrystallized vadose pisoids. A supratidal to terrestrial origin of these intervals is indicated by their position within the sedimentary succession (Fig. 15). They are mainly found above microbial laminites and commonly are erosionally overlain by intraclast conglomerates and oolites at the base of the overlying cycle.

Platform configuration. The reconstruction of the platform during the entire Upper Cambrian is hampered by the absence of a distinct platform margin; however, if we look at coeval successions around Laurentia, then some conclusions with respect to platform organization can be drawn. Read (1989) described facies and platform architecture of the Upper Cambrian successions in the Appalachians. There, cycles containing abundant thrombolites and, in general, restricted facies are present on the middle and inner shelf. Demicco (1985) developed a model for Upper Cambrian sections in the northern Appalachians, in which tidal flats prograded from the platform rim in a landward direction. He also documented the migration over tens of kilometers of oolite shoals from the margin into the back-barrier lagoon. In the Canadian Appalachians, thick oolite shoals were deposited along the platform margin, and numerous microbial buildups grew between the shoals (James et al., 1989). All along the eastern margin of Laurentia, there was a belt of nearshore siliciclastic deposits between the barren craton and the carbonate platform, the intrashelf basin or the inner detrital belt (Fig. 9).

In the Zonda and La Flecha Formations, the peritidal complex is present across the entire eastern basin and there is only little evidence of major facies changes. The section in the Guandacol subbasin shows more subtidal facies and was probably slightly deeper than the remainder of the platform (Fig. 14). Neither a siliciclastic belt nor the platform margin are exposed. The peritidal complex may have been a broad barrier on the outer platform (Read, 1985).

Sedimentary succession of La Flecha Formation. Sections of the La Flecha Formation are present in all three subbasins (Fig. 3); however, the sedimentary successions differ from subbasin to subbasin. In the San Juan subbasin, two successions are recognized within the La Flecha Formation. The lower one starts with sediments of the intraclast grainstone-oolite-thrombolite association that pass upsection into intertidal and supratidal deposits with abundant caliche horizons toward the top. The upper succession is more subtidally influenced and seems to be less cyclically developed.

In the Guandacol subbasin, the La Flecha Formation consists of an alternation of limestones and dolostones. The entire succession shows a well-developed cyclicity (Cañas, 1995a). Cycles span the low-energy subtidal to supratidal environments.

In the Talacasto subbasin, outcrops of the La Flecha Formation are centered around Cerro La Silla, south of Jáchal. All sections show a bipartite subdivision. The lower unit (Agua Negra Member of Baldis et al., 1981a) is a succession of peritidal cycles with abundant stromatolites and thrombolites. This basal succession is similar to the type section. The upper part (Los Diaguitas Member of Baldis et al., 1981a), however, shows a different development. The member is very evident in the field because its thick-bedded white rocks contrast with the underlying and overlying gray rocks. Typically, the succession is composed of small-scale cycles of light gray to white dolosparites and dolomudstones. Dolosparites commonly show calcite as a replacement of gypsum nodules, whereas in the dolomudstones, pseudomorphs after gypsum prevail. LLH stromatolites, thrombolites, and intraclast grainstones or oolites are present only to a very minor extent.

Evolution and sequence stratigraphy of the Zonda and La Flecha Formations. The basal succession within the Zonda Formation is regarded as a sedimentologic continuation of the upper part of the La Laja Formation (Fig. 7). The uppermost package of oolites of the La Laja Formation gradually gives way to intertidal and supratidal rocks. The top of this succession is marked by a transgressive surface, which brings subtidal mudstones onto supratidal laminites. This surface (sequence boundary 6; Fig. 7), which is present in all sections, is interpreted to be a type 2 sequence boundary because there is no significant basinward shift in facies, nor are there indications of large-scale subaerial exposure or karstification.

In the San Juan and the Talacasto subbasins, units 2 and 3 of the Zonda Formation form one major shallowing-upward succession. No shelf-margin wedge is developed; consequently, the transgressive surface coincides with the sequence boundary (Fig. 12). The thick subtidal sediments above the transgressive surface represent a transgressive systems tract. Their transition into the highstand systems tract is expressed by the change to thinner bedded units with a more pronounced cyclicity and the return to formation of stromatolites and microbial laminites. In the La Flecha section, several breccia beds are observed below the contact to the La Flecha Formation, and are interpreted to be evaporite dissolution beds. These horizons are present on top of an overall shallowing-upward succession and indicate that the sediments had been subaerially exposed and that the rocks were fractured during dissolution of the evaporites. Prolonged periods of subaerial exposure are also indicated by mudcracks and flat-pebble conglomerates. Within this shallowing-upward succession, the culminating event is the prolonged exposure of the uppermost strata, and consequently, the top of the uppermost breccia of the Zonda Formation can be interpreted as a type 1 sequence boundary. This boundary (sequence boundary 7; Fig. 7) also coincides with a flooding surface above which thrombolites, oolites, and intraclast conglomerates indicate shallow-subtidal environments.

Two sequences are recognized within the La Flecha Formation. The lower one begins with small-scale cycles (here taken as synonyms of parasequences) in which sediments of shallow subtidal and lower intertidal environments are abundant. Upsection, intertidal and supratidal rocks are more important and there is an increase in caliche rocks (Fig. 16). Within this lower succession, cycle thickness tends to decrease, mainly at the expense of thrombolitic and stromatolitic rocks, resulting in a progradational stacking pattern. The sequence boundary (sequence boundary 8) is marked by the disappearance of caliche horizons and the return to more subtidally influenced parasequences and the return of thrombolites. In this upper succession, caliche horizons are rare (Fig. 16). Stacking patterns do not reveal a clear pattern of being progradational or retrogradational. The upper boundary of this sequence (boundary 9) will be described in the context of the La Silla Formation.

Cañas (1995a), constructing Fisher plots for the La Angostura section in the Guandacol subbasin (*Crepicephalus* to *Dunderbergia* [?] zone) from more than 50 shallowing-upward cycles, found evidence of a sequence boundary at the transition from the middle to the upper third of the exposed succession. The sequence boundary is located in the uppermost *Crepicephalus* zone. Both sequences, below and above the sequence boundary, are third-order sequences. Lithologically, the bed above the sequence boundary shows lithoclasts with coarse quartz grains.

Together with abundant microbial laminites and mudcracks in this interval, the terrigenous input is taken as evidence of a type 1 sequence boundary. If this interpretation is correct, there ought to be a correlative surface in the other subbasins.

One possibility is that the sequence boundary in the Guandacol subbasin correlates to the sequence boundary within the

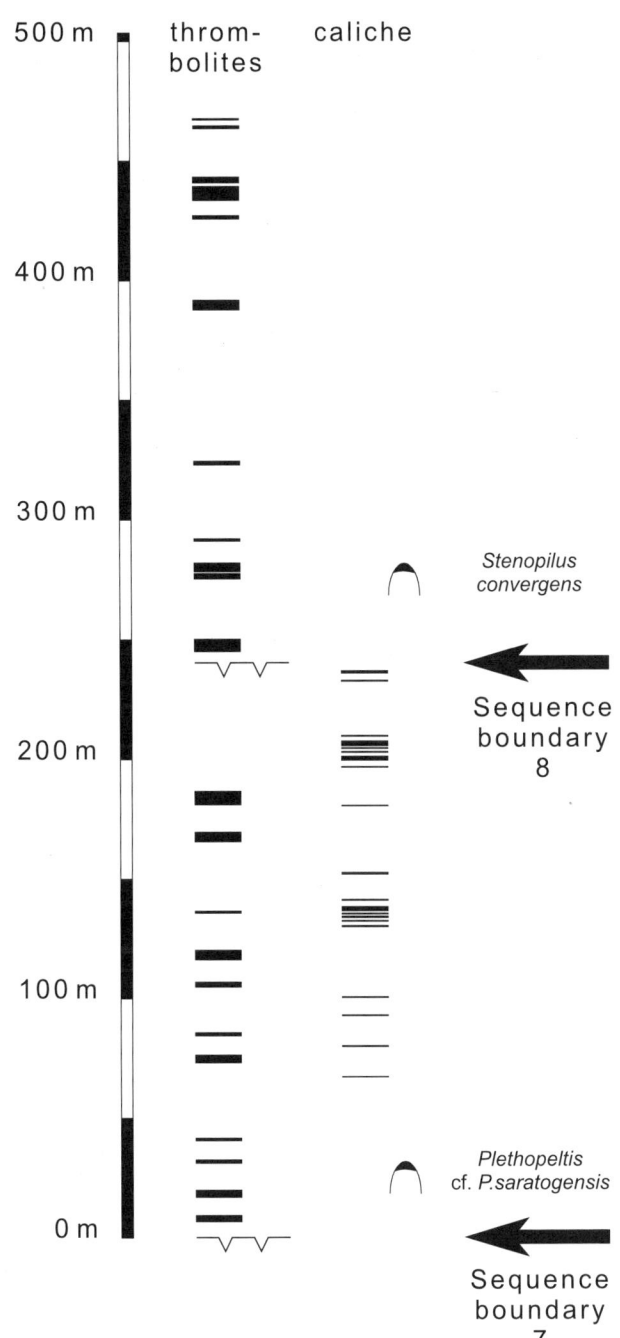

Figure 16. Distribution of thrombolites and caliche horizons in type section of La Flecha Formation. Increasing abundance of caliche toward middle of section is taken as evidence for sequence boundary. Trilobite symbols mark biostratigraphic control. See text for discussion.

Zonda Formation (boundary 6). This would imply that most of the strata, both below and above the sequence boundary, in the La Angostura section are correlative and coeval to the Zonda Formation.

An alternative interpretation is that the type 1 sequence boundary that separates the Zonda Formation from the La Flecha Formation in the San Juan subbasin (boundary 7) is the same type 1 sequence boundary that is present in the Guandacol subbasin. This assumption implies a major hiatus between the Zonda Formation and the La Flecha Formation in the San Juan subbasin which would support the interpretation of the formational boundary as a type 1 sequence. This also implies that the rocks above the sequence boundary in the La Angostura section have no coeval representatives in the San Juan subbasin. Could this mean that in this scenario these rocks represent a kind of lowstand deposit?

Biostratigraphically, both scenarios are viable: whatever correlation will eventually be shown to be correct, it has no bearing on platform configuration. The study by Cañas (1995a) showed that the rocks of the Guandacol subbasin represent slightly deeper facies than both the Zonda and La Flecha Formations.

The evolution of the Upper Cambrian strata in the Argentine Precordillera shows four major successions, the lowermost originating in the Middle Cambrian (top of La Laja Formation). All cycles are shallowing upwards and are bounded at the top by sequence boundaries. According to their thicknesses and the average time involved in each sequence, they represent third-order sequences (Fig. 7).

La Silla Formation

The La Silla Formation comprises a thick succession of mostly medium- to thick-bedded, bluish-gray limestones with minor amounts of dolomite and dolomitic limestones. Biostratigraphic data are still sparse (e.g., Vaccari, 1994; Lehnert, 1995a) and are only available from the Jáchal area. There, the La Silla Formation begins during the latest Cambrian (Keller et al., 1994) and ends in the latest Tremadoc.

The La Silla Formation was introduced (Keller et al., 1994) as a formal unit within the carbonate platform deposits of the Precordillera. The type section was chosen at Cerro La Silla southeast of Jáchal (Fig. 3), where the lower and the upper boundaries are well exposed. There, the upper part of the La Flecha Formation consists of cyclically arranged dolomites with abundant stromatolites and gypsiferous mudstones. The boundary with the La Silla Formation is marked by 20 cm of dark gray silty dolomite. Above this bed light gray grainstones are present, in which several white dolomitic microbial mounds are developed. Above the silty dolomite, almost all the chert and the majority of the stromatolites disappear. In addition, the dominant rock type is limestone, whereas in the underlying La Flecha Formation dolomites prevail.

In the La Flecha section, a similar transition is observed. The upper part of the La Flecha Formation consists of an alternation of lime mudstones, dolomites, and silicified stromatolites. Above the contact, only gray limestones are present; chert is virtually absent. About 5 m above the base the first nautiloids and gastropods are observed. The overlying succession is highly cyclical (Lehnert and Cooper, 1995, personal commun.), but hitherto has not been studied in detail.

Three main areas with sedimentary rocks of the La Silla Formation are recognized (Fig. 3). In the San Juan subbasin, sections of the La Silla Formation have been measured at Las Lajas, Los Berros, and Puesto de los Potrerillos (Fig. 17, see oversized plate at the back of the volume). Within the Talacasto subbasin, two areas are distinguished: sections in the western thrust sheets (Talacasto [west], Gualilán, Pachaco; Figs. 3 and 17), where only the upper succession of the La Silla Formation is exposed; and the northern sections around Jáchal (Cerro La Silla, Cerro La Chilca, Cerro Viejo de San Roque; Fig. 17). Correlations between the sections in the north and in the south are difficult, mainly because sections in the south are incomplete and because biostratigraphic data there are completely absent.

Lithofacies. The La Silla Formation is composed of a variety of rocks which were deposited in a peritidal environment. Boundstones are very abundant in all sections. In the type section, several white microbial mounds are found near the base of the formation. These mounds have a height of as much as 50 cm; the bases are as wide as 80 cm. No internal structures are visible. Peloidal grainstones onlap the mounds, but irregular erosional contacts have also been observed.

Microbial laminites (5 in Fig. 18) form prominent horizons to 2 m thick. In places, they show mudcracks and in situ brecciation near the top of the beds. Other organic structures are thrombolite mounds, which in the Cerro La Silla section have coalesced to form a 2-m-thick massive thrombolite horizon (Fig. 17). LLH and SH stromatolites are rare in the succession; however, they are easily detected because many of them are silicified. The majority of these rocks shows fenestral fabrics, which vary in size from microscopic to 2 cm. Internal (geopetal) sediment is rare and most of the fenestrae are filled with white blocky calcite cement.

Mudstones are very abundant and consist of pure lime mud. In the middle part of the succession, many of the mudstones are moderately to strongly bioturbated (6 in Fig. 18). In other mudstones, thin-section analysis reveals that much of the mud is of microbial origin and that structure grumeleuse (Cayeux, 1935) is abundant. Hence, many of the mudstones are better classified as microbial boundstones. Mudstones and microbial boundstones show abundant fenestral fabrics.

Wackestones and packstones with gastropods and nautiloids are also present (1 in Fig. 18). Their matrix is muddy or, in places, consists of abundant peloids. Other bioclasts include trilobites, crinoids, sponges, and algae (2 in Fig. 18).

Peloidal grainstone (4 in Fig. 18) is a very common rock type in the La Silla Formation; oolites are less abundant. Ooids are commonly replaced by dolospar and only ghost structures are preserved. The size of the ooids varies between 0.4 and 0.7 mm.

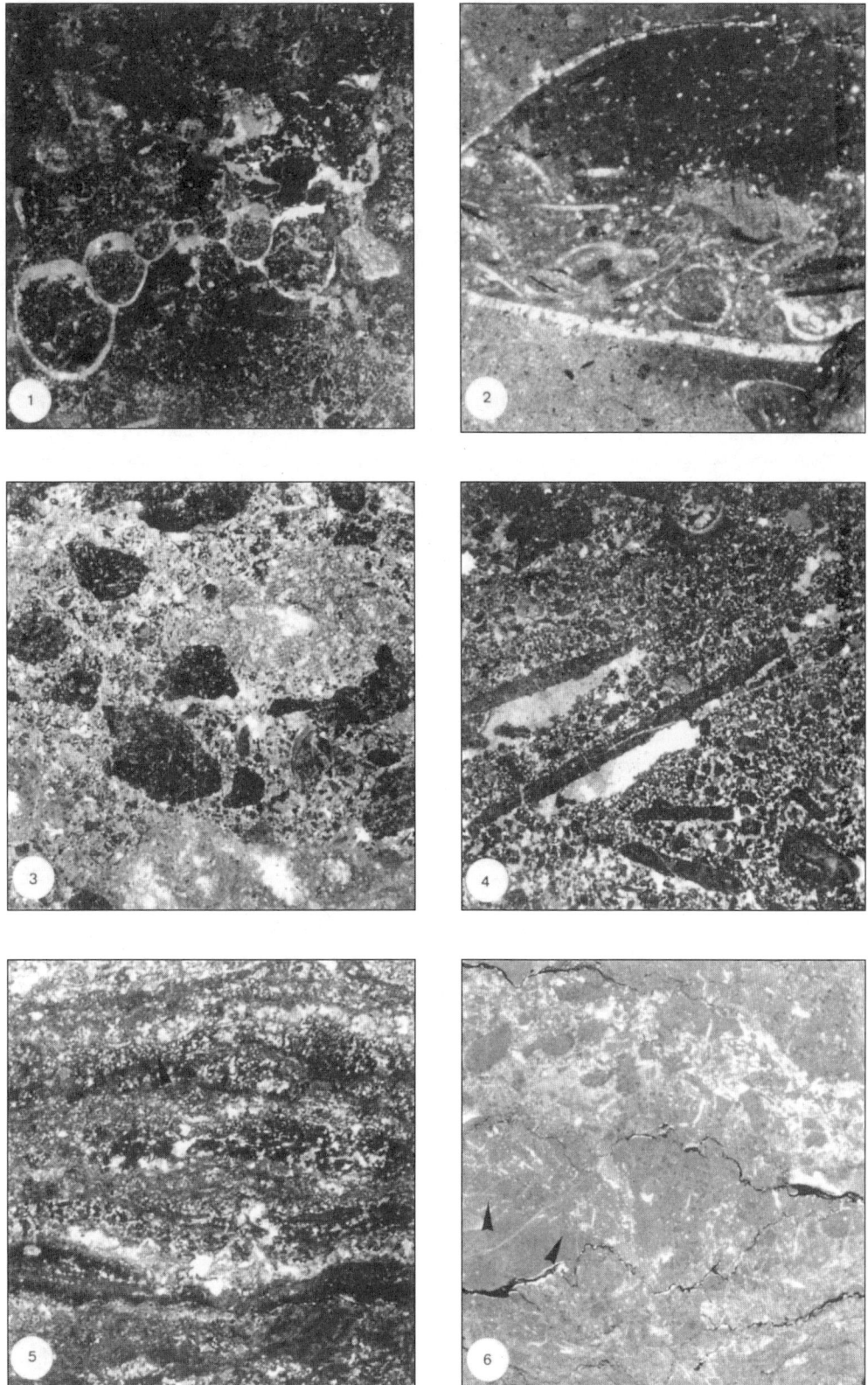

Figure 18. Facies of La Silla Formation. 1: Gastropod wackestone with internal fill of peloidal sediment and geopetal cements. Wackestone association. Magnification is 4.5. 2: Large gastropod shell with graded fill of bioclastic packstone. Bioclasts are mainly gastropods and trilobites. Magnification is 4.5. 3: Poorly sorted intraclast conglomerate. Clasts show variety of microfacies. Matrix is peloidal grainstone with a few bioclasts. Magnification is 4.5. 4: Cross-bedded intraclast conglomerate with abundant flat-pebbles and shelter porosity. Large pebbles float in matrix of peloidal grainstone. Magnification is 4.5. 5: Microbial laminite. In places, well-preserved *Girvanella* (arrow) are observed that encrust microbial layers. Magnification is 4.5. 6: Wackestone with sponge-fragment (arrows) is overlain by thin intraclast layer. Entire sediment is strongly biomottled. Magnification is 4.5.

Fragments of trilobites and gastropods as well as peloids served as nuclei for the radially concentric ooids. Grapestones or coated grains are also present. In many oolite beds almost all ooids are diagenetically altered to microquartz or chalcedony. Individual beds rarely exceed 50 cm in thickness and many of them show cross-bedding. In places, onlap of oolite sands onto thrombolite mounds has been observed.

Intraclast grainstones (3 in Fig. 18) and intraclast conglomerates are irregularly distributed throughout the sections. Some of them fill small channels. In places, intraclasts within the conglomerates are imbricated (4 in Fig. 18). Many of the beds show ripple-drift cross-laminations. The intraclasts are mainly dark gray to brown mudstones (6 in Fig. 18), peloidal grainstone, or reworked microbial mats. In addition, some oncoids have been observed (6 in Fig. 18). In a few beds there are lithoclasts that are entirely composed of oolitic rocks.

Flat-pebble conglomerates and breccias show large tabular clasts in a peloidal-grainstone matrix. These conglomerates are mostly monomict and the clasts are either peloidal grainstone or mudstones and microbial boundstone.

Facies associations and depositional environment. Biologically mottled mudstones and wackestones (wackestone association) were formed in a subtidal environment where carbonate mud was produced and deposited (Fig. 19). Slow sedimentation rates are indicated by the presence of numerous hardgrounds and bedding-parallel trace fossils. The scarceness of ripples together with the rare and low-diversity fauna indicate stagnant waters in which the absence of wave and current activity led to unfavorable (hypersaline?) living conditions. Garrett (1977) described a similar environment from the interior of the Bahamian platforms, where a low-diversity fauna with few individuals is responsible for an almost total biological mottling of the muds. Wackestones and packstones with abundant nautiloids and gastropods indicate a temporary improvement of environmental conditions, which also locally enabled receptaculitids and sponges to settle.

The peloidal-grainstone association has its classic counterpart in the muddy sand lithofacies of the Bahamas (Newell et al., 1959), which occupy shallow subtidal to lower intertidal environments (Fig. 19). Storm and tide reworking of early lithified sediments provided a source for the abundant intraclast layers and the intraclast-peloidal grainstones.

The oolitic rocks of the La Silla Formation (oolite association) were discussed in detail by Cañas (1995a, 1995b), who concluded that the oolitic grainstones resemble accumulation in sand shoals, which is corroborated by the frequent cross-bedding in the rocks. Although some of the oolites may represent tidal-bar facies (herringbone cross-stratification), the majority of the ooid sands were deposited as laterally extensive sheets following major storm events. A distinction between both types is not possible, because this distinction is based on the orientation of the sands with respect to the platform margin (Ball, 1967; Halley et al., 1983). In both sand types, tidal activity is the primary agent in sediment distribution and formation of bed forms. A secondary overprint is due to storms, which are responsible for the formation of channels and spill-over lobes (Perkins and Enos, 1968; Hine, 1977) as well as for important sediment transport toward the platform margin and into the basin. Storms are also responsible for the reworking of semilithified muds, which are incorporated into the tempestites as intraclasts.

Other constituents of the oolite association are thrombolites and microbial mounds, which grew in subtidal areas and were onlapped by the carbonate sands. The giant stromatolites described by Dill et al. (1986) from tidal channels in the Exuma Sound (Bahamas) are a recent counterpart.

The microbial-boundstone association with its laminites, abundant mudstones, and its associated sedimentary structures is interpreted to be of higher intertidal to supratidal origin (Fig. 19). This is shown by the growth forms of the laminites and stromatolites, mudcracks, birdseyes, and evaporite pseudomorphs. Thin layers of grainstone and packstone, mostly with intraclasts, are the products of storms that transported subtidal material onto the intertidal and supratidal flats. In this environment, flat-pebble conglomerates and breccias resulted from desiccation and reworking of carbonate layers.

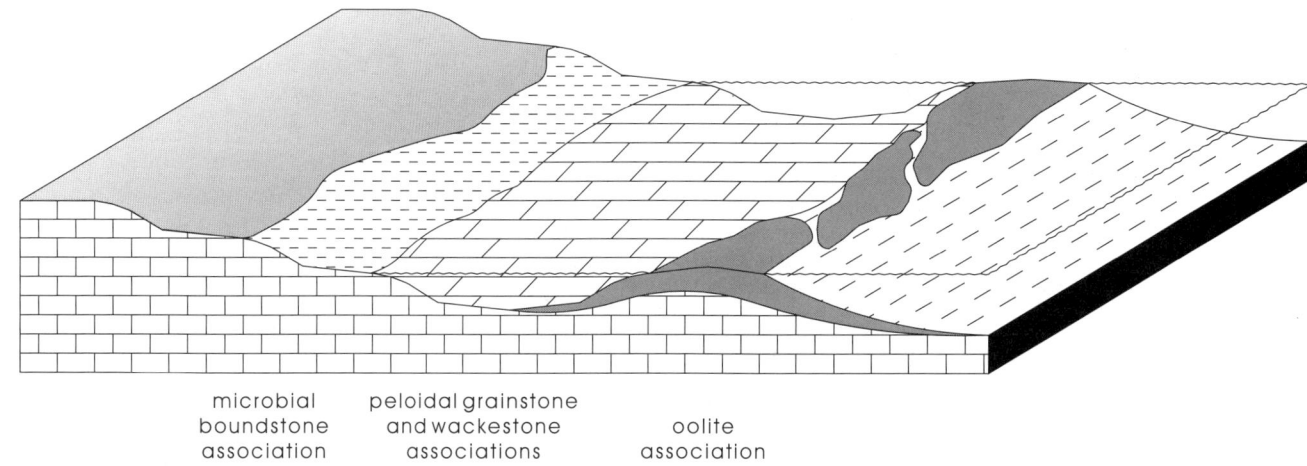

Figure 19. Depositional environments and facies associations of La Silla Formation. Legend as in Figure 14.

Platform configuration. The sections of the La Silla Formation are all very similar in their facies succession (Fig. 20), which is dominated by restricted subtidal and peritidal facies. The oolite association is present in all sections of the Talacasto subbasin and in the sections of the San Juan subbasin, but thick oolite accumulations developed only locally (e.g., Cerro La Chilca; Figs. 17 and 20). The larger oolite sands were effective enough to protect the platform interior from most wave and tidal currents and hence created restrictive conditions on the platform interior. Sediment accumulation in the intertidal and supratidal environments led to the formation of microbial boundstones and storm-induced mud layers with mudcracks and evaporites.

The absence of extended subtidal flats and a distinct slope make it difficult to interpret the platform configuration. In the context of the carbonate ramp (see Appendix 1) model (Aigner, 1985; Read, 1985; Burchette and Wright, 1992), only two facies belts are recognized: the oolite sand bodies of the high-energy barrier zone and back-barrier, protected lagoonal and peritidal environments. Cañas (1995a, 1995b) considered a rimmed-shelf model (see Appendix 1) the most likely for the deposition of the La Silla Formation with the oolites being margin-related accumulations. This model is attractive in that it explains the presence of deeper water facies as related to a lagoonal environment in back-barrier settings. However, in the outcrop area of the La Silla Formation, there are no indications of a nearby shelf break and the oolites are present all across the platform. The westernmost outcrops of the La Silla Formation at Gualilán, Talacasto, and Pachaco (Fig. 21, see oversized plate at back of volume), where

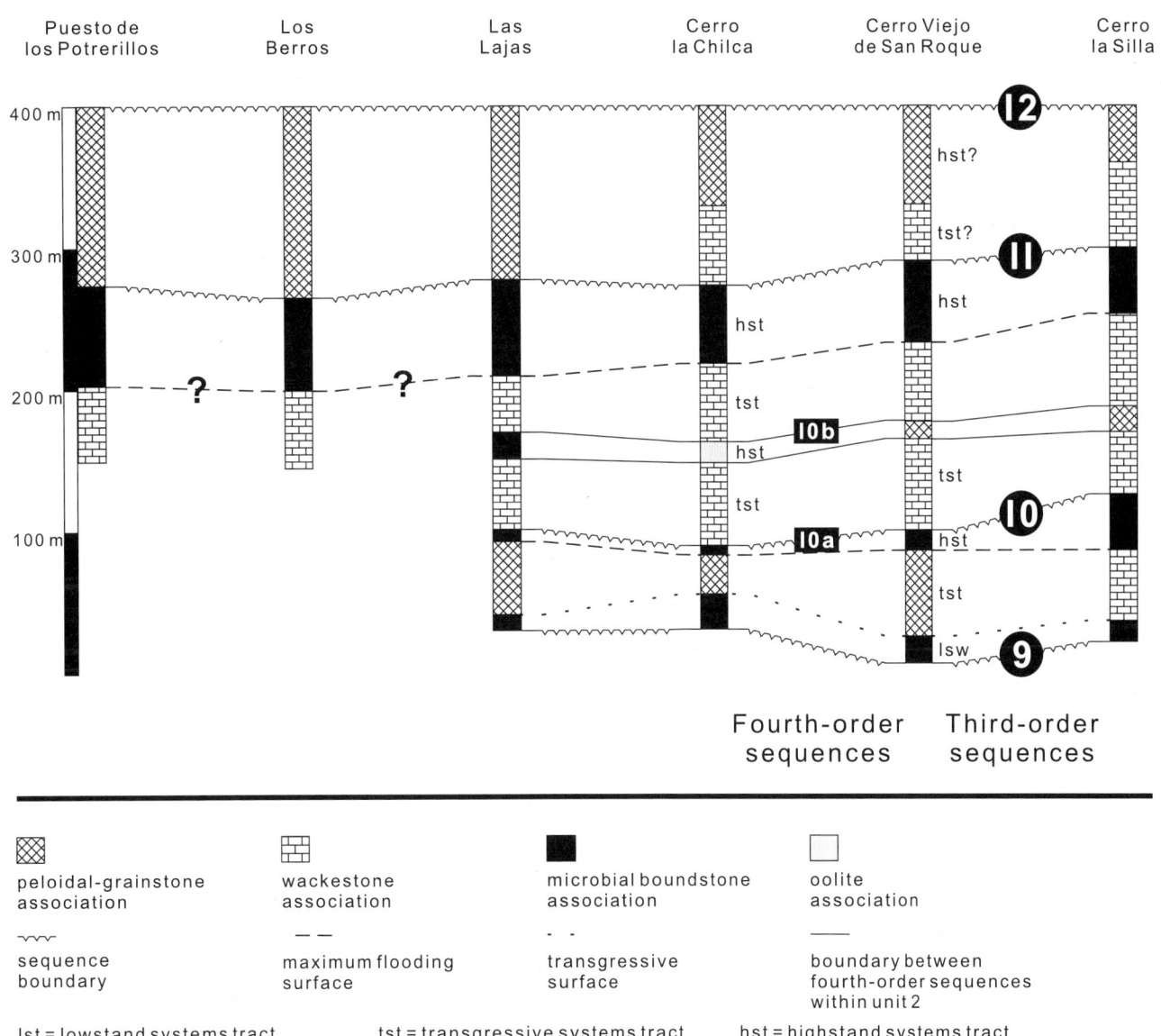

Figure 20. Sequence-stratigraphic interpretation of La Silla Formation. Third-order sequence numbers are those of entire carbonate platform succession.

only the upper part of the formation is exposed, still show the wackestone and peloidal-grainstone associations. No microbial boundstones or other indicators of an intertidal environment are present there. If this is taken as evidence of a slightly deeper environment, then a ramp-like configuration of the sedimentary environment of the La Silla Formation seems to be a viable alternative.

Sedimentary succession of the La Silla Formation

Talacasto subbasin. Near Jáchal, the La Silla Formation consist of three units (sequences 9 through 11; Figs. 17 and 20) easily distinguished in the field. The middle unit is dark gray and sandwiched between the lower and the upper units, which both show light gray and white rocks. This subdivision is also reflected in the sediment composition. Unit 1 begins with 20–25 m of peloidal grainstones, dolomitic limestones, and microbial laminites (Fig. 20). Upsection, peloidal grainstones with minor intercalations of microbial laminites form the bulk of unit 1. The rocks are thin bedded but rapidly pass to massively bedded successions upsection. Toward the top of unit 1, which is made up of the microbial-boundstone association, chert becomes increasingly abundant and commonly masks the microbial structures. The thickness of this basal unit is around 100 m in the Cerro La Silla and Cerro Viejo de San Roque sections, but less than 60 m in the Cerro La Chilca section (Fig. 17; see also Fig. 20).

Unit 2 starts with dark gray rocks of the wackestone association with intercalations of peloidal grainstone. Bioturbation and biologically mottled limestones are much more common than in the other units. Upsection, more and more intraclast conglomerates and microbial laminites are intercalated with peloidal grainstones. In the upper part, microbial laminites are dominant. In addition, many of the rocks are dolomitic. In the Cerro La Silla section, the top of this unit is marked by a thin breccia bed with chert clasts cutting erosionally into the underlying rocks. This upper part of unit 2 is strongly silicified; beds of pure chert more than 2 m thick have been observed (Fig. 17).

Unit 2 is composed of two shallowing-upward successions. The basal succession starts with mudstones and peloidal grainstones. In its upper part, there are some microbial laminites and a prominent thrombolite bed more than 1 m thick. Silicification is very intense in the thrombolites and the microbial laminites. The change from mudstones and grainstones toward organogenic rocks is accompanied by a change toward light colors. In the Cerro La Chilca section, a major oolite package forms the upper part of the lower shallowing-upward succession.

The upper succession within unit 2 shows a comparable evolution of facies and a similar change of color. The mudstones and wackestones in the lower part of this cycle show a more diverse fauna than observed in the other units. Sponges and rare receptaculitids have been observed besides trilobites, gastropods, and *Hyolithes*.

Unit 3 is an alternation of gray and light gray sediments of the wackestone association and intraclast grainstones, which upsection pass into white and gray dolomitic rocks of the peloidal-grainstone association (Fig. 20). Some of the sediments are of microbial origin.

San Juan subbasin. A correlation of the three main units of the Talacasto subbasin to the Las Lajas section (Fig. 3) is possible, and even the subdivision of the second unit into two shallowing-upward successions is recognized (Fig. 20). In the Las Lajas section, the lower two units each end with prominent microbial laminites (Fig. 17). The upper unit consists predominantly of the peloidal-grainstone association (Fig. 20). Farther south, in the Los Berros and the Puesto de los Potrerillos sections (Fig. 3), only two units are preserved below the San Juan Formation. On the basis of lithostratigraphic correlation and the sedimentary succession, the two units in the south are correlated with units 2 and 3 in the Las Lajas section and, consequently, with units 2 and 3 of the sections in the north. In the Los Berros and Puesto de los Potrerillos sections, no subordinate successions are recognized within the units.

The lowermost exposed rocks of unit 2 are dolomitic limestones. Muddy textures prevail; however, grainstones and packstones with intraclasts, nautiloids, and scarce trilobites are present. In places, oolitic limestones are developed. Toward the top of the unit, microbial laminites become more abundant, alternating with fenestral mudstones. In both sections, unit 2 ends with a thick microbial laminite (Fig. 17).

The upper unit (unit 3) shows a basal succession of the wackestone association with peloids and few algae, which upsection pass into peloidal grainstones with gastropods and scarce trilobites. The uppermost interval of this unit is composed of peloidal grainstones and mudstones with abundant fenestral fabrics, mainly birdseyes, and is also present in the more outboard sections at Gualilán and Pachaco. Most of the rocks are thick bedded and gray. In the southern sections, this succession is about 125–130 m thick.

Evolution and sequence stratigraphy of the La Silla Formation. The three units present in the La Silla Formation each record a shallowing-upward trend. Although the components of these successions vary slightly, the framework of these shallowing-upward successions consists of a basal subtidal unit, mainly represented by the wackestone association. The successions then show an interval of packstones and peloidal grainstones with increasing amounts of intraclasts, which reflect deposition under more turbulent conditions. Together with ripples and small channels these rocks indicate a shallow subtidal to lower intertidal environment, where waves constantly winnow and rework the sediments.

The upper part of each of the lower two units is composed of abundant microbial laminites and fewer stromatolites, which are the sites of important chert accumulations. Although shallow subtidal and lower intertidal sediments are present, the majority of the upper part of both units formed in intertidal and supratidal environments, as indicated by the low-relief boundstones and accompanying mudcracks.

The uppermost succession (unit 3) differs in that its upper

part consists of mudstones with abundant birdseyes and some thrombolites. Birdseye formation is attributed to intertidal (Shinn, 1983b) or supratidal environments (Bathurst, 1971). As unit 3 lacks any other evidence of supratidal deposits, the thick birdseye limestones are interpreted to have formed in the intertidal realm. Hence this unit lacks a well-developed supratidal cap. The boundaries of the units reflect abrupt changes in relative sea level. They are depositional sequences according to sequence stratigraphy view (Sarg, 1988; Handford and Loucks, 1993; Fig. 20). Sedimentation of the La Silla Formation began above a distinctive surface, which in the type section is characterized by the presence of dark silty dolomites. The change from peritidal dolomites of the La Flecha Formation with abundant chert and evaporites toward grainstones with mud mounds marks the flooding of the peritidal platform and the onset of another depositional regime. Consequently, the boundary between the two formations is regarded as a sequence boundary (sequence boundary 9 of the carbonate platform succession; Figs. 7 and 20). The silty dolomites in the Cerro La Silla section are accompanied by large dolomitic extraclasts in the overlying rocks, which required a prolonged time for lithification, erosion, and reworking. A similar horizon is present in the Puesto de los Potrerillos section. This indicates that the entire platform was affected by a sea-level drop and the accompanying basinward shift of facies. Consequently, it is likely that the boundary is a type 1 sequence boundary. Above the sequence boundary, there is a cyclical interval with dolomitic limestones and microbial laminites (Fig. 20). The individual parasequences shoal upward from peloidal grainstone or wackestone toward laminites. The entire parasequence set is interpreted to represent the lowstand deposits above the sequence boundary. Because they are connected to a type 1 sequence boundary, they form a lowstand systems tract (Van Wagoner et al., 1988). In this context, the silt of the silty dolomites may have been washed in during the time of absolute sea-level lowstand.

The transgressive surface is located on top of the last microbial laminite of the underlying lowstand systems tract (Fig. 20). Above it, the basal sequence in the La Silla Formation (sequence 9 of the carbonate platform succession) shows peloidal grainstones at the base that pass upsection into mudstones and microbial laminites or, as in the Las Lajas section, into oolitic grainstones with few laminites. The subtle change from peloidal grainstones to microbial laminites is interpreted to reflect the change from the trangressive systems tract toward the highstand systems tract.

Unit 2 of the La Silla Formation consists of two shallowing-upward successions (sequences 10a and 10b; Fig. 20). The basal one starts with subtidal rocks as the result of a retrogradational pulse and is topped by a thick package of oolitic grainstones in the Cerro La Chilca section. Elsewhere, peloidal grainstones, microbial laminites, or thrombolites mark the top of the succession which in its entity is regarded as a fourth-order sequence within the second unit (sequence 10). The flooding on top of this fourth-order sequence is easily recognized and marks the next retrogradational pulse with the return to subtidal conditions, and a locally abundant fauna including sponges and receptaculitids. Continued flooding, the rate of deposition more or less equal to accommodation, maintained subtidal conditions at the base of sequence 10. Finally, however, carbonate production outpaced relative sea-level rise, resulting in a progradational pattern in the sediments. The change from aggradation to progradation marks the limit between the transgressive systems tract and the highstand systems tract and consequently is the maximum flooding surface (Fig. 20). In the middle sequence of the La Silla Formation (sequence 10), the highstand systems tract is dominated by dolomitic limestone with abundant fenestral fabrics and thick microbial laminites (Fig. 17). The contact to the overlying unit three is sharp and represents a transgressive, marine flooding surface along which there is locally evidence of subaerial exposure (abundant mudcracks and a breccia bed in the Cerro La Silla section; Fig. 17). Biostratigraphic data in this part of the La Silla Formation are very sparse and do not allow the recognition of a possible hiatus, so the breccia bed and the accumulation of mudcracks at this surface are the only arguments to consider the sequence boundary a type 1 boundary. From a regional perspective, it is much more likely that the sequence boundary (sequence boundary 11 of the carbonate platform succession; Fig. 20) represents a type 2 boundary, because there is no well-defined basinward shift in facies or areally exposure or karstification.

The third sequence in the La Silla Formation is less well defined. Flooding above the sequence boundary led to the return of a normal-marine subtidal environment in the northern sections, whereas the southern sections are dominated by the accumulation of peloidal muds. There is hardly any variation visible in the Puesto de los Potrerillos or Los Berros sections. In contrast, in the north there is a tendency toward more oolitic grainstone and thrombolitic mounds toward the sequence boundary that marks the top of the La Silla Formation. In general, this upper sequence seems to be dominated by aggradation; only in its uppermost interval is there some expression of progradation. The uppermost sequence within the La Silla Formation lacks a distinct supratidal cap.

San Juan Formation

The San Juan limestones have been known since the pioneer work of Kayser (1876) and Stelzner (1873). At that time, the limestones were attributed to the "Infra-Silurian". Kobayashi (1937) proved that the upper fossiliferous part of the entire carbonate succession was of Ordovician age. However, it was not until recently that a formal definition of the San Juan Formation was given (Keller et al., 1994). The lower boundary coincides with the sudden appearance of a fully developed and very diverse marine fauna, which is commonly accompanied by a lithologic change from thick-bedded to thin-bedded and platy limestones. Hitherto, the base of the San Juan Formation has only been dated in the type section at Cerro La Silla (Keller et al., 1994; Lehnert, 1995a, 1998c). There, conodonts indicate a late Tremadocian age for the base of the San Juan Formation.

The San Juan Formation is conformably overlain by a variety of formations (Gualcamayo Formation, Los Azules Formation, Las Aguaditas Formation; Fig. 4). In some places, a characteristic package of dark lime mudstones, marlstones, and shales is present at the transition (transfacies calcáreo-pelíticas of Baldis and Beresi, 1981).

From a regional point of view, five groups of sections can be distinguished. In the San Juan subbasin sections are present (Fig. 3) in the Sierra de Villicum (Don Braulio) and Sierra Chica de Zonda (Las Lajas, Los Berros, Puesto de los Potrerillos). In the Talacasto subbasin, the sections around Jáchal in the north (Cerro La Silla, Cerro Viejo de San Roque, Cerro La Chilca, Agua Hedionda, Niquivil; Fig. 3) are discussed as the northern sections; sections along the Río San Juan (Río Sassito, Río Sasso, Tambolar, Pachaco; Fig. 3) as the southern sections; and sections in the western thrust sheets north of the Río San Juan (Talacasto, Gualilán, Las Chacritas; Fig. 3) as the western sections. There are sections north of the Río Jáchal toward Guandacol that were well documented and interpreted by Cañas (1995a, 1995b). The sections measured in the limestones of the San Juan Formation and their lithologic correlations are documented in Figure 21.

Time frame of deposition. Although detailed paleontologic work on the San Juan Formation began with Kobayashi (1937), there are only a few sections that are biostratigraphically well dated and zoned. There are only isolated data from most sections and they do not allow a precise biostratigraphic correlation to other sections.

The first biostratigraphic subdivision of a section of the San Juan Formation dates back to Serpagli (1974), who found five characteristic conodont assemblages (faunas A–E; Fig. 22) in a section along the Río San Juan. One of the biggest shortcomings of this paper is that there is no mention from which locality the samples were obtained. The map shown in the paper (Serpagli, 1974, p. 4) indicates a position between the Pachaco and the Tambolar sections. This has led to confusion, e.g., when dating the top of the formation. Lehnert (1995a), considering the Pachaco locality (Fig. 3) as the original described section, pointed out the discrepancies between his data from the Pachaco section and those of Serpagli (1974). However, if it is assumed that the samples were originally obtained from the Tambolar section (Fig. 3), the Serpagli data fit those of Lehnert (1995a, 1995b) and thicknesses given in Serpagli (1974), and those reported herein are almost identical (Fig. 22). In addition, the interpretation of the biostratigraphic data in connection with the sedimentary record of the Tambolar section fits well into the regional picture. Consequently, in this paper the correlation of the Tambolar section to the other sections of the San Juan Formation, as documented in Figure 21, is based on the assumption that Serpagli's (1974) samples are from the Tambolar section.

Another subdivision of the San Juan Formation into conodont assemblage zones was given by Lehnert (1990, 1993) for the Niquivil section (Fig. 21). These zones are widely applicable and proved very helpful in dating various parts of the formation (e.g., Lehnert, 1995a; Lehnert and Keller, 1993b; Keller and Lehnert, 1998).

A brachiopod zonation of the Cerro La Silla section was established by Herrera and Benedetto (1991), who subdivided the San Juan Formation into five brachiopod zones. These zones span the late Tremadoc through early Llanvirn. The Niquivil section (Fig. 3) is located in the same range as the type section at Cerro La Silla (Fig. 3), and because this section has been studied in detail (Lehnert, 1995a) there is a good correlation of brachiopod zonation and conodont biostratigraphy in this part of the Precordillera.

The top of the San Juan Formation has been dated in various localities (Ortega et al., 1985; Sarmiento, 1986, 1990; Albanesi, 1991; Lehnert, 1995a, 1995b). These studies revealed a marked difference in age between the sections around Guandacol (Fig. 3) and those to the south of the Río Jáchal. Near Guandacol, there are sections where the black shales of the Gualcamayo Formation yielded graptolites of the late Arenig *Isograptus victoriae* zone (Ortega et al., 1985). From a section a bit farther south (Cerro Potrerillo), Albanesi (1991) described conodonts and graptolites from the Arenig-Llanvirn boundary. South of the Río Jáchal, however, the top of the San Juan Formation and the transition into the overlying formations is always of early Llanvirnian age (*Eoplacognathus suecicus* zone: Sarmiento, 1986, 1990; Lehnert 1995a, 1995b).

Lithofacies. The San Juan Formation is the most widespread of all carbonate successions and shows the most variations in facies and lithology, perpendicular as well as parallel to depositional strike. It is composed of a variety of carbonate rocks that cover the entire range of carbonate textures.

Birdseye mudstones are thick bedded and light gray. Within the reef succession they are thick bedded, but near the base of the formation at Cerro La Silla they are thin bedded. There, mudstones with abundant pseudomorphs after gypsum have been observed alternating with the birdseye limestones.

Whole-fossil mudstones and wackestones (Wilson, 1975) are present in medium- to thick-bedded intervals and show a diverse fauna with brachiopods, trilobites, crinoids, gastropods, nautiloids, ostracods, and sponge spicules. Among the algae *Nuia* and *Girvanella* are the most prominent; in places, almost monomictic *Nuia* wackestones were deposited. However, in the upper reef mound interval, calcareous algae are much more important than *Girvanella* or *Nuia*. The latter is rare in these horizons. The color of the mudstones and wackestones is gray and dark gray.

Thin-bedded and platy black mudstones and wackestones (6 in Fig. 23) are present at the top of the formation, where they alternate with black shales. These rocks, which constitute the "transfacies calcareo-pelitica" of Baldis and Beresi (1981), mainly contain deep-water trilobites and conodonts.

Nodular wackestones (5 in Fig. 23) are present in the lower third of the sections and near the top of the formation (Fig. 21). They contain the same fauna and flora as the whole-fossil wackestones. In the upper interval, abundant large *Maclurites* together with *Lituites* sp. and orthocones more than 2 m long have been observed (e.g., Talacasto west; Fig. 3). In the San Juan subbasin,

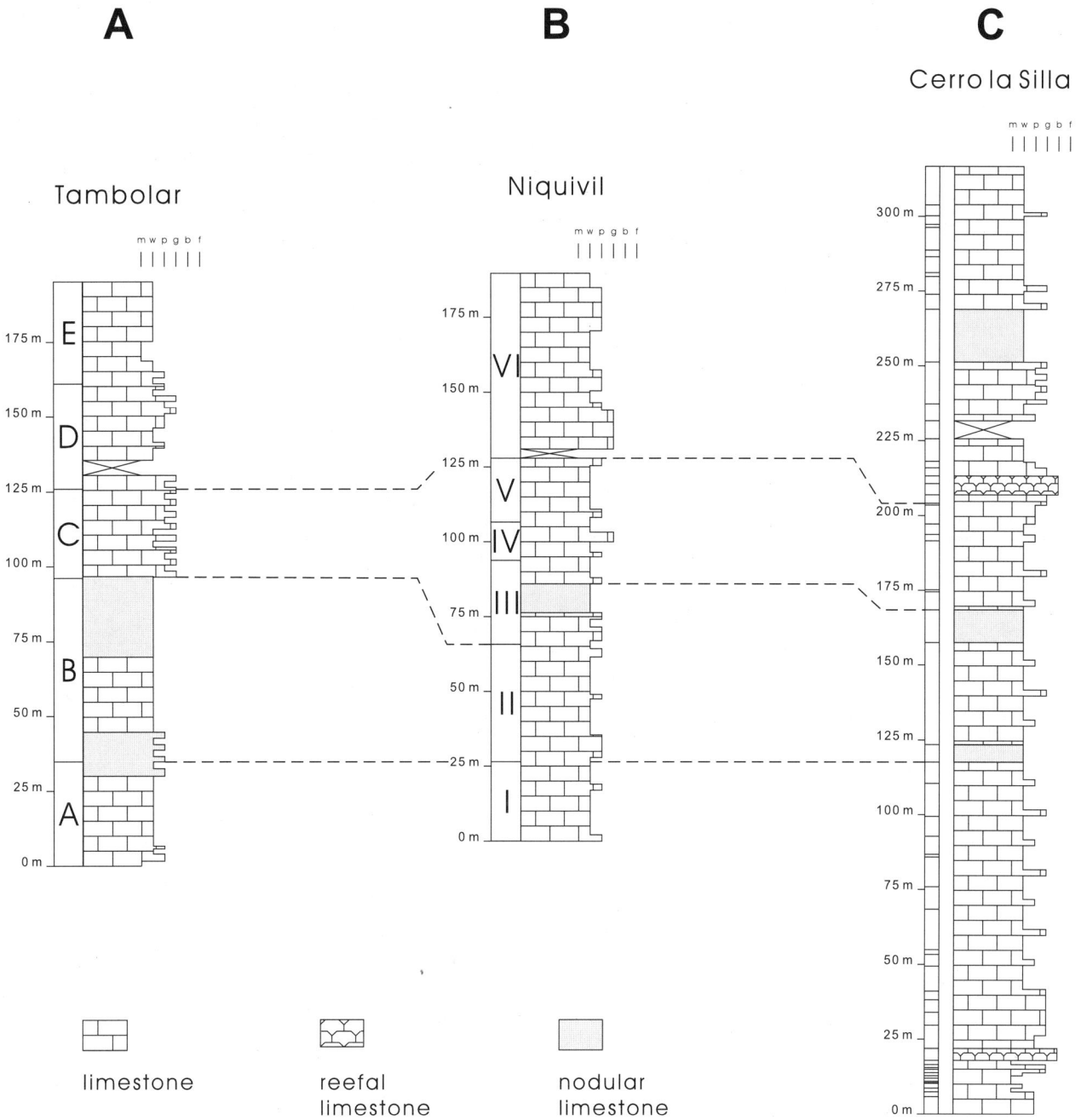

Figure 22. Biostratigraphic subdivisions of San Juan Formation and their correlations to type section at Cerro La Silla (C). B shows zonation of Niquivil section (modified from Lehnert 1990, 1993). A is Serpagli's (1974) subdivision plotted against measured section at Tambolar. Limits between faunas A to E are not fixed, but are best fits. b = boundstone, f = framestone, g = grainstone, m = mudstone, p = packstone, w = wackestone.

the nodular interval in the lower third of the sections is dominated by yellow, red, and green colors. The nodular texture is either caused by intense bioturbation or pressure solution. Diagenesis together with an elevated primary clay content make these rocks transitional to marly limestones. At Cerro La Chilca, there is a prominent hardground developed with vertical burrows penetrating several centimeters into the uppermost bed of the San Juan Formation.

Nodular packstones together with some wackestones and marlstones are present at the top of the limestone succession in Las Chacritas section. These rocks are not typical of the San Juan Formation. They are dark gray and thin to medium bedded, and some of the beds are graded. The individual beds are often separated by dark gray to black shale partings. The fauna consists of abundant sponges or their remnants (spiculites) and fragments of small trilobites, brachiopods, crinoids, and *Nuia*. Ostracods and calcareous algae are present only to a minor extent. The dark matrix is often peloidal and locally shows small intraclasts. In some horizons, bioturbation is pervasive and responsible for the destruction of primary sedimentary structures as well as, at least in part, the fragmentation of the bioclasts.

Bioclastic wackestones (1 in Fig. 24) and packstones are abundant in all sections. Their fauna and flora are similar to those of the mudstones and wackestones; however, many of the individuals are broken or abraded. In addition, sponge fragments (3 in Fig. 24) and peloids may be present. Many of the packstones are dominated by one faunal element; i.e., crinoidal packstone or trilobite packstone are developed, among others. With increasing abundance of lithoclasts, there is a transition into bioclastic-lithoclastic packstones (4 in Fig. 24). The lithology of the lithoclasts is often that of the underlying beds (intraclasts; 5 in Fig. 24) but (dolomitic) mudstones (extraclasts) have also been observed.

In some sections, oncolitic packstones prevail (6 in Fig. 24). The size of the oncoids varies between 1 and 3 cm. The bed thickness of the bioclastic wackestones and packstones changes from thin bedded to thick bedded. Many of the thin beds show an erosional base, and some are graded and/or cross-bedded.

Calcisiltites are rare and present only in thin beds toward the top of the formation. Traction-induced bedding is responsible for a thin lamination or low-angle cross-bedding. In places, these beds are graded.

Among the grainstones three groups are distinguished. Bioclastic grainstones show broken or abraded clasts of a variety of fossils. Intraclasts are present locally and where they are abundant, intraclast grainstones and intraclast-bioclast grainstones are developed. Some of the components show micrite envelopes. Pelmatozoan grainstones are commonly cross-bedded and well sorted. They are present as continuous beds or as small channel fills.

Rudstones are restricted to the reef-bearing horizons where they fill erosional channels between the mounds or onlap them. The fauna is dominated by fragments of sponges, receptaculitids, and, in the upper horizon, rare stromatoporoids. These clasts are observed in a matrix of bioclastic or crinoidal grainstone.

Boundstones are present in the two reef-mound horizons and their lateral equivalents. They were described in detail by Cañas and Carrera (1993), Cañas and Keller (1993), Keller and Bordonaro (1993), and by Keller and Flügel (1996). In the lower reef-mound horizon the boundstones are part of a complex reef rock association. The mound horizon is composed of numerous bioherms, which are found as isolated mounds or coalesce to form a three-dimensional network of mounds (3 in Fig. 23). The main components are sponges, the receptaculitid *Calathium* and *Girvanella* (4 in Fig. 23). Microbial mats and *Girvanella* encrust rock fragments and organisms to form solid bindstones. A detailed description of these facies was given by Cañas and Carrera (1993).

Large isolated bioherms are present in the upper reef interval (2 in Fig. 23). They consist of a solid framework of stromatoporoids (Keller and Flügel, 1996). Some of these framestones still contain internal wackestone with abundant algae and brachiopods. The mounds are onlapped by pelmatozoan grainstone. Some of the mounds are covered with packstones in which gastropods (*Maclurites*) and nautiloids are concentrated. Thick biostromes are also observed in the upper reef interval. Some of them are composed entirely of stromatoporoids with a variety of growth forms. Large domical individuals are embedded within massive grainstone in the Don Braulio section (1 in Fig. 23). High columnar forms are present in massive wackestone and packstone in the Puesto de los Potrerillos section. There, some of the carbonate mud may have been passively baffled by the stromatoporoids; however, the influence of current activity is clearly visible in many places. Other biostromes are a mixture of stromatoporoid and sponge-algal associations (Cañas and Keller, 1993). The latter are very similar to the mounds in the lower reef horizon. The mounds form small patch reefs, which in places are eroded to form rudstone. Rudstone and grainstone form the matrix for these small reefs.

Sponge biostromes are present in the Talacasto area (e.g., Beresi, 1986b; section Talacasto east; Fig. 3) and at Cerro La Chilca and include thick-bedded wackestones, packstones, and grainstones. The fauna is dominated by a variety of sponges that seem to have trapped the sediment.

Facies associations and depositional environment. A widespread facies association of the San Juan Formation is the wackestone-intraclast packstone association (Fig. 25). Wackestones and rare mudstones with well-preserved fossils represent the background sedimentation on the carbonate platform. The diversity and abundance of fossils indicate normal marine conditions within the photic zone. This background sedimentation was frequently interrupted by storms, which stirred up the sediments, destroyed habitats and colonies of animals, and resedimented this mixture to form tempestites. Similarly, the wackestone-oncolitic packstone association (Fig. 25) documents background sedimentation with storm events. However, the source area for the tempestites differed and was probably a belt of nearby oncolite shoals, which today are nowhere exposed.

There are clear proximal-distal trends in the tempestites. In the sections of the San Juan subbasin (Fig. 3) tempestites are rare. They are very abundant in the sections around Jáchal and from there toward the west and southwest. In the northern sections of the Talacasto subbasin (Fig. 3) tempestites are thicker and more often show erosional features at the base, grading, and cross-bedding. Farther to the west, these sediments in general are thinner and have a more regular base. Cross-bedding there is rarely present. There are also compositional variations across the platform. In the eastern sections, abundant oncolite beds are present that are absent in the west. In addition, intraclasts become less abundant toward the west. In the Pachaco section (Fig. 3), no intraclasts or oncoids have been observed (Fig. 21). The tempestites there are thin layers (1–10 cm) of bioclastic wackestone and grainstone.

Figure 23. Field aspects of San Juan Formation. 1: Large isolated stromatoporoid in main biostrome of Don Braulio section. Dark layers consist of pure chert. Hammer is 32 cm long. 2: Framestone of stromatoporoids with laminar growth forms. Upper reef interval, Las Lajas section. Hammer tip is 5 cm long. 3: Lower reef-mound horizon at Cerro La Silla. Irregular surfaces are depositional surfaces of individual sponge-algal mounds. Inner ramp environment. 4: Outcrop of mound facies with abundant sponges and receptaculitids. Large sponge in upper right corner is ~5 cm long. Cerro La Silla section. 5: Lower nodular limestone interval in Cerro La Silla section. Note shale partings between limestones. Succession is essentially free of tempestites and marks major deepening event. Deep-ramp deposit. Hammer is 32 cm long. 6: Deepening succession at Cerro La Chilca. Stratigraphic top is to right; succession shown is ~2 m thick. Note decreasing thickness of limestone beds and increasing contents of shale interbeds toward top. This succession is typical of "transfacies" and is part of drowning succession.

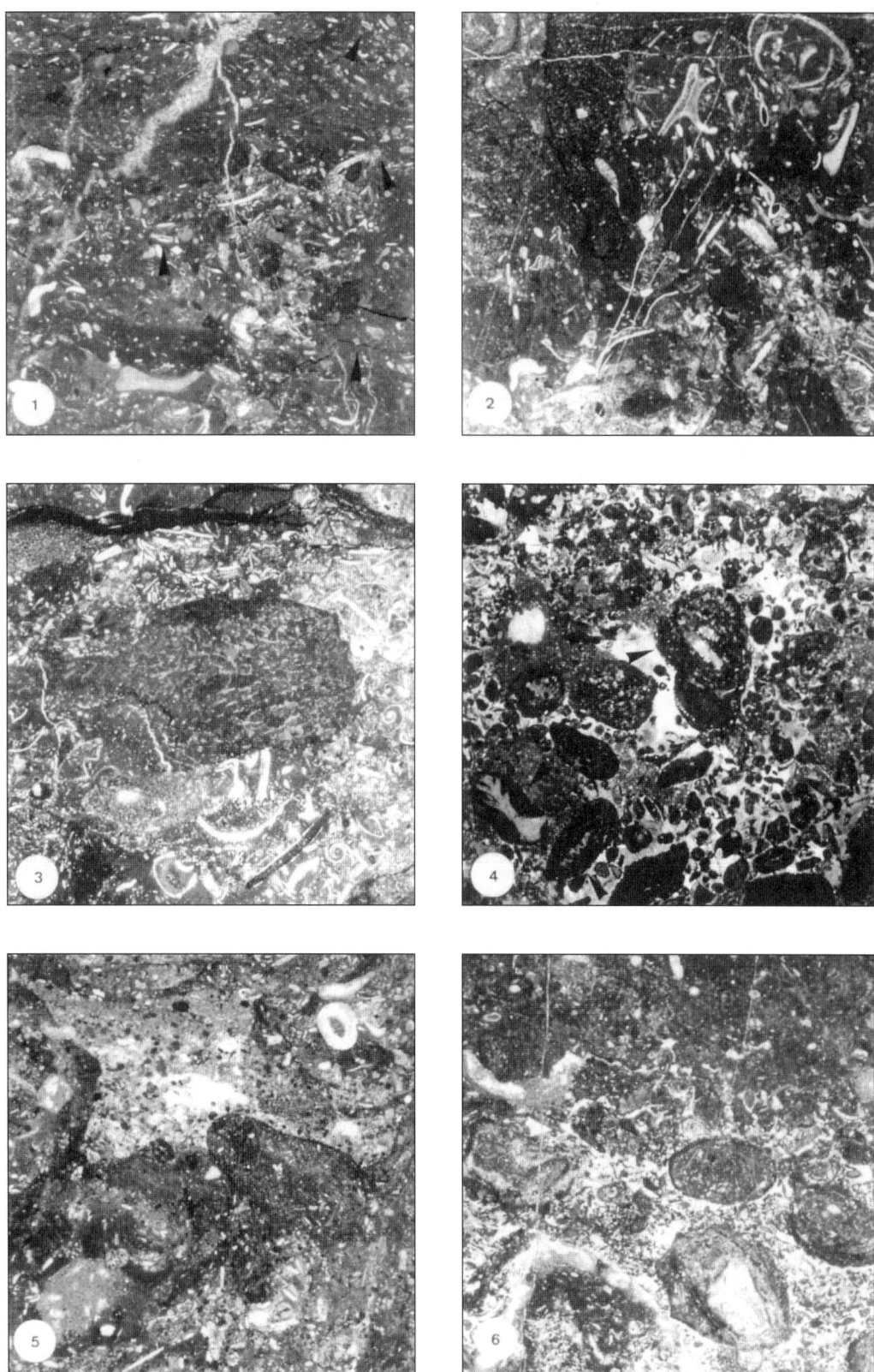

Figure 24. Facies of San Juan Formation. 1: *Nuia* wackestone-packstone. Most of small white bioclasts are cross sections of possible alga *Nuia* (arrows). In addition, rounded crinoidal fragments and other bioclasts are present. Mid-ramp deposit. Magnification is 4.5. 2: Bioclastic wackestone-packstone with gastropods, sponge fragments, and many other bioclasts. Many of these beds have tempestitic origin and were bioturbated after deposition. Mid-ramp deposit. Magnification is 4.5. 3: Bioclastic packstone of upper reef interval. Note abundant gastropods and large sponge fragment. Distal inner ramp environment. Magnification is 4.5. 4: Tempestitic packstone-grainstone with abundant intraclasts and oncoids (arrow). Note bioclasts, which served as nuclei for oncoids. Mid-ramp deposit. Magnification is 4.5. 5: Biomottled packstone with bioclasts and isolated intraclasts. Sediment is poorly sorted and contains crinoids, sponge spicules, *Nuia*, and ore particles. Magnification is 4.5. 6: Graded oncolitic tempestite with transition from grainstone to wackestone. Magnification is 4.5.

The nodular-wackestone association (Fig. 25) is dominated by whole-fossil wackestones and few mudstones. Tempestites are absent in this association. The rocks were deposited well below storm-wave base, but still within the photic zone. This is shown by the composition of the abundant and diverse fauna and the algae present. Brachiopods, pelmatozoans, and trilobites are frequent; reef-building organisms that contribute to the reef-mound horizons are absent. A similar facies was described from the Guandacol area (Cañas, 1995b) and interpreted in the same way.

The nodular-packstone association, only found in the Las Chacritas section, was deposited in the same general environment. Rare graded beds and shallow-water bioclasts indicate that some of the material was transported to a deeper environment, probably during storms. The intense bioturbation is responsible for the textural homogenization of sediment and, in addition, points to reduced sedimentation rates. This is corroborated by shale partings and thin shale interbeds in this association.

The reef and reef-mound facies association (Fig. 25) is the most complex and diverse group of rocks. The sections in the San Juan subbasin and the northern sections of the Talacasto subbasin around Jáchal (with the exception of Cerro La Chilca) are composed mainly of grain-supported rocks and boundstones (Fig. 21). The lower reef mound horizon at Cerro La Silla and its lateral equivalents at Agua Hedionda, Cerro Viejo de San Roque, and Cerro La Chilca are dominated by a sponge-algae-*Calathium* community, which was described in detail by Cañas and Carrera (1993) and Cañas (1995a, 1995b). The abundance of fragmented reef organisms within the mounds as well as in the neighboring sediment points to considerable turbulence in the shallow subtidal environment. This is also indicated by deep channels that cut into the mounds and that are filled with rudstone. In addition, the intermound sediment is also mainly coarse-grained packstone, grainstone, and rudstone, indicating that this shallow subtidal area was within the range of currents and tides (Fig. 25).

The upper interval also records a high-energy environment. Rigid stromatoporoid patch reefs are surrounded by well-sorted pelmatozoan grainstone with abundant tidal channels and ripple-drift cross-bedding. Lateral variations are biostromes in which the stromatoporoids are found in a grainstone matrix (Don Braulio section), but also in a wackestone-packstone matrix (Puesto de los Potrerillos section). This indicates a variety of subenvironments. A very protected environment, for example, is indicated in the Puesto de los Potrerillos section by a thin mudstone horizon with abundant birdseyes (birdseye mudstone association). This horizon forms the base for a level with small (<1 m) patch reefs composed of sponges, algae, and stromatoporoidal crusts. In the Los Berros section, a thick interval with birdseye mudstone is present (Fig. 21). Together with a few horizons with birdseyes and gypsum below the lower reef mound horizon in the Cerro La Silla section, these rocks are the only vestiges of an intertidal to supratidal environment (cf. Shinn, 1983b). No indications of subaerial exposure have been found (see also Cañas, 1995a). A back-barrier (lagoonal) environment present in protected areas of the proximal inner ramp is the most plausible interpretation (Fig. 25).

The lateral equivalent of the upper reef horizon is a packstone-grainstone association with intercalated biostromal boundstones. This association is present in all those sections where there are no true bioherms (Fig. 21). It is also present in those sections of the Talacasto subbasin in which the reef interval is preserved beneath the various younger erosional surfaces. Where present in the Talacasto subbasin, the interval is characterized by an alternation of packstones and grainstones, but wackestones are also found. The environment was slightly deeper and less agitated than that of the reef facies proper, but still within the range of tides and currents (Fig. 25). Isolated stromatoporoids and small colonies are only found in grainstones indicating that the low domical or bulbous organisms (growth forms according to Keller and Flügel, 1996) were sensitive to mud pollution. Within biostromal boundstones, sponges are present, many of them in

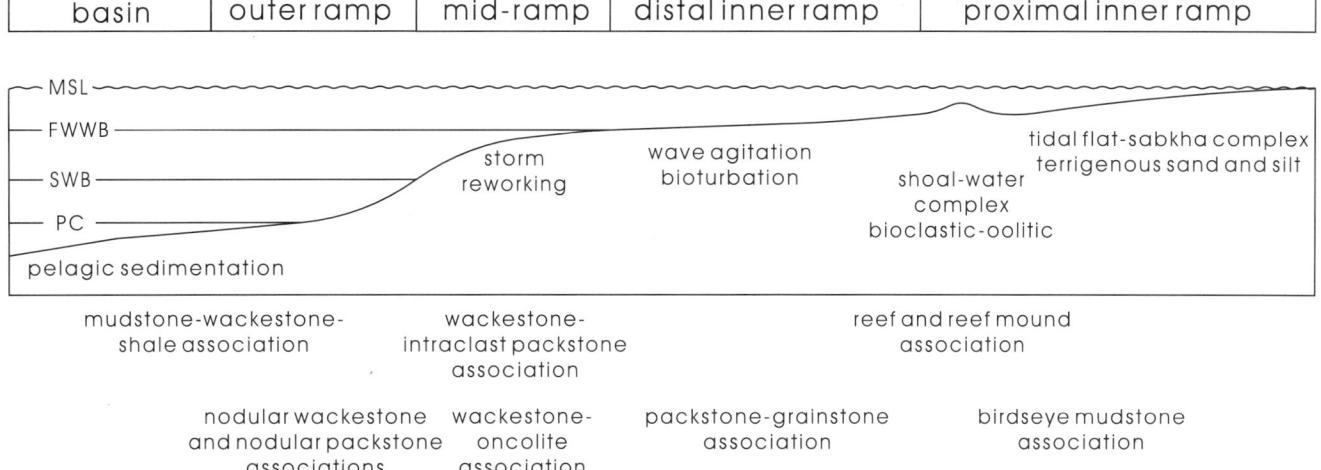

Figure 25. Facies associations of San Juan Formation and their attribution to depositional environments of carbonate ramp (modified from Burchette and Wright, 1992). MSL = mean sea level, FWWB = fair weather wave base, SWB = storm wave base, PC = pycnocline.

life position. In contrast to the stromatoporoids, the sponges grew in a mud-dominated environment and probably acted as bafflers.

The mudstone-wackestone-shale association is preserved in a few sections on top of the San Juan Formation, e.g., in the Cerro La Chilca section. The association is composed of dark to black mudstones and wackestones with intercalations of black shale beds. The latter may be as thick as 7 cm. The rocks were deposited under very quiet conditions well below storm wave base (Fig. 25). Rare calcisiltite beds within this association are interpreted to represent distal tempestites (Aigner, 1985). The limestones are typical deep-water sediments (Wilson, 1969) and a hemipelagic origin is invoked. Deposition probably took place near the boundary between oxic and anoxic bottom conditions (Cañas, 1995a), shown by the rapid change between trilobite-rich layers and anoxic mudstones without ichnofauna. This association is the "transfacies calcareo-pelítica" of Baldis and Beresi (1981), which marks the drowning (see Appendix 1) of the platform.

An additional facies within this association was described by Cañas (1995a). Between the Guandacol and Gualcamayo rivers, fine- to medium-bedded bioclastic packstones and grainstones are present. They are poorly sorted, show shelter porosity, and in places are graded. The grainstones are also found as small channel fills. The fauna is a mixture of autochthonous deep-water trilobites and platform-derived bioclasts, including algae. This facies is interpreted to represent distal tempestites deposited below storm wave base.

Platform configuration. No oolites or vestiges of a tidal-flat environment are preserved. The overwhelming majority of the sediments that constitute the San Juan Formation are of subtidal origin. The vast storm-influenced subtidal flats become progessively deeper toward the west, but there are no indications of a slope. The progressive deepening is most accentuated in the Guandacol area (Cañas, 1995a, 1995b).

The absence of a distinct slope is strong evidence that the San Juan Formation was deposited on a homoclinal ramp (Ahr, 1973; Read, 1985; Burchette and Wright, 1992). A similar model was developed by Cañas (1995b) for the area between Cerro La Silla and Guandacol. In the San Juan Formation, the inner ramp above fair-weather wave base is represented by the reef facies with its mounds and grainstones (Fig. 25). The packstone-grainstone association was deposited on the seaward side of the reefs. The presence of carbonate mud indicates a slightly deeper part of the inner ramp toward the transition to the mid ramp. Except for intervals of birdseye limestones, lagoonal deposits or the tidal flat are absent.

The middle ramp is the zone between fair-weather wave base and storm wave base. Tempestites are common in this segment and proximal-distal trends are usually recognized (Aigner, 1982, 1985; Burchette and Wright, 1992). In the San Juan Formation, the wackestone-intraclast-packstone and the wackestone-oncolite associations were deposited on the middle part of the ramp (Fig. 25).

The outer ramp shows much less evidence of storm influence because it is located below storm wave base. In the San Juan Formation, both nodular wackestone intervals as well as the nodular packstone association are outer ramp deposits (Fig. 25).

Adjacent to the outer ramp is the basin. Burchette and Wright (1992) addressed the problem of recognizing truly basinal facies and concluded that the attribution of sediments to a basinal setting is not always conclusive and that a combination of factors has to be used. The deep-water character of the mudstone-wackestone-shale association was previously stressed. Organic-rich mudstones and their transition into graptolite shales together with the absence of coarse-grained detrital material indicate a basinal environment. However, the tempestite facies within this association (Cañas, 1995a) are more typical of the outer ramp; hence the entire interval represents environments from the outer ramp to the basin (Fig. 25).

For paleogeographic reconstructions it is important to note that sediments that might have been deposited on the landward side of the barrier system are almost absent and that an important part of a complete platform is missing.

Sedimentary succession of the San Juan Formation. The discussion of the sedimentary succession and the lateral variations of the San Juan Formation is hampered by the fact that many sections are incomplete. Some of the sections are incomplete at the base (especially those along the Río San Juan) because thrust faults have cut up to the base of the San Juan Formation (see Von Gosen, 1992). Many sections, however, are incomplete toward their top, which is caused by at least four different phases of erosion (intra-Ordovician, uppermost Ordovician, Carboniferous, and Tertiary).

San Juan subbasin. The sections in the San Juan subbasin all start with mid-ramp deposits (Figs. 21 and 26). They contain a varied fauna and flora with abundant *Nuia*. The rocks are thin bedded and gray and correspond to the "Miembro de calizas lajosas" of Beresi and Bordonaro (1984). Upsection, distinct variegated nodular limestones (Miembro de calizas y margas varicolores) are present. No tempestitic packstones or grainstones have been observed in this outer-ramp succession. The nodular-wackestone association has a constant thickness, that varies between 25 and 35 m. Isolated conodont data (Beresi et al., 1987) from the Las Lajas section indicate the *Oepikodus evae* zone (conodont assemblage zone II) and the *O. evae/ O. intermedius* zone (conodont assemblage zone III of Lehnert, 1993) for the nodular-wackestone association. The latter is overlain by platy packstones and grainstones (mid-ramp) that form the base for the first major reef accumulation in the sections of the San Juan subbasin. The inner-ramp reef succession shows a variety of reef types and environments, among which are bioherms and biostromes as well as a lagoonal facies with birdseyes (Los Berros section). The reef interval is more than 150 m thick, but its top is only exposed in the Don Braulio and Puesto de los Potrerillos sections (Fig. 26). There, a sudden change from predominantly crinoidal grainstones to thick-bedded, in places marly, wackestones is observed. In the Don Braulio section, the uppermost beds of this outer-ramp nodular-wackestone association are overlain by the black graptolitic shales of the Gualcamayo Formation. The upper successions in both the Don Braulio and the Puesto de los Potrerillos sections are almost identical and in both sections the top is placed in the *Eoplacognathus suecicus* zone; therefore, it seems justified to combine the data from both sections for the interpretation of platform development and the discussion of the geodynamic evolution (Fig. 26).

Northern sections of the Talacasto subbasin. Around Jáchal, the San Juan Formation is characterized by a lower interval of about 20 m thickness in which several small-scale shallowing-upward successions are found between the top of the La Silla Formation and the lower reef-mound horizon. They are characterized by wackestones that pass upward into grainstones and finally into fenestral mudstones. Many of these fenestrae seem to have been gypsum or anhydrite. These horizons, interpreted to have been deposited on the proximal and protected inner ramp (Fig. 25), form the base for the lower reef-mound horizon. In the Agua Hedionda and Cerro La Chilca sections, the succession corresponding to the reef interval is characterized by coarse detrital limestones (grainstone and rudstone; Fig. 21), and no mounds have been observed there. Mid-ramp wackestones with a varying amount of packstones and intraclast packstones overlie the limestones.

Higher up in the succession, there are two distinct intervals of the nodular-wackestone association with an intermediate succession of non-nodular wackestone and packstones (Fig. 21). In the sections studied, the base of this deep-ramp interval coincides with the base of the *O. evae* zone. It is not clear whether there is only one thick amalgamated nodular interval in the Cerro La Chilca section, or whether there is an upper interval hidden in the zone without outcrops.

Above the deep-ramp nodular limestones, wackestones with intercalations of packstones (middle ramp) are present that mimic those underneath the nodular limestones (Fig. 26). The mid-ramp rocks are abruptly overlain by thick-bedded grainstones and rudstones that form the foundation for the upper reef-mound horizons. In the Cerro La Chilca section, this succession is represented by thick-bedded grainstones and rudstones. In Figure 25, a distinction is made between the inner-ramp association of grainstone and packstone and the reef facies proper, which is also attributed to the inner ramp.

The upper reef horizon, which is not as thick as in the sections of the San Juan subbasin, gradually passes into mid-ramp wackestones. The boundary between both was drawn at a level where the abundant coarse-grained tempestites disappear and oncoids are absent.

The upper part of the Cerro La Chilca section (Fig. 26) shows thick-bedded, in places nodular wackestones (outer ramp) followed by dark platy hemipelagic mudstones and wackestones (transfacies). Overlying these are black shales and siltstones of the Los Azules Formation (6 in Fig. 23). A similar transition is observed at Cerro del Fuerte (Fig. 3; also called Quebrada de Los Gatos), a bit north of Cerro La Silla (Cañas, 1995a).

Western sections of the Talacasto subbasin. The western sections are characterized by a rather monotonous succession of facies. In these sections, rocks of the wackestone-intraclast-packstone association (mid ramp) prevail (Fig. 26). In some of the intercalated tempestites oncoids are concentrated. In the Gualilán section, there is a lower interval of deep-ramp nodular limestones (Fig. 26), which from its stratigraphic position does not correspond to the main interval of nodular limestones in the northern sections.

The main interval of nodular limestones (outer-ramp deposits) is found higher up in the succession. A situation corresponding to the northern sections is found in the Las Chacritas section, where two nodular intervals are separated by wackestones and some packstones (Fig. 26). Upsection, mid-ramp sediments are observed.

The upper part of the western sections is composed of thick-bedded to very thick-bedded limestones of the inner-ramp packstones-grainstones association (Fig. 26). Among the organisms, there are sponges and stromatoporoids. This interval correlates to the upper reef interval and the composition of the sediments points to a distal inner-ramp environment, seaward of the reefs. In the Talacasto section, outer-ramp wackestones form the top of the exposed succession, whereas in the Las Chacritas section (also called Las Tunas section) about 60 m of nodular packstones are present (Fig. 21). A sharp surface marks the contact between the thick-bedded interval and the overlying outer-ramp limestones. According to Espisua (1968), these rocks, the age of which is not quite clear, represent the upper member of the San Juan Formation. Following Astini (1991) and Keller et al. (1993b), they are part of the Los Azules Formation (see later discussion).

Southern sections of the Talacasto subbasin. The sections along the Río San Juan are the most difficult to interpret. In the Río Sassito and Río Sasso sections the basal thrust is within the lower part of the San Juan Formation. The base of the section at Puerto Tambolar is not accessible, whereas in the Pachaco section the thrust is found in the upper part of the La Silla Formation. There, however, Ordovician erosion cut down as deep as the upper *O. intermedius* zone (conodont assemblage zone IV of Lehnert, 1993), whereas in the other sections the top of the preserved succession of the San Juan Formation is significantly younger. The lower part of the Tambolar and Pachaco sections is composed mainly of wackestones with abundant intercalations of packstones and grainstones (middle ramp). In the Tambolar section, outer-ramp nodular limestones are found near the base of the *O. evae* zone (Fig. 26). This interval is hardly recognized in the Pachaco section, where it might coincide with a mudstone-wackestone succession that lacks tempestites. In the Tambolar section, the remainder of the succession is made up of distal mid-ramp deposits, whereas at Pachaco the sediments are attributed to the deep ramp (Fig. 26).

Guandacol subbasin. The interpretation of the Guandacol sections is based mainly on the descriptions given by Cañas (1995a). Unfortunately, no thicknesses are given in that paper. The succession around Guandacol starts with mid-ramp deposits

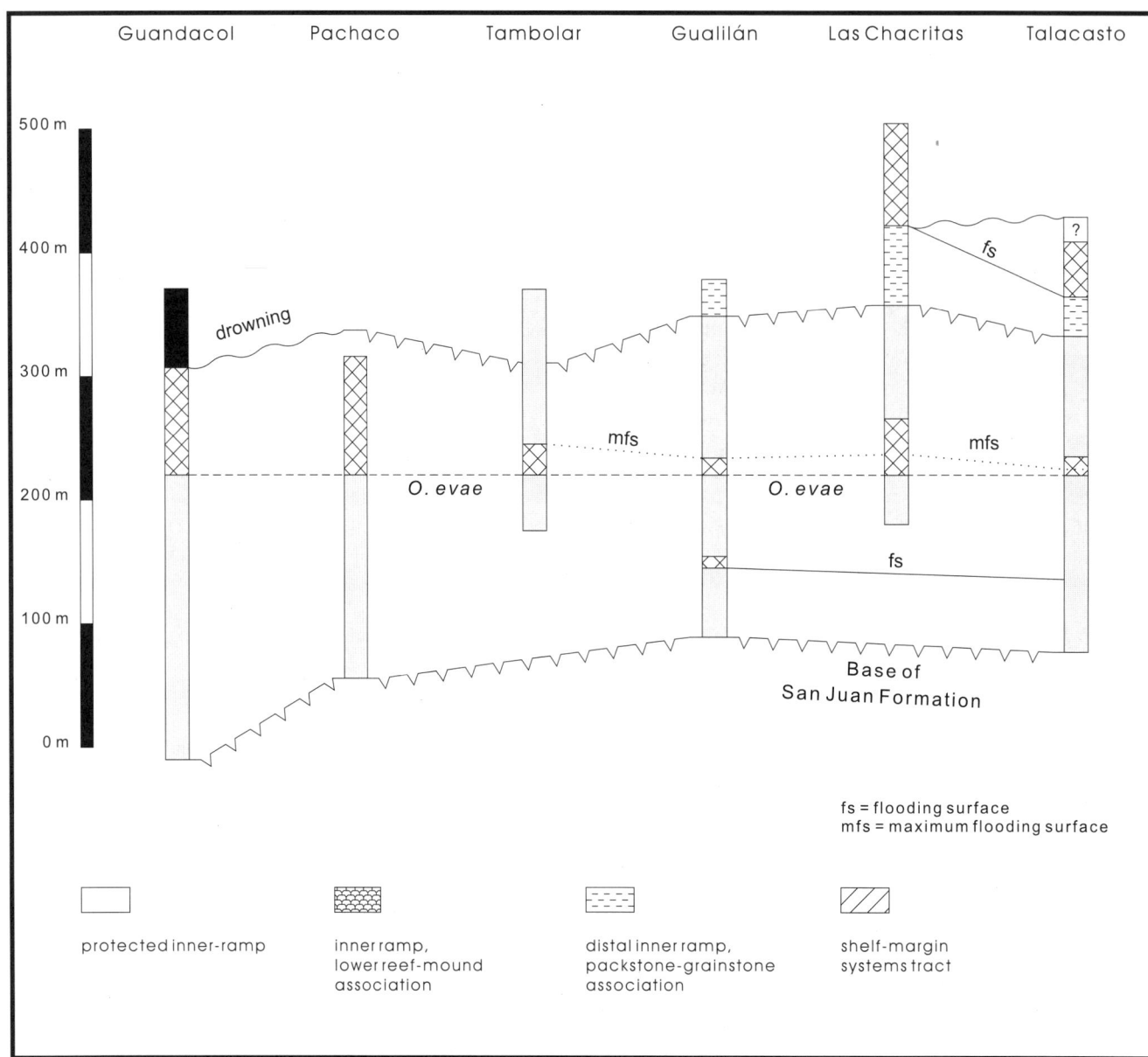

Figure 26 (this and opposite page). Depositional environments and ramp successions of San Juan Formation and their sequence stratigraphic interpretation. Note highly irregular thicknesses and succession of flooding surfaces below drowning unconformity. Reference line is base of *O. evae* conodont zone.

much like those found in the western sections. As in most other sections, near the base of the *O. evae* zone there is a deepening and the transition to deep-ramp nodular limestones (Fig. 26). Near the base of conodont assemblage zone VI ("*Parapanderodus*" *nogamii–P. gracilis–Ansella jemtlandica* zone) and approximately coeval with the onset of reef growth in the other sections, an interval dominated by proximal tempestites with black shale intercalations is described. Upsection, it rapidly gives way to black shales and lime mudstones of the Gualcamayo Formation.

Evolution and sequence stratigraphy of the San Juan Formation. Deposition of the San Juan Formation began near the Tremadoc-Arenig boundary (*Paltodus deltifer* zone). The incursion of a fully marine fauna and the deposition of wackestones indicate a flooding of the vast peritidal deposits of the upper La Silla Formation. No erosional or subaerial exposure features have been observed at the formation boundary; however, the dramatic change in fauna and depositional facies qualifies this boundary as a sequence boundary (Fig. 26). The conventional view of sequence stratigraphy requires the presence of a shelf-margin wedge or a lowstand systems tract between the sequence boundary and the transgressive systems tract (e.g., Sarg, 1988). The most common position for the low-

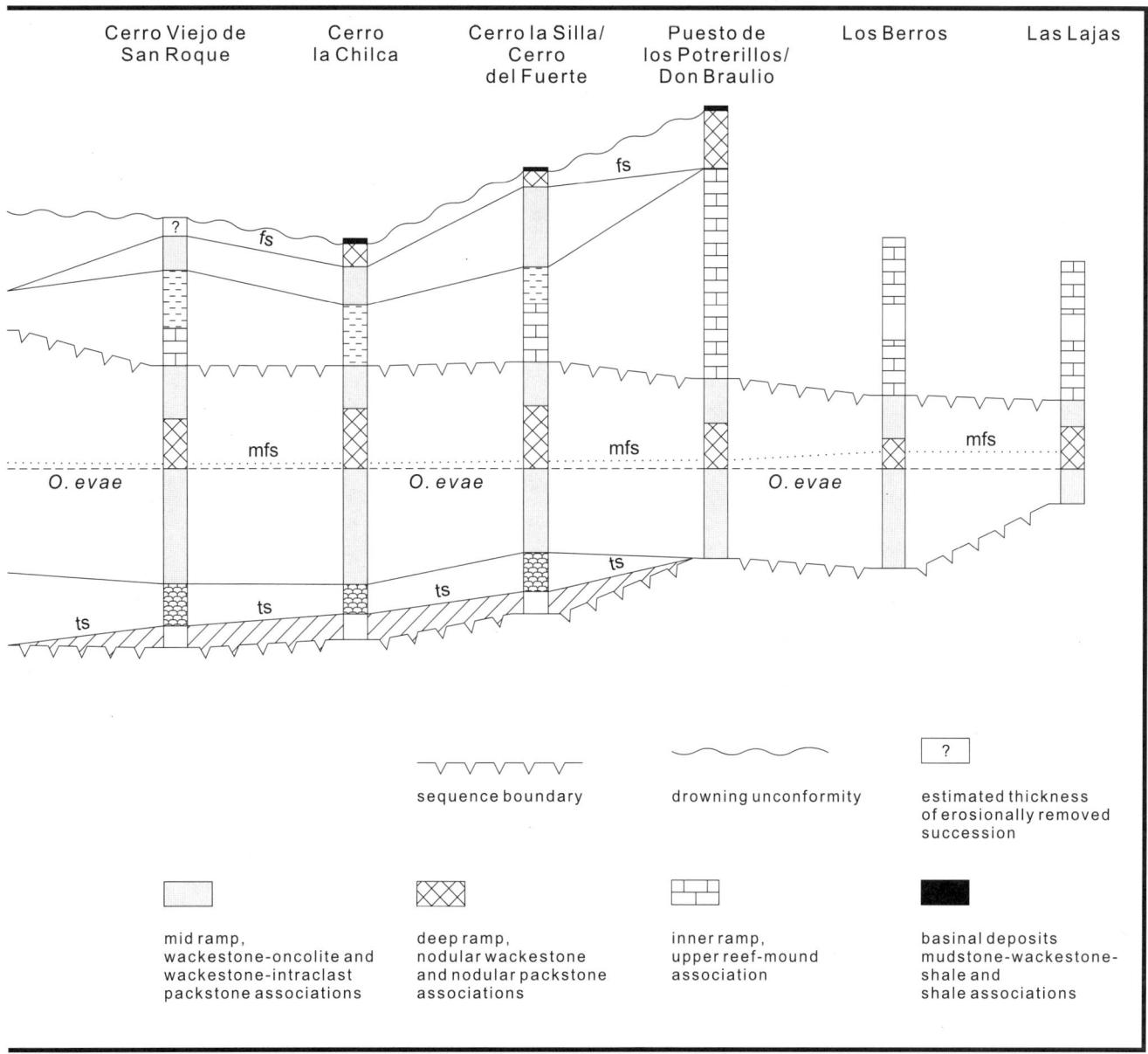

stand systems tract is at the outer edge of the platform or off the platform.

In the northern sections around Jáchal several small-scale shallowing-upward successions are found between the top of the La Silla Formation and the lower reef-mound horizon. Each of these cycles forms a parasequence, and the parasequences together form a progradational parasequence set. Cañas (1995a) interpreted these deposits to represent the lowstand systems tract. However, the presence of this parasequence set well inboard of the presumed platform margin of the upper La Silla Formation suggests that these cycles are a shelf-margin wedge (Fig. 26). The presence of a shelf-margin wedge in addition to the absence of subaerial exposure classify the sequence boundary between the La Silla Formation and the San Juan Formation as a type 2 sequence boundary. The prominent surface below the lower reef-mound horizons forms the transgressive surface, and the overlying sediments represent the transgressive systems tract (Fig. 26). In the sections of the San Juan subbasin, the parasequence set between the La Silla Formation and the San Juan Formation is not developed, and no facies that might represent a lowstand deposit are found (Fig. 26). Instead, the dominant mid-ramp wackestones without apparent cyclicity in the lower part of those sections represent the transgressive systems tract, and the sequence boundary and the flooding surface merge to one surface. Consequently, the shelf-margin systems tract wedges out by onlap toward the San Juan subbasin (Fig. 26).

In the western and southern sections of the Talacasto subbasin there is a much more uniform development of facies across the La Silla–San Juan formational boundary. One of the characteristics of a shelf-margin wedge is its thinning by downlap in a basinal direction (Van Wagoner et al., 1988). Hence it is plausible to assume that the shelf-margin systems tract was restricted to the more inboard sections and that it has only a very weak expression in the western sediments. Although the transition from the La Silla Formation to the San Juan Formation is much more subtle in the west, there is still no true correlative conformity recognized (Fig. 26).

In the northern sections of the Talacasto subbasin, the succession from the base of the lower reef mounds toward the upper reef horizon is interpreted as one sequence with one transgressive systems tract and one highstand systems tract (Cañas, 1995a). The maximum flooding surface is placed within the higher part of the *Prioniodus elegans–O. communis* zone (conodont assemblage zone I). According to this interpretation, the subsequent highstand comprises the *O. evae* zone (zone II) through the lowermost part of the *P. nogamii–P. gracilis–A. jemtlandica* zone (zone VI).

However, the transition from the lower reef-mound horizon toward the overlying mid-ramp deposits indicates a deepening of the depositional environment, which is accompanied by a change in the conodont faunas from predominantly hyaline warm-water forms to temperate water forms (Lehnert, 1995a). The occurrence of the different conodont groups in the Argentine Precordillera is mainly depth-related (Lehnert and Keller, 1993a) and supports the sedimentologic indications of a deepening at this level, which happened during the basal *Acodus deltatus–Paroistodus proteus* zone (Lehnert, 1995a). Consequently, the transition from the inner-ramp reef-mound interval to mid-ramp limestones marks a marine flooding surface (Van Wagoner et al., 1988; Fig. 26). A deepening event is also recorded in the Gualilán section, where there is a thin interval of nodular deep-ramp limestones.

Near the base of the *O. evae* zone (conodont assemblage zone II), there is another marked deepening event (Fig. 26), which is recognized in almost all sections by the presence of strongly bioturbated and often variegated nodular limestones. This succession of deep-ramp facies, which is almost 50 m thick in the Cerro La Silla section, is also characterized by a conodont fauna dominated by cold-water or deep-water forms (Lehnert, 1995a) and comprises the conodont assemblage zones II (*O. evae* zone) and III (*O. evae–O. communis* zone) of Lehnert (1993). The boundary between the underlying mid-ramp deposits and the deep-ramp limestones is sharp and, according to the few data available, the transition is synchronous over the entire ramp, including the Guandacol area. The limit between the mid-ramp and deep-ramp associations is here interpreted as another marine flooding surface (Fig. 26), which again is supported by the change in the conodont faunas. Cañas (1995a) placed the maximum flooding surface in this part of the section. As outlined here, the nodular wackestone interval is actually a triplet of nodular limestone–wackestone/packstone–nodular limestones, and hence shows the first pulses toward a renewed shallowing just above the lower nodular limestones. This is taken as good evidence of the change from retrogradation toward aggradation in the carbonate-producing system and consequently, the maximum flooding surface is placed on top of the lower nodular deep-ramp limestones (Fig. 26). Although this lower interval south of the Río Jáchal does not fit the classical condensed interval at the top of the transgressive systems tract (Loutit et al., 1988), it may be taken as a proximal equivalent of it. A better candidate for the condensed interval is present in coeval strata in the Guandacol area, where several hardgrounds are observed and glauconite is abundant in the sediments.

The succession above the lower nodular limestones indicates a successive shallowing of the depositional environment interrupted in some sections by the second nodular-limestone interval. However, the increasing amount of packstones together with the reappearance of sponges, calcareous algae, and stromatoporoids suggests the evolution of an early highstand systems tract. In the rocks, this development is documented by the transition from deep-ramp nodular limestones toward sediments of the mid-ramp wackestone-intraclast packstone association (Fig. 26).

There is no evolutionary pattern observed in the highstand systems tract, which might indicate a late highstand (thinning of beds, shallowing, progradational patterns). Instead, there is an abrupt change from mid-ramp deposits to shallow-water packstones and rudstones with abundant reef structures. This abrupt change marks a basinward shift in facies, which in siliciclastic systems indicates the nearby presence of a sequence boundary (e.g., Posamentier and Vail, 1988). Because carbonate ramps act much like siliciclastic systems (Burchette and Wright, 1992), the contact underneath the reef interval is suspected as being the sequence boundary. No subaerial exposure phenomena have been observed along the contact; however, in the lower beds of the reef interval, *Cordylodus* cf. *angulatus* was found (Lehnert, 1990, 1993), which is restricted to the latest Cambrian and the Tremadocian. There is strong paleontologic evidence of a major erosional event, probably in a more cratonward position, prior to the deposition of the reef interval. Hence, there must have been an area where a marked drop in sea-level led to erosion down to levels where this species had lived. In addition, there are indications in the conodont faunas that might point to a hiatus in the upper part of conodont assemblage zone V or between zones V and VI (Lehnert, 1997, personal commun.). Especially the absence of any species of *Baltoniodus* might indicate a hiatus comprising much of the *B. navis–B. triangularis* zone (Lehnert, 1995a). The absence of subaerial exposure features in the preserved parts of the ramp, the possible existence of a hiatus, and the sudden change in the character of the sediments at the base of the upper reef interval are typical of a type 2 sequence boundary, a conclusion also drawn by Cañas (1995a). In the sediments of the Guandacol area, the shift of facies across the sequence boundary led to the deposition of a thick package of proximal tempestites in an environment that otherwise was dominated by black shales.

Cañas (1995a) interpreted much of the reef interval to represent the corresponding lowstand systems tract, which is incompatible with the principles of sequence stratigraphy. A sea-level lowstand associated with a type 2 sequence boundary results in the deposition of a shelf-margin wedge (Posamentier et al., 1988). Here the idea is proposed that reef growth was related to the relatively rapid sea-level rise typical of the beginning of the trangressive systems tract (Van Wagoner et al., 1988). One of the main arguments is reef growth. As demonstrated by Keller and Flügel (1996), the principal reef mounds composed of *Zondarella* had synoptic relief and are dominated by domical growth forms. No vertical zonation is present and there are no indications that reef termination was caused by subaerial exposure. Hence it seems that the stromatoporoid mounds grew in an environment that did not allow them to outpace steadily rising sea level. Such a situation is in better agreement with a transgressive systems tract than with a lowstand systems tract or a shelf-margin systems tract. The development of reefs and reef mounds within a trangressive systems tract seems to be much more common than within a lowstand systems tract (Burchette and Wright, 1992), although this generalization is not an argument in favor of the scenario described here. The transgressive systems tract directly overlies the sequence boundary, and the transgressive surface and the sequence boundary have merged to one surface (Fig. 26).

The interpretation of the subsequent evolution of the San Juan Formation with its transition into the overlying rocks is much more complicated than the previous events and first requires a description of the contact and the overlying successions.

ROCKS OVERLYING THE SAN JUAN FORMATION

Conformable and Unconformable Successions

The contact between the San Juan Formation and the overlying units is exposed in many sections. Two different situations are present:

1. The San Juan limestones gradually pass into younger rocks. The transition from the San Juan Formation into the overlying conformable successions has various expressions in the rocks. Because of this nonuniform character, these transitional beds have received a variety of names or were assigned to different formations (Fig. 4). The transitional nature was recognized by Baldis and Beresi (1981), who introduced the term "transfacies calcareo-pelítica" for these beds.

2. An erosional contact separates the San Juan Formation from younger units. Erosion on top of the San Juan Formation occurred during at least four different events. Besides the continuing Holocene erosion, the youngest, Tertiary erosion affected mainly sections in the San Juan subbasin along the eastern margin of the Argentine Precordillera. This latest phase of erosion affected the carbonate platform rocks in the same area as did the pre-Late Carboniferous erosion. The effects of the latter are seen where the Tertiary events did not totally remove the Carboniferous strata. The unconformity between the San Juan Formation and the Upper Carboniferous rocks can be traced north as far as the Agua Hedionda section. However, there is a fundamental difference between the sections in the San Juan subbasin and the area around Agua Hedionda. In the former, Carboniferous glacial deposits cover steeply tilted strata, whereas in the north, the contact is paraconformable (Buggisch et al., 1994a). This is the effect of a different structural history of the San Juan subbasin and the Talacasto subbasin during pre-Late Carboniferous time (Ortiz and Zambrano, 1981). The Agua Hedionda section is the only section in the Talacasto subbasin where Carboniferous sediments are found on top of the carbonate platform rocks.

In many parts of the Talacasto subbasin, a major unconformity separates the San Juan Formation from younger deposits (Fig. 4). The unconformity is well marked by a chert-pebble conglomerate at the base of the overlying sedimentary rocks (Tambolar and La Chilca Formations). This conglomerate is composed of a reddish or brownish, in places greenish shale matrix with small clasts of chert. Clast size is on the order of a few millimeters to 5 cm; the thickness of the unit varies between a few centimeters and about 20 cm. Graptolites found just above the conglomerate indicate a latest Ashgillian age for the deposits of the La Chilca Formation (Peralta, 1990).

Río Sassito section. With the discovery of Caradocian carbonates (Lehnert 1995a, 1995b) in the Río Sassito section (Fig. 3) it became clear that the pre-Silurian unconformity represents at least two different events. The effects and the amount of erosion of the individual events, however, are difficult to distinguish because in most sections both unconformities merge to form one surface. A clue to the importance of the individual events is the section at Río Sassito. There, Llandeilian siliciclastic strata and Caradocian carbonates overlie the San Juan Formation (Lehnert 1995b). In this section, the top of the latter was dated as the lower part of conodont assemblage zone VI. The Caradocian carbonates are overlain by the pre-Silurian conglomerate. The San Juan Formation and the Río Sassito succession are separated by a chert-pebble conglomerate very similar to the younger one at the base of the Silurian. Astini and Cañas (1995) described a basal conglomerate with clasts more than 1 m across. This basal conglomerate does not exist. The "clasts" described are blocks that resulted from fracturing of the San Juan limestones by several faults.

The Río Sassito succession consists of a lower siliciclastic succession and an upper calcareous interval. The upper succession has a thickness of about 15 m, and was interpreted as the remnants of an isolated or pelagic carbonate platform (see Appendix 1; Lehnert 1995a,1995b; Keller and Lehnert, 1998). Conodonts found in the limestones indicate a Caradocian age, but at present it cannot be excluded that parts of the carbonates belong to the Ashgill. The importance of this outcrop is many fold: (1) it demonstrates that carbonate production was still possible during the Late Ordovician; (2) carbonate production was possible despite the fact that in many adjacent areas siliciclastic strata were deposited; and (3) similar carbonates may have been the source for reworked Caradocian conodonts found in debris-flow deposits at San Isidro.

Cerro La Chilca section. In this section, which is the standard section for the upper part of the San Juan Formation (Keller et al., 1994), thick-bedded limestones of the nodular-wackestone association are present in the upper part of the formation (Fig. 21). The top of this succession is marked by a prominent hardground with abundant vertical burrows. Above this hardground, about 4.5 m of the mudstones-wackestones-shale association are exposed (6 in Fig. 23) which upsection rapidly give way to black and dark gray silty shales. In this section, the shales are assigned to the Los Azules Formation (Furque, 1983).

The top of the San Juan Formation *definitely* belongs to the *E. suecicus* zone (Lehnert, 1995a), whereas the overlying carbonates are *tentatively* assigned to the same zone (Lehnert, 1995a).

Las Chacritas section. This section was originally described by Espisua (1968), who subdivided the San Juan Formation into two members. There is no doubt that the lower member described by Espisua (1968) is part of the San Juan Formation. The top of this member belongs to the packstone-grainstone association with intercalated isolated mounds (inner ramp; Fig. 26). Upsection, there is a sharp contact above which more than 60 m of deep-ramp nodular limestones (upper member of Espisua, 1968) were deposited (Figs. 21 and 26). Along the contact, which is not well exposed, neither indications of hardground formation nor other indicators of a break in sedimentation have been observed.

The packstone-grainstone association yielded conodonts of the *E. suecicus* zone (Lehnert, 1995a); however, the younger carbonates are undated at present. Conodonts found do not allow an assignment to the *E. suecicus* or to the following *Pygodus serra* zone.

Las Aguaditas section. Although the contact between the San Juan Formation and the overlying rocks is faulted in places, it is clearly visible that the mudstone-wackestone-shale association is in sedimentary contact with the top of the limestones. In their discussion of the Las Aguaditas section, Keller et al. (1993b) included this association into the Las Aguaditas Formation because the latter is composed of dark mudstones, wackestones, and shales with intercalations of mass-flow deposits. There is no doubt that in this section the "transfacies" and the rocks of the Las Aguaditas Formation are genetically related; they are indistinguishable in the field. Rocks below and above the contact between the San Juan Formation and the Las Aguaditas Formation are dated both with graptolites and conodonts (Brussa, 1994; Lehnert, 1995a). The contact is placed in the *E. suecicus* zone.

Guandacol subbasin. A sedimentologic study of the transition between the San Juan Formation and the Gualcamayo Formation in the Guandacol area was given by Cañas (1995a). There, more than 50 m of dark gray nodular mudstones and wackestones are intercalated between mid-ramp limestones and the transitional beds that form the base of the Gualcamayo Formation (Fig. 26). From the description and interpretation given by Cañas (1995a), these limestones are much like the deep-ramp nodular packstones in the Las Chacritas section.

Biostratigraphic data from the Guandacol sections indicate that deposition of graptolitic black shales was initiated much earlier than south of the Río Jáchal. Graptolites from black shales just above the limestones belong to the *Isograptus victoriae* zone (Ortega et al., 1985), whereas the conodonts (Hünicken and Sarmiento, 1982, 1985) are representatives of the *O. evae* zone. Both fossil zones belong to the interval around the early-late Arenig transition.

At Cerro Potrerillo (Fig. 3) the transition was dated by Albanesi (1991), who described conodonts from the Arenig-Llanvirn boundary interval. Hence this section seems to represent an intermediate position between the Guandacol area and the sections south of the Río Jáchal.

Discussion. The transition from the San Juan Formation into the overlying units is one of the most critical intervals in the early Paleozoic evolution of the Argentine Precordillera for several reasons. (1) It marks the end of large-scale carbonate platform evolution, although carbonate production continued locally. (2) The transition into the younger rocks is not uniform; a variety of sedimentary successions is present on top of the San Juan Formation (Fig. 4). (3) The transition is not coeval. (4) The noncoeval transition is used as an argument in favor of a diachronous top of the San Juan Formation (Hünicken, 1985; Benedetto et al., 1986; Beresi, 1988). (5) The early Llanvirn was a time of major eustatic sea-level rise (Fortey, 1984; Ross and Ross, 1995) and is thought to be responsible for the transgression and the accompanying deposition of black shales (Beresi, 1990; Heredia and Beresi, 1995). In contrast, there have been also tectonic interpretations for the termination of the carbonate platform (Astini, 1991, 1993; Keller et al., 1993b).

Except for the Guandacol subbasin, the final countdown for the termination of regional carbonate production started near the Arenig-Llanvirn boundary, i.e., with the transition from conodont assemblage zone VI of Lehnert (1993) toward the *Histiodella sinuosa* zone of the North American subdivision. This transition marks an abrupt deepening event on top of the packstone-grainstone association or the upper reef horizon (Fig. 26). With the deepening, limestones of the deep-ramp nodular-wackestone association were deposited. Relative sea-level rise continued into the *Histiodella kristinae–E. suecicus* zone, and the carbonate producing system was barely able to catch up with relative sea-level rise. This is shown in the Cerro La Chilca section, where a well-developed hardground is present. Another major rapid deepening event led to the final drowning of the platform and for a short time mainly deep-water limestones and shales were deposited, blanketing the carbonate platform. The corresponding flooding surface is a very sharp surface and nothing is seen in terms of a transition. In most areas, the "transfacies" forms the base of the overlying succession; however, in the Las Chacritas section, deep-ramp nodular limestones are present above a sharp contact (Fig. 26). In all areas south of the Río Jáchal, this surface represents the drowning unconformity (see Appendix 1).

In the Guandacol subbasin, similar events are observed; however, they are considerably older there. The available biostratigraphic data indicate that the onset of basinal deposition is coeval to the development of the upper reef horizons, i.e., during a marked shallowing of the rest of the platform.

Don Braulio Section

The Don Braulio section (Fig. 3) is the only section in the San Juan subbasin in which Middle and Upper Ordovician rocks are preserved. In addition, the Silurian rocks on top of the Ordovician show a development quite different from that of the other parts of the Argentine Precordillera. Four formations are distinguished above the San Juan Formation: in ascending order, these are the Gualcamayo, La Cantera, Don Braulio, and Mogotes Negros or Rinconada Formations (Fig. 4).

Gualcamayo Formation

In the Don Braulio section, rocks of the nodular-wackestone association of the San Juan Formation overlie the upper reef-mound interval. Upsection, there is a sharp surface (1 in Fig. 27) above which the first black shales with intercalations of dark mudstones, the "transfacies", are observed. Peralta (1995) included these beds in the Gualcamayo Formation as its lower member. The ages of both the top of the San Juan Formation and the lower member of the Gualcamayo Formation are well established and, according to conodont determinations, both belong to the *E. suecicus* zone (Sarmiento, 1986; Sarmiento and Rábano, 1992). An early Llanvirn age is also indicated by the graptolites *Paraglossograptus tentaculatus* and *Glyptograptus austrodentatus austrodentatus* (Peralta, 1995).

Lithofacies and the sedimentary succession of the Gualcamayo Formation. The lower member of the Gualcamayo Formation is composed of the mudstone-wackestone-shale association (1 in Fig. 27).

The middle member is composed predominantly of black graptolite shales, which upsection gradually pass into variegated shales. Overlying these are diamictites (see Appendix 1), and conglomerates are present which traditionally were assigned to the La Cantera Formation. Peralta (1995), however, separated the strata beneath the main conglomerate horizon from the La Cantera Formation and included them in the Gualcamayo Formation as its upper member. This is based on graptolite biostratigraphy, which suggests a hiatus at the base of the main conglomerate. Although there is clear field evidence of truncation of different stratigraphic levels beneath the main conglomerate, the abrupt change in depositional style at the base of the lowermost diamictite is here taken as the boundary between the Gualcamayo Formation and the La Cantera Formation.

La Cantera Formation

The La Cantera Formation is an entirely siliciclastic succession (Fig. 28) and its basal conglomerates are an outstanding feature. Clast size varies from a few centimeters to >30 cm. The clasts are dominantly sandstones and to a minor extent metaquartzites. Quartz pebbles are abundant in places. Graptolite-bearing shale clasts are as long as 2 m and show green and black colors.

Among the calcareous clasts there are light gray limestones of the San Juan Formation and dark to black lime mudstones with a typical deep-water microfacies. In addition, there are orange to brown sandy limestones (2 in Fig. 27), in places calcareous sandstones, with an abundant fauna of thick-shelled gastropods, brachiopods, and some trilobites. Very few granite pebbles are present, but no other plutonic clasts have been observed.

Predominantly clast-supported conglomerates are present. Components are moderately to well rounded, and their size varies between coarse sand and large cobbles. The matrix, where present, is sandy or arkosic.

The conglomerates show highly variable geometries. Narrow and steep channels contrast with broad and shallow channels. In places, individual conglomerate bodies are more than 100 m long and exhibit sheet-like geometries. Amalgamation of individual conglomerate beds and normal and inverse grading within the beds is frequently observed.

Diamictites contain pebbles and boulders in a green silty and sandy matrix. The clast spectrum of the diamictites is identical to that of the conglomerates.

Fine- to coarse-grained sandstones, in places pebbly sandstones, are a major constituent of the La Cantera Formation (Fig. 28). In the coarser beds amalgamation is visible. These beds commonly are closely connected to conglomerates by grading from fine to coarse or vice versa. Sublitharenites dominate, but subarkoses and arkoses are also present in significant amounts in the La Cantera Formation.

The medium- to fine-grained sandstones are thick to thin bedded and show brown to yellow-brown colors. Many of the beds are graded and have a rippled top, but others are structureless. In addition, pinch-and-swell structures and flute casts have been observed. The latter indicate a constant east–west directed transport. Erosion at the base of the beds commonly incorporated small mud chips of the underlying beds into the sandstones.

Siltstones, silty shales, and pure clay shales are medium to thin bedded. The rocks are gray, brown, or yellow. Plane-parallel ripple lamination is present in some of the siltstones.

Many of the sandstones and siltstones are combined to form turbidites. Following the classic Bouma subdivision, T_{a-c} and T_{a-d} turbidites are observed in coarse-grained sandstones grading upward into siltstones. Medium- and fine-grained sandstones together with siltstones are mainly arranged in T_{b-c} or T_{b-d} successions. T_{c-d} patterns are developed only locally.

Sedimentary succession of the La Cantera Formation. The section measured in the La Cantera Formation (Fig. 28) is the type section described by Furque and Cuerda (1979) and Baldis et al. (1982). Four units are recognized in the type section (LC1–LC4). The lowermost unit (unit 1, ~24 m thick) corresponds to the upper member of the Gualcamayo Formation as described by Peralta (1995), but is here included in the La Cantera Formation as originally defined. This lower unit is a succes-

Figure 27. Don Braulio section. 1: Transition of San Juan limestones into Gualcamayo Formation (top). In contrast to succession at Cerro La Chilca, there are only few limestone beds intercalated in basal black shales. Thick limestone bed is ~15 cm thick. 2: Limestone clast in La Cantera Formation. These fossiliferous sandy limestones yielded late Llanvirnian and earliest Llandeilian fossils. In addition, metaquartzite clast and clasts of vein quartz are present. Quartz clasts are ~1 cm across. 3: Olistolith of reworked limestone conglomerate in Rinconada Formation. In addition to limestones, metaquartzite clasts and sandstone clasts are present within conglomeratic olistolith. View is 27 × 18 cm. 4: Huge olistolith (10 m high) entirely composed of conglomerates. In places, original bedding of conglomerates is still visible (arrow). 5: Olistolith consisting of alternation of quartzarenites, siltstones, and shales. Within limits of Precordillera there is no known succession that might have served as source for this olistolith.

sion of thin conglomerates, sandstones, diamictites, siltstones, and shales. It is overlain with a marked unconformity by the main conglomerate of the La Cantera Formation. Laterally, this conglomerate cuts down deeply into the underlying strata, in places into the Gualcamayo Formation. The unconformity is paleontologically characterized by a hiatus (Peralta, 1995). Unit 1 still belongs to the Llanvirn sensu lato or the basal Llandeilo, whereas graptolites from above the conglomerate indicate a Llandeilian age.

Units 2 to 4 are overall fining-upward and thinning-upward successions (Fig. 28). At the base, unit 2 comprises the main conglomerates of the La Cantera Formation, which upsection gradually pass into coarse-grained sandstones and finally into finer grained sandstones. Siltstones and shales are only minor components of the succession. In the model of Mutti and Ricchi Lucci (1972), the conglomerates represent facies A and B and, together with the abundant T_{a-b} to T_{a-d} turbidites, are indicators of a proximal environment. Unit 2 is ~37 m thick.

Unit 3 (~33 m) begins with a prominent sandstone bed above the siltstones and shales on top of unit 2. Unit 3 consists mainly of T_{b-c} to T_{c-d} turbidites with intercalated siltstones and shales. Flute casts are abundant in this unit.

A thick package of sandstones marks the base of unit 4, which is composed of fine-grained sandstones, siltstones, and shales (~47m). Flute casts and slump structures are present in several beds. Although a T_{a-d} turbidite bed is present, the majority of the beds exhibit T_{c-d} successions. The siltstone and shale content is significantly higher than in the underlying units. Graptolites found in this unit correspond to the *Nemagraptus gracilis* zone (Sánchez et al., 1991) of the basal Caradoc.

Discussion. The turbiditic nature of the La Cantera Formation was recognized by Peralta (1986). Astini (1991) presented a model for the deposition of the conglomerates and coarse-grained sandstones that closely follows the fan-delta model of Mutti and Ricchi Lucci (1972). Although there is no doubt that most of the conglomerates are channeled, their vertical and lateral amalgamation to laterally extensive sheets at the base of unit 2 indicates that the depositional environment was not fed by a point source, but rather that the conglomerates were deposited along a line source. Hence, a slope apron configuration is favored here.

Unit 1 reflects the sudden input of coarse detritus above the black shales of the Gualcamayo Formation. It lacks an obvious internal arrangement of facies similar to that in the overlying units. The most important erosional event is documented in the main conglomerate horizon not only paleontologically, but also by the clast content. Dark limestone pebbles assigned to the Gualcamayo Formation or the Las Aguaditas Formation yielded upper Llanvirnian (Albanesi and Benedetto, 1992) and early Llandeilian faunas (*Eoplacognathus lindstroemi*; Albanesi et al., 1995), indicating that sedimentation on top of the carbonate platform had been continuous until the early Llandeilian. Unit 1 records the first change in depositional style; however, the most important erosional event was connected with the deposition of the main conglomerate. This erosion led to the local removal of unit 1 and to erosion of the uppermost layers of the underlying Gualcamayo Formation or its equivalents. After this event, the rapid decrease in grain size and bed thickness within the individual units, but also as an overall pattern, indicates a rise in relative sea level. This relatively rapid rise was interrupted only by two minor pulses of coarser clastics. Whether this overall pattern was caused by eustatic events or by tectonics is hard to evaluate; however, as discussed later, a tectonic control is most likely.

No indicators of transport directions have been found in the conglomerates, but flute casts in the overlying sandstones and siltstones prove an eastern source area. If it is assumed that the conglomerates were also derived from the east, the noncarbonate clast content reflects a source area to the east where a siliciclastic sedimentary succession was present. One major component of this succession must have been sandstones, because they are the dominant lithotype in the conglomerates. Another major constituent of the succession was metaquartzite. The metaquartizes may be stratigraphic equivalents of the sandstones, eroded from a deeper crustal level. Alternatively, they may represent a stratigraphically older sedimentary succession that underwent metamorphism prior to the deposition of the sandstones.

Granitoids were also present in the source area. Their debris (quartz and feldspar, mainly K-feldspar) is found in the arkoses and subarkosic sandstones. Isolated granite intrusions of Early Ordovician age have been described from the western Sierras Pampeanas (Rapela et al., 1998). These might have been the source for the granitic pebbles observed.

Don Braulio Formation

The Don Braulio Formation was originally defined and subdivided into three members by Baldis et al. (1982); it is a rather heterogenous unit, composed almost entirely of siliciclastic rocks. On the basis of a detailed sedimentological study, Astini (1991) described two units within the Don Braulio Formation, a subdivision that is followed here (DB1–DB2).

Lithofacies. Diamictites and pebbly sandstones are an important lithofacies of the Don Braulio Formation (Fig. 28). Pebble- to boulder-sized clasts occur in a matrix of gray shale or fine sandstone. Many of the clasts are striated or faceted. Sandstones and metaquartzites are the main clast types, but igneous rocks are also present in the clast population.

A conglomerate layer separates the lower from the upper member. It exhibits features (lateral and vertical amalgamation) similar to the main conglomerate in the La Cantera Formation, although the Don Braulio conglomerate is thinner.

Sandy limestones and calcareous sandstones are present in thin beds or lenticular bodies. These rocks contain an abundant fauna of brachiopods, trilobites, and bryozoans. Many of the faunal elements are typical of the *Hirnantia* fauna (Benedetto, 1985, 1986). In weathered outcrops these sediments are brown to orange.

Green to yellowish-brown shales alternating with a few gray siltstones are another important rock type. Bioturbation is localized and a few trilobites have been found in the shales. Within the siltstones a horizon is present in which chert pebbles are observed. The pebbles are as much as 4 cm across and float in the silty matrix. In places, the pebbles are densely packed.

Iron oolites (Fig. 28) form thin orange to red horizons that alternate with siltstones. The oolite packages attain a maximum thickness of about 1 m. Internal stratification features have not been observed. The ooids are densely packed, and their size varies between 0.2 and 1 mm.

Sedimentary succession of the Don Braulio Formation. The lower unit of the Don Braulio Formation is composed predominantly of diamictites and conglomerates (Fig. 28). Faceted and striated clasts are glacial features first described by Peralta and Carter (1990a), and are related to the Late Ordovician glaciation of Gondwana (Buggisch and Astini, 1993). Above the basal diamictite (~17 m thick), a bed is present that consists mainly of sandstone but laterally interfingers with coarse conglomerates. This horizon is separated by an intervening diamictite from a second conglomerate that marks the base of the overlying upper unit of the Don Braulio Formation.

In the section measured (Fig. 28), calcareous sandstones with the *Hirnantia* fauna immediately overlie the conglomerate and are overlain by silt and clay shales containing fragments of graptolites. Hitherto, this conglomerate was placed at the top of the lower unit of the Don Braulio Formation (Baldis et al., 1982; Peralta, 1990, Astini, 1991). However, if this conglomerate is traced laterally toward the north, it successively cuts out older strata: it first cuts out the upper diamictite interval, then the conglomerate "cannibalizes" the lower sandstone-conglomerate succession. The conglomerate then cuts into and through the main diamictite and finally rests upon the uppermost sedimentary rocks of the La Cantera Formation. Erosion prior to the deposition of this conglomerate created a considerable relief. In the section measured, the *Hirnantia*-bearing horizon rests directly on the conglomerate; however, if traced to the north, increasing amounts of green shales are present between the conglomerate and the calcareous sandstones until 22 m of shale separate the two horizons.

This erosional unconformity, covered and marked by the conglomerate, is the compelling reason to include the conglomerate into the upper unit as its basal part. The overlying marine shales record a rapid relative rise in sea level with the corresponding onlap of sediments onto the erosional morphology. Upsection, siltstones dominate and finally sandy intercalations are present, indicating an overall shallowing of the environment.

A silty interval with chert pebbles is present toward the top of the formation; higher up, iron oolites with siltstones and shales represent the topmost beds of the Don Braulio Formation. Shaly interbeds within the oolite succession, the upper member of Baldis et al. (1982) and Peralta (1990), are dated as earliest Silurian (Peralta, 1986).

Discussion and sequence stratigraphy of the Don Braulio Formation. A sequence-stratigraphic analysis must focus on the two important sequence boundaries within the Don Braulio Formation. The sequence boundary separating the Don Braulio Formation from the La Cantera Formation is important because of the major hiatus involved, whereas the younger sequence boundary is characterized by an impressive erosional relief. The lower sequence, which predominantly consists of diamictite, does not reveal features that might be used in attributing the succession to a distinct systems tract. The upper member, however, contains some characteristics that help in a sequence-stratigraphic interpretation. The erosional unconformity and the associated conglomerates are the expression of a major drop in sea level accompanied by a basinward shift of coastal onlap. Hence the conglomerates most likely filled an incised valley system, characterizing a type 1 sequence boundary, which subsequently was flooded. The transgressive surface is probably represented by the onset of shale deposition above the conglomerates. The maximum flooding surface is located close to the horizon of sandy limestones and calcareous sandstones containing the *Hirnantia* fauna. This horizon is the first horizon recognized in all sections because it locally rests directly on the conglomerate. Hence sea level rise had reached a position in which the erosional relief was entirely flooded. A faunistic analysis of these fossiliferous horizons (Sánchez et al., 1991) revealed differences in faunal composition and age over short distances, horizontally as well as vertically. It is here assumed that these differences are, in fact, the result of differential onlap of the sediment onto a preexisting erosional surface. This is corroborated by the faunas, which are composed of shallow-water associations near the base of the fossiliferous interval and deeper water communities higher up in the succession. The change in composition of the faunas is observed in <20 m of section (Sánchez et al., 1991).

Consequently, the transgressive systems tract of the upper sequence within the Don Braulio Formation is represented by the green shales and their varying thickness. They are separated from the overlying highstand systems tract by a maximum flooding surface positioned near the fossiliferous interval. The highstand systems tract includes the iron oolites and the chert pebble conglomerate as an expression of a return to shallower water environments. There is no doubt that the upper sequence boundary is related to the Late Ordovician glaciation. However, as discussed later, the effects of eustasy are matched by the effects of continental breakup and the onset of drifting. Tectonics played a major role in the formation of sequences during the Ordovician-Silurian boundary interval.

Rinconada Formation

In the Sierra de Villicum (Fig. 3), a thick succession of Silurian rocks is present, which has been described as wildflysch (Amos, 1954; Borello, 1969) and was called the Mogotes Negros Formation (Furque and Cuerda, 1979). Similar deposits in the Sierra Chica de Zonda are known as Rinconada Formation (Amos, 1954). Both formations are sedimentologically identical and coeval (Peralta, 1990). Consequently, only the term Rinconada Formation is used in this paper (Fig. 4), because it corresponds to the older definition (Amos, 1954); the term Mogotes

Negros Formation is superfluous and should be abandoned. In the Sierra de Villicum, the Rinconada Formation conformably overlies the Don Braulio Formation, but in the Sierra Chica de Zonda it is on top of the San Juan Formation (Fig. 4).

The thickness of the strongly deformed sedimentary rocks is difficult to establish. Conservative estimates for the succession at La Rinconada show about 1000 m (Von Gosen et al., 1995), whereas the 4000 m reported at Don Braulio (Peralta, 1990; Loske, 1992) are probably caused by repetition of section along cryptic thrusts.

Lithofacies. The Rinconada Formation consists of a matrix of gray to dark gray, in places greenish shales and siltstones that host centimeter to kilometer sized olistoliths. In addition, fine-grained sandstones, as well as a few diamictites and conglomerates, are part of the autochthonous succession. The conglomerates occur in broad, shallow channels. Most of the clasts were derived from the San Juan Formation (3 in Fig. 27), but greenish metaquartzites, a few basalt clasts, and even fewer granitoid clasts have also been observed. Similar conglomerates from the Don Braulio section were described by Loske (1992), who also described iron-oolite clasts. In a matrix of siltstones, diamictites show large (10–20 cm across), well-rounded pebbles of a variety of lithologies. Most important are intermediate to acidic volcanics and striated clasts of different lithologies. The latter have been interpreted to be reworked from older glacial deposits of the Don Braulio Formation (Von Gosen et al., 1995).

Slumps and slides are abundant in the rocks and were used to determine transport directions (Von Gosen et al., 1995). Northwest and southeast-directed transport is documented in the sediments at La Rinconada.

Among the olistoliths, there is a marked difference is size and composition between the Don Braulio section and the outcrops along the eastern border of the Sierra Chica de Zonda. In the latter, limestone slabs of the San Juan Formation dominate, their size varying between ~100 m and several thousand meters. In the Don Braulio section, they are not more than 300 m across. There, a variety of other olistoliths has been observed, which merits special attention. Clasts composed of iron oolite are present near the base of the formation. These clasts are clearly derived from the underlying iron oolite within the Don Braulio Formation. Another olistolith, several tens of meters across, consists of a well-bedded succession of white and brown quartz arenites and siltstones (5 in Fig. 27). The facies within the olistolith has no known counterpart in the Ordovician succession of the eastern Argentine Precordillera. Most important are olistoliths of conglomerates, which are rarely more than 20 m across (4 in Fig. 27). The internal fabric of the reworked blocks is mostly clast supported; in places long-axes orientations have been observed. Among the clasts within the olistoliths there are greenish metaquartzites, sandstones, and siltstones, which together form the bulk of the conglomerates. The igneous rocks present were described in detail by Von Gosen et al. (1995), who distinguished acidic plutonic rocks and basic to intermediate volcanic rocks. Carbonate components are mainly derived from the San Juan Formation; however, in the La Rinconada section silicified oolites are present, which have not been observed in the carbonate platform rocks. The cobbles are composed of well-sorted ooids. These ooids are characterized by an extremely regular and concentric arrangement of the individual ooid layers. In addition, the thickness of the individual layers is very uniform. The size of the individual components (>2 mm) together with their uniform internal structure might indicate that the ooids are in fact pisoids. However, cobbles in general do not reveal pisolite textures.

At La Rinconada, a few conglomerate blocks have been observed that are composed entirely of small limestone clasts (6–8 cm across). Most of the clasts represent facies of the San Juan Formation, but dark lime mudstones similar to those of the "transfacies" have also been found. Conodonts described from these pebbles (Lehnert 1995a, 1995b) are not younger than the early Llanvirn, hence the conglomerate is composed of rocks of the San Juan Formation and the basal Gualcamayo or Las Aguaditas Formations.

At the same locality, black shales with Ordovician graptolites are present (Peralta and Uliarte, 1986), but there is strong evidence that the fossils were found in allochthonous blocks (Peralta, 1990; Von Gosen et al., 1995). Graptolites and facies prove that these blocks were derived from the Gualcamayo Formation.

In the sections at Don Braulio and at La Rinconada, an inverse stratigraphy of the clasts is observed. In the Don Braulio section, this is true for the base of the succession, where the lowermost clasts are iron oolites derived from the immediately underlying beds of the Don Braulio Formation. Upsection, conglomerate clasts are present that were derived from the La Cantera Formation and, finally, olistoliths of the San Juan Formation were deposited. Above this interval, however, conglomerate olistoliths dominate. At La Rinconada, an olistolith of black shales resembling the Gualcamayo Formation is present near the base. Upsection, mainly carbonate slabs are intercalated; according to conodonts (Lehnert, 1995), these are also inversely arranged.

Discussion. The Rinconada Formation records sediment accumulation in a rapidly subsiding basin. This is indicated by its thickness, abundant slump and slide structures, and the presence of large olistoliths and thick olistostromes. The inversely arranged olistoliths in the Don Braulio section testify to rapid subsidence and/or uplift of the basin shoulders. The giant carbonate slabs at La Rinconada can have been derived only locally by pure gravitational processes.

Nothing is known about the original extension of the basin and its continuation toward the east. However, it seems that the basin in which the Rinconada Formation accumulated, was originally north-south trending, bounding the Sierra de Villicum to the east, and that the basin continued south along the eastern border of the Sierra Chica de Zonda. Toward the west, in the Talacasto subbasin, siliciclastic platform sediments were deposited coevally and uninterrupted from the Late Ordovician into the Devonian. Therefore, the sediments of the Rinconada Formation and the clasts and olistoliths contained in it cannot have been derived from a western source area beyond the Sierra de Villicum and Sierra Chica de Zonda.

The basin-forming mechanism is interpreted to have been large-scale extensional tectonics (Loske, 1992; Von Gosen et al., 1995) and the basin, along its western margin, must have been bounded by east-dipping faults.

An important feature of the Rinconada Formation is its clast and olistolith contents. In the Don Braulio section, the clasts and olistoliths record erosion through the Don Braulio Formation and down into the San Juan Formation. Clasts and olistoliths must have been locally derived, i.e., from the present-day Sierra de Villicum. However, the conglomerate olistoliths above the inversely-arranged succession and the siliciclastic olistolith described earlier (5 in Fig. 27) were probably derived from a different (eastern?) source area. The siliciclastic olistolith has no known in-situ counterpart exposed in the Argentine Precordillera. The conglomerate olistoliths are compositionally different from the La Cantera conglomerates. In addition, they are volumetrically too abundant to be solely derived from the La Cantera Formation. They either represent a lateral equivalent of the La Cantera Formation that, during deposition of the La Cantera Formation had a different source area, or the conglomerate olistoliths represent an independent (younger or older) depositional system unrelated to the La Cantera Formation and nowhere exposed within the actual limits of the Precordillera.

The presence of Gualcamayo blocks and of presumably reworked (Von Gosen et al., 1995) striated clasts in the La Rinconada section indicates that both the Gualcamayo Formation and the Don Braulio Formation originally were much more widespread deposits than their present-day outcrop distribution indicates.

CONTINENTAL MARGIN SEDIMENTS AND ALLOCHTHONOUS CAMBRIAN AND ORDOVICIAN ROCKS

La Cruz Limestones

Middle and Upper Cambrian limestone olistoliths are present in all sections along the Ordovician continental margin within the Los Sombreros Formation (Keller, 1995; Bordonaro and Banchig, 1996). The largest slab is found in the Los Túneles section (Fig. 3) and is ~1500 m long and ~300 m thick (Fig. 29). Other occurrences include the Cambrian of San Isidro (Borrello, 1971; Pina et al., 1986), which only recently has been shown to be in an allochthonous position (Bordonaro et al., 1993). Many of these olistoliths show a characteristic deep-water facies described in the following. Different names have been applied to these rocks: Caliza La Cruz or La Cruz Formation (Borrello, 1971), Cerro Solitario Formation (Borrello, 1971), and Estancia San Martín Formation (Pina et al., 1986). A compilation of the most important outcrops of these limestones was made by Bordonaro and Banchig (1996). It is here proposed to include all Middle and Upper Cambrian olistoliths in deep-water facies into the La Cruz Limestones, which requires a redefinition of this unit. One of the problems of formal nomenclature is the absence of reference with respect to giant olistoliths. The olistoliths described in the following fulfill most criteria to assign them formational status (Codigo Argentino de Estratigrafía): they can be mapped on a scale of 1:25,000, the outcrops are laterally continuous, and they have a distinct lithology. However, the La Cruz Limestones are redefined as an informal unit, because they lack distinct stratigraphic boundaries owing to their allochthonous origin.

La Cruz Limestones: Redefinition and lithology. The La Cruz Limestones are a succession of deep-water limestones, marlstones, and shales with a characteristic deep-water or open-marine fauna composed of agnostid trilobites and sponge spicules.

Derivatio nominis and type section. The name is derived from the Quebrada de la Cruz in the San Isidro area west of Mendoza city (Fig. 30). The section in the Quebrada de la Cruz is the original type section of Borrello (1971). There, the La Cruz Limestones are present as huge olistoliths within the Ordovician continental-margin succession (Los Sombreros Formation).

Boundaries. The lower boundary is unknown, because hitherto no olistolith has been found in which underlying rocks with a different lithology are exposed.

The upper boundary is exposed in the Los Túneles section. There, within an olistolith, Lower Ordovician slope sedimentary rocks overlie Middle Cambrian agnostid-bearing rocks with an erosional unconformity.

Age. Many ages have been obtained from the La Cruz Limestones (Benedetto and Vaccari, 1992; Bordonaro, 1985; Bordonaro and Baldis, 1987; Bordonaro and Banchig, 1990; Borrello, 1971; Cuerda et al., 1985b; Heredia, 1987, 1995; Heredia and Bordonaro, 1986; Vaccari, 1987). The most recent discussions of the olistoliths and their trilobites were by Bordonaro and Liñan (1994) and Bordonaro and Banchig (1996). The oldest olistoliths yield trilobites of the *Glossopleura* zone (early Middle Cambrian), whereas the youngest olistoliths belong to the *Saukia* chron of the latest Cambrian.

Thickness. A minimum of 300 m is preserved in the Los Túneles megaolistolith.

Distribution. Locally derived olistoliths of the La Cruz Limestones are present in all sections of the continental-margin facies (Keller, 1995). Hence they were originally a laterally extensive unit that supplied olistoliths to the sections at San Isidro, Los Sombreros, Ojos de Agua, Quebrada de los Ratones, and the Los Túneles section near Jáchal (Fig. 3). The limestones are also present east of the Quebrada Santa Elena and along the road between the Pampa de Yalguaraz and the Puesto Santa Clara de Abajo (Fig. 3) within the "Facies Alojamiento" of the Villavicencio Group of Harrington (1971). In addition, it seems that almost all the other limestones within this "Facies Alojamiento" resemble the La Cruz Limestones (Harrington, 1971; p. 24).

Discussion. The La Cruz Limestones represent a widespread facies in the Argentine Precordillera, but they are only present as allochthonous blocks in the Los Sombreros Formation. The olistoliths give a wide range of ages for the sediments, which were

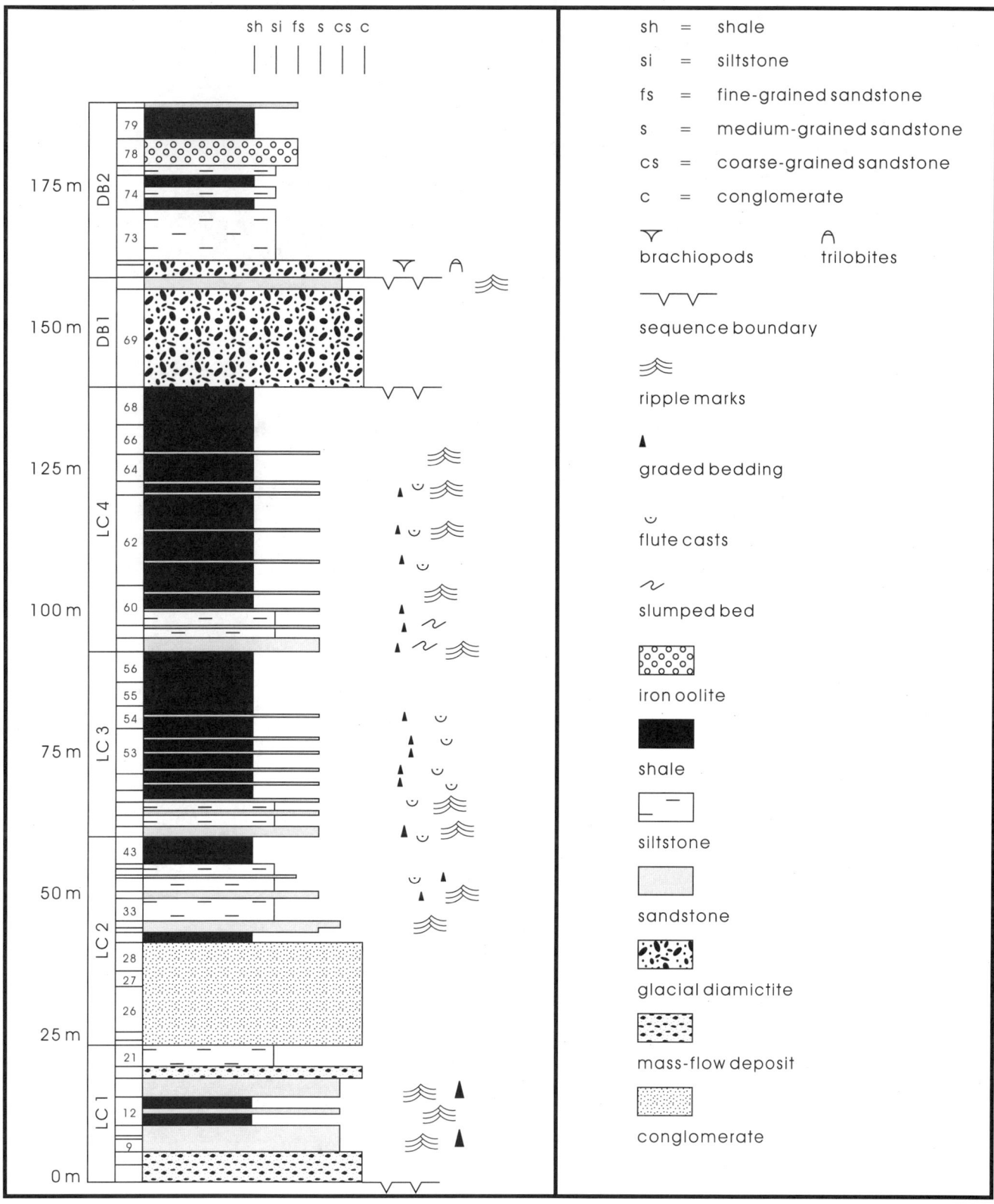

Figure 28. Sedimentary succession of La Cantera and Don Braulio Formations in Don Braulio section. LC 1–LC 4 are lithologic units within La Cantera Formation. DB 1 and DB 2 are lithologic units within Don Braulio Formation. See text for details.

Figure 29. Aspects of Los Sombreros and Empozada Formations. 1: Chancellorid wackestone of Lower Cambrian olistolith at Los Túneles locality. (× 10). 2: Los Túneles megaolistolith. Light gray carbonate slab (arrow) is >1 km long and > 325 m thick. Olistolith is composed of La Cruz Limestones and is embedded in Los Sombreros Formation. 3: Olenellid trilobite coquina of Lower Cambrian olistolith at Los Túneles locality (× 10). 4: Boulder-bearing shale of Los Sombreros Formation at San Isidro. Boulder consists of brown sandstone. In purely descriptive sense, these sedimentary rocks are true diamictites. Backpack for scale. 5: Diamictite in Empozada Formation at San Isidro. Fine-grained sandstones and siltstones of platform environment host abundant limestone clasts and chert pebbles. Although sediments have possible Ashgillian age, there is no final clue as to (glacial?) origin of these sediments. 6: Load casts are abundant in sedimentary rocks of Empozada Formation. Cast is about 5 cm across.

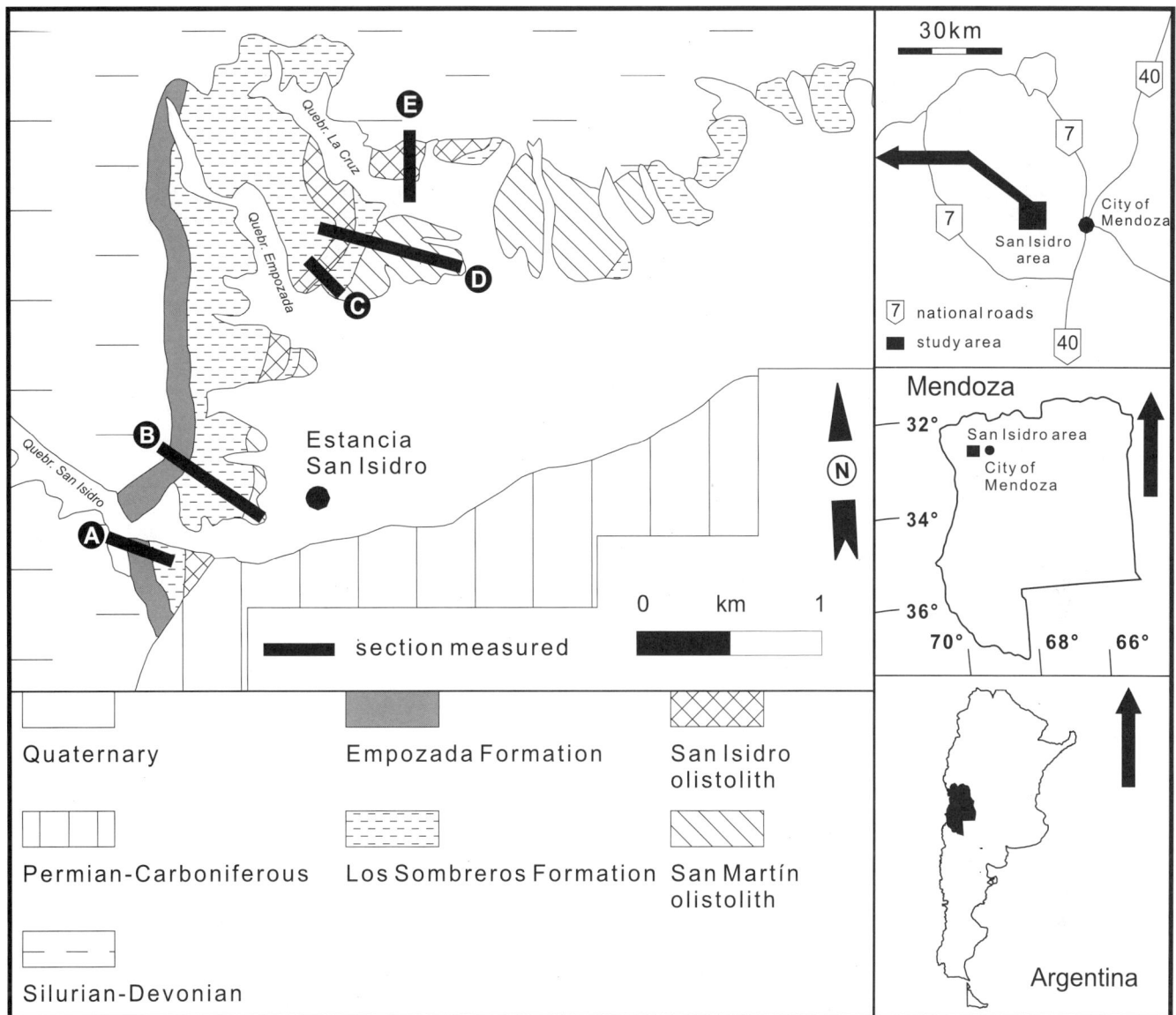

Figure 30. Cambrian-Ordovician outcrops at classical San Isidro locality. Section A is proposed type section for Empozada Formation. Sections B and D form composite section shown in Figure 21. Sections C and E were measured through San Isidro olistolith which represents facies of La Laja Formation (modified from Keller et al., 1993a).

deposited probably in a deep outer platform setting. A variety of names was applied to these rocks, mainly based on their differing ages. For example, Borrello (1971) included the rocks at Cerro Pelado into his La Cruz Formation, but the former later were redefined as the El Relincho Formation by Heredia (1990, 1996). With the proof of more and more olistoliths with similar lithologies but even a wider variety of ages, it became clear that all these rocks originally were part of the same depositional system; therefore there is no need to create an increasing number of formational names. The name La Cruz Limestone for all these deposits was chosen because the type section is the most easily accessible and the San Isidro area is a classical locality for the study of Middle and Upper Cambrian limestones. With the redefinition given here, the names Solitario Formation and San Martín Formation are superfluous, and should be abandoned; however, the individual olistoliths may be referred to as, e.g., San Martín olistolith and Solitario olistolith.

Lithofacies. Carbonates are the dominant lithology. Mudstones, wackestones, and, in places, packstones contain trilobites, phosphatic brachiopods, and sponge spicules. Silt-sized quartz and pyrite nodules are common. Bioturbation is observed only very locally. The sedimentary rocks are thin to medium bedded, and dark gray to black. Dark brown to yellow-brownish colors are observed where the matrix is recrystallized to microspar. Following the microfacies concept (see Appendix 1) of Wilson (1975) and Flügel (1982), spiculitic mudstones (SMF 1), pelagic mudstones (SMF 3), and microbioclastic calcisiltites (SMF 2) are present. The microbioclastic calcisiltites are composed of fine carbonate detritus, peloids, and small abraded bioclasts, among which are trilobites and sponge spicules. In places, very small white calcispheres of unknown origin have been observed. Quartz silt, equigranular to the other components, is always present. Many of the beds are very thinly stratified, and some of them are graded and show traction-induced bedding. Small-scale scouring at the base of the beds has been observed in places.

Dark marlstones and brown and black shales are present but only locally abundant (e.g., in the San Martín olistolith). The shales are fissile. The fauna in the marlstones and shales is the same as in the limestones. In all the rocks bioturbation is absent, and the rocks show a very regular, in places slightly undulose, bedding. Although in many cases the sediments are tectonically strongly deformed, slumping can still be observed in several horizons.

Interpretation and depositional environment. Sediments of the La Cruz Limestones reflect a deep-water depositional environment. This is shown by three microfacies (SMF 1–3) that are characteristic of a slope or basinal setting (Wilson, 1969) and their association with marlstones and black shales. Other features that corroborate such an interpretation are the dark color of the sediments, their thin lamination, micrograding and microscouring, and the lack of bioturbation. In addition, the agnostid trilobite fauna (Bordonaro, 1990b; Bordonaro and Liñan, 1994) and conodonts indicate a fairly deep, open-marine environment.

In many Paleozoic deep-water environments the lime mud is almost exclusively derived from adjacent platforms (peri-platform ooze; Schlager and James, 1978). The main depositional processes are settling out of suspension and sedimentation from turbidity currents (Scholle et al., 1983). In the La Cruz Limestones, the graded beds of calcisiltite with their small scours are interpreted to represent distal turbidites. In contrast, most of the mudstones, spiculites, shales, and marlstones are autochthonous deposits.

Although there is no doubt that the La Cruz Limestones reflect a deep-water environment, the original site of deposition, i.e., slope or basin, is more difficult to assess, because of the allochthonous nature of the outcrops. One characteristic of the La Cruz Limestones is the absence of debris-flow deposits, conglomerates, and megabreccias, which are much more common on slopes than in the basins (Cook and Mullins, 1983). This may point to a basinal setting of the La Cruz Limestones, but as stated by McIlreath and James (1984) and Burchette and Wright (1992), a carbonate platform may pass into deeper water without a marked break at the platform margin. In such a ramp-like configuration, the La Cruz Limestones probably represent the deep-ramp and basinal environment together.

Los Sombreros Formation

The Los Sombreros Formation was introduced by Cuerda et al. (1983) for a thick succession of intermixed carbonates, shales, breccias, and conglomerates, to which they assigned an Ordovician age. Cuerda et al. (1985b, 1986) reported Cambrian and Ordovician fossils from these rocks. The recognition that the Cambrian and Lower Ordovician rocks are allochthonous blocks (olistoliths) within the Los Sombreros Formation (Bordonaro, 1990a; Benedetto and Vaccari, 1992) and the distribution of the Los Sombreros Formation all along the western margin of the eastern basin led to almost identical redefinitions (Benedetto and Vaccari, 1992; Banchig and Bordonaro, 1994). At present, the Los Sombreros Formation is regarded as an Ordovician slope succession comprising upper Arenigian through Caradocian sedimentary rocks (Banchig and Bordonaro, 1994). In this paper, the outcrops of the slope facies at San Isidro, which hitherto have been included in the Empozada Formation, are also assigned to the Los Sombreros Formation.

San Isidro section. Green and black shales are the dominant lithology in the San Isidro section (Figs. 30 and 31). They contain a variety of mass-flow deposits. Conglomerates are composed of clasts of sandstone, shale, alum shale (see Appendix 1), siliceous shale, chert, and limestones with a variety of textures. The conglomerates are predominantly clast supported; the maximum clast size is ~5 cm. The conglomerates are laterally discontinuous and show channel-fill geometries. Some of them are vaguely graded and there is a transition toward current-induced stratification in the upper parts of the beds. Amalgamation of conglomerates is visible in a few outcrops. Channels may also be filled with coarse- to medium-grained graded sandstones with ripple-drift cross-bedding.

Several breccia beds are present in the lower part of the section. They are as much as 50 cm thick and can be laterally traced for about 40 m until they thin and disappear. There is a clear difference between carbonate and noncarbonate clasts. Carbonate clasts are mostly tabular and in places are imbricated. They are composed of lime mudstone and wackestone and some calcisiltites. The noncarbonate clasts, which are smaller and much more spherical, include quartz, chert, and siliceous shale.

Diamictites are a prominent feature. They contain centimeter- to meter-sized clasts of different lithologies (4 in Fig. 29). Several beds of arkosic sandstones are present in the lower part of the succession. They are graded and have an erosional base.

In general, the San Isidro section (Fig. 31) shows the greatest variability among the clasts and among the olistoliths of all sections of the Los Sombreros Formation. Many of the rocks have been described as the lower part of the Empozada Formation (Gallardo et al., 1988; Gallardo and Heredia, 1995) or as part of the Cambrian of San Isidro (Keller et al., 1993a).

The Cambrian of San Isidro is a thick succession of carbonates with intercalated packages of shales. This succession hitherto was thought to be Early to early Middle Cambrian and Late Cambrian in age. Several formations were established; in ascending order, the Estancia San Martín Formation, San Isidro Formation, and the La Cruz Formation, (Borrello, 1971; Pina et al., 1986). However, it was demonstrated (Bordonaro, 1992; Bordonaro et al., 1993) that all three "formations" are huge olistoliths within the Ordovician slope facies of the Los Sombreros Formation, separated by the shale intervals (Fig. 31). Trilobites and conodonts are known from both the San Isidro olistolith and the La Cruz olistolith (Bordonaro, 1985; Heredia, 1987; Heredia and Bordonaro, 1986). However, no macrofossils have been described from the Estancia San Martín olistolith. Ordovician olistoliths are rare at San Isidro, and there is only one documented occurrence (Bordonaro and Peralta, 1987).

Estancia San Martín olistolith. The Estancia San Martín olistolith is composed of the La Cruz Limestones. Because no age-diagnostic macrofossils have been reported from this olistolith to date, the assignment of the Estancia San Martín Formation to the Cambrian is now mainly based on the occurrence of abundant sponge spicules, which seem to be typical of the Cambrian (Beresi and Rigby, 1994).

San Isidro olistolith. The San Isidro olistolith is the best dated (Bordonaro, 1989, 1992). The trilobites are mainly platform types and are characteristic of open-marine conditions (Bordonaro et al., 1993). The trilobites are very similar to those of the coeval La Laja Formation.

Packstones and grainstones are the most important lithology of the San Isidro olistolith (Fig. 31). The main bioclasts are trilobite fragments, but hyolithids, pelmatozoans, and brachiopods have also been observed. Other components include ooids, abundant oncoids, and detrital quartz grains. Glauconite is abundant in some horizons and is present either as a replacement of components or as diagenetic filling of intergrain spaces (Keller et al., 1993a). Wackestones contain trilobites and brachiopods that are better preserved than in the packstones. Sponge spicules are also present. The rocks are often nodular and strongly bioturbated. Marlstones and siltstones are present to a minor extent and have a fauna similar to that of the other rocks. In places, ripple-drift cross-lamination has been observed.

Two facies associations are recognized that are also present in the coeval La Laja Formation; i.e., the packstone-grainstone association and the siltstone association. In addition, there are the wackestones and marlstones, which can be assigned to the wackestone-intraclast packstone association or the mudstone-wackestone association of the La Laja Formation. Consequently, the sedimentary rocks of the San Isidro olistolith do not have to be reinterpreted here. Instead, identical rocks and almost identical trilobites to the coeval strata of the La Laja Formation are more than strong evidence that the San Isidro olistolith was part of the La Laja depositional environment. The recognition of an olistolith of the La Laja Formation at San Isidro implies that the La Laja Formation extended from the surroundings of San Juan to at least the San Isidro area.

Ordovician olistolith. Graptolite black shales are present at the southern margin of the Quebrada de San Isidro (Fig. 30). They were dated as Arenig (Bordonaro and Peralta, 1987) and were thought to be in an autochthonous position. However, the findings of Middle Ordovician graptolites in levels stratigraphically beneath these shales (Bordonaro et al., 1993) indicate that the graptolite shales are also an olistolith. One important conclusion of the presence of this olistolith is that in the San Isidro area a deep-water environment existed during the Arenig. Sedimentation might have taken place in a similar environment as, e.g., in the Guandacol area, where black shales were deposited from the early-late Arenig boundary onward.

Los Sombreros section. The type section (Fig. 32) shows various mass-flow and rockfall deposits (Fig. 33) that yielded fossils spanning the Middle Cambrian through Early Ordovician. This led to some confusion about the stratigraphic range of some of these fossils. Bercowski and Fernandez (1988) found *Nuia* in several limestones in the upper part of the succession. On the basis of its presence in horizons stratigraphically below beds containing Middle Cambrian trilobite *Tonkinella stephensis*, they concluded that *Nuia* might have a stratigraphic range from the Middle Cambrian into the Ordovician. With the realization that the limestones in the Los Sombreros Formation are of allochthonous origin, this interpretation has to be reconsidered. The clasts with *Nuia* are part of a thick megabreccia deposit, and according to their microfacies the clasts can easily be attributed to the San Juan limestones. Besides trilobites, the first *Cordylodus proavus* fauna from South America was described from one of the Cambrian deep-water limestone olistoliths (Lehnert 1994).

Several megabreccias are present in the upper part of the section (1, 2 in Fig. 33). In these deposits, La Flecha type rocks and clasts of the La Silla Formation and the San Juan Formation are found associated with clasts of sandstones and turbidites (3, 6 in Fig. 33). Another interesting sedimentary rock is fault-graded beds (Seilacher, 1969) in which small vertical faults propagate upward and die out toward the top of the horizon (5 in Fig. 33).

Ojos de Agua section. Sandstones and conglomerates are important constituents in this section (Fig. 34). The maximum clast size is ~60 cm. Among the clasts, garnet-bearing gneisses, schists, and granitoids have been observed (Banchig et al., 1990a). Many of the conglomerates fill erosional channels and are laterally discontinuous. They are often graded and pass into coarse-grained sandstones. The sandstones have erosional features and are graded. Amalgamation was observed in several beds. Bouma sequences T_a to T_b or T_a to T_c are common in these deposits. Thick cross-bedded sandstones and sandstone without internal sedimentary structures are also present. At the base of the section, huge boulders are exposed that consist entirely of conglomerates, but their connection to the section is not clear. In addition, thick carbonate megabreccia bodies have been described from this section (Banchig et al., 1990b).

Los Ratones section. The outstanding feature of these outcrops (Fig. 3) is an apparent channel fill, in which basement boulders are present. Among the angular boulders, metaquartzites, gneisses, and granitoids have been observed. In addition, a huge olistolith of the La Cruz Limestones is present (Fig. 35).

Los Túneles section. This section shows a uniform succession of slates in which one major olistostrome with a variety of olistoliths is present (Fig. 36). Among these is the largest olistolith found in the Los Sombreros Formation. It is here termed the Los Túneles megaolistolith. It is more than 1500 m long and

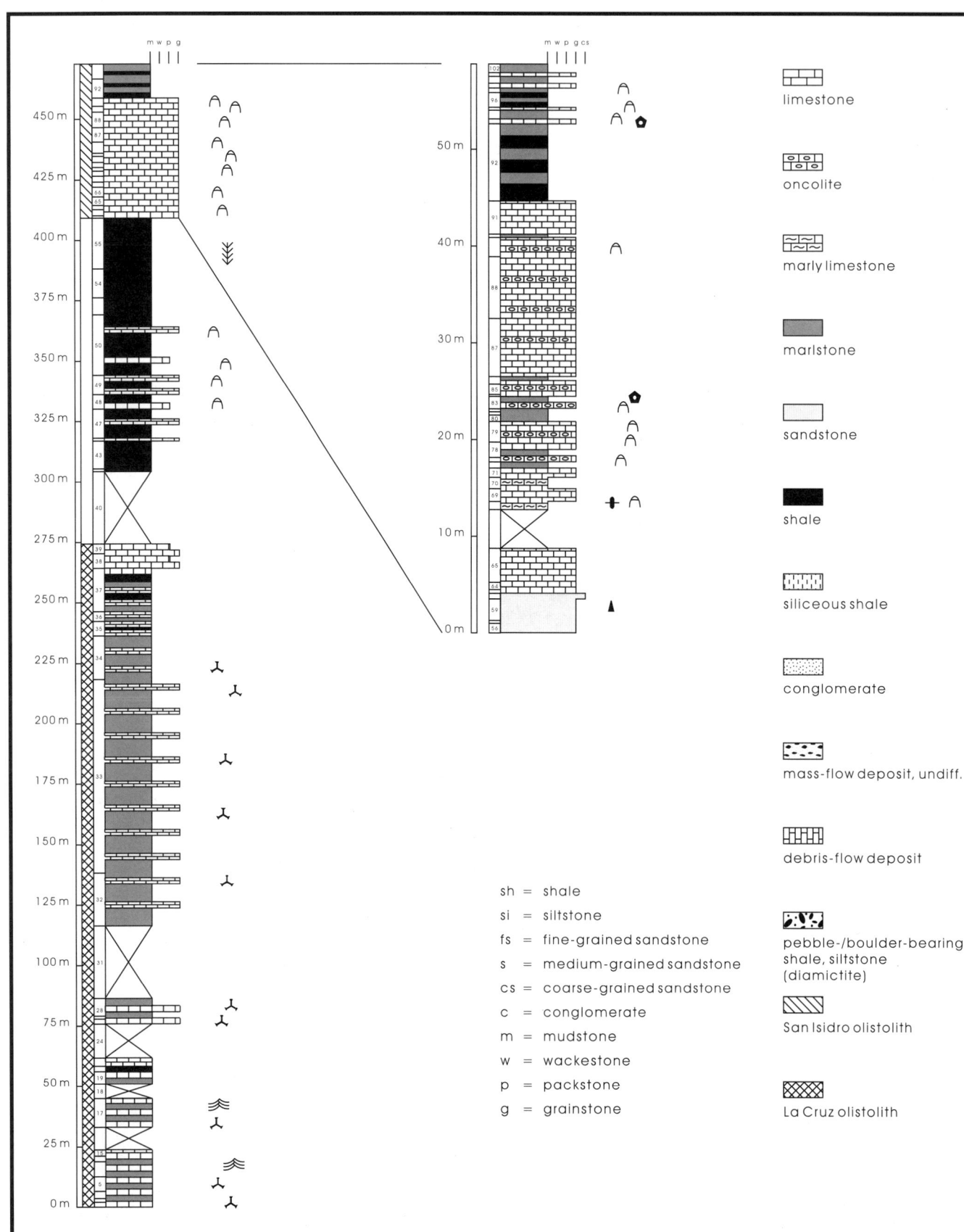

Figure 31 (this and opposite page). Section of Los Sombreros Formation at San Isidro.

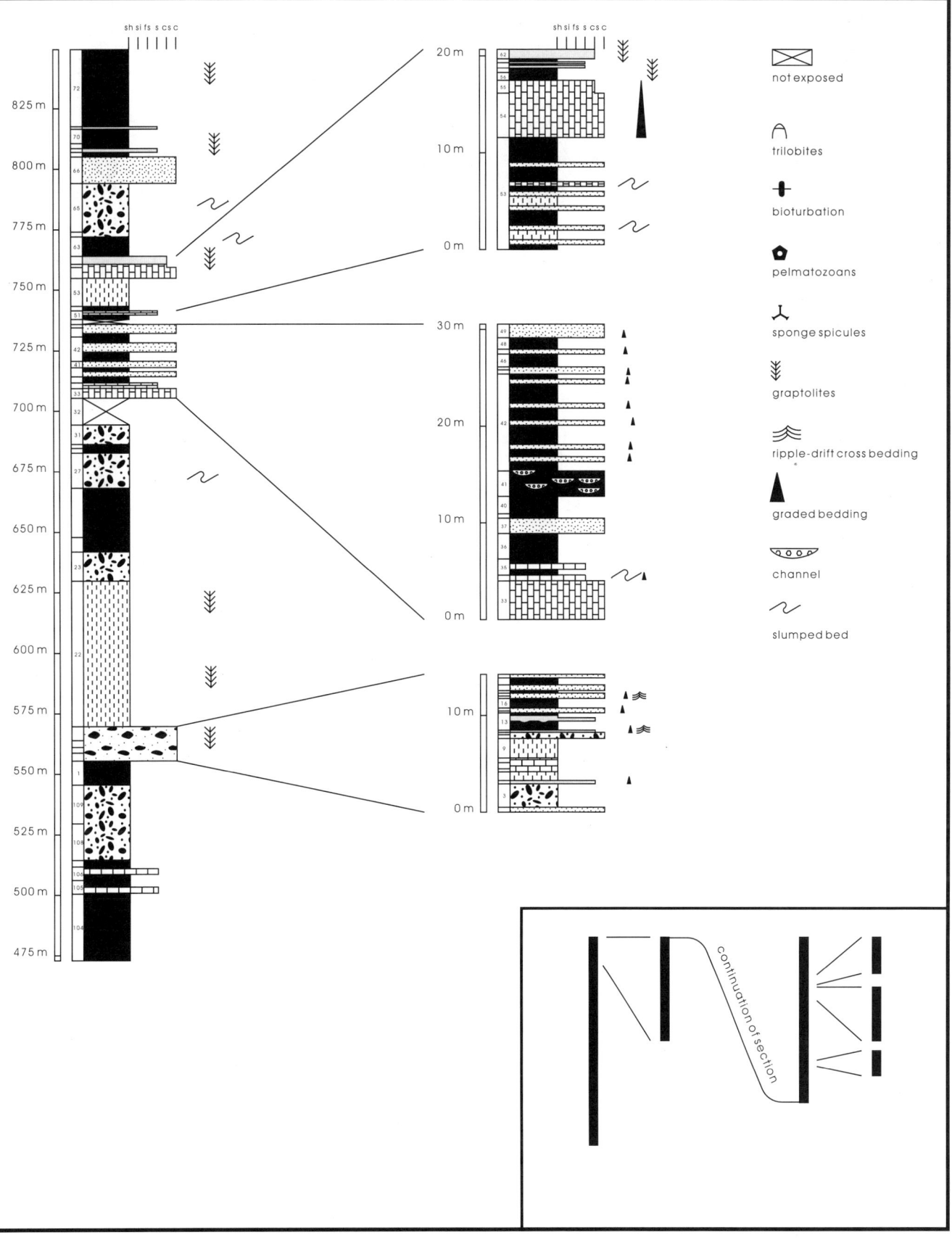

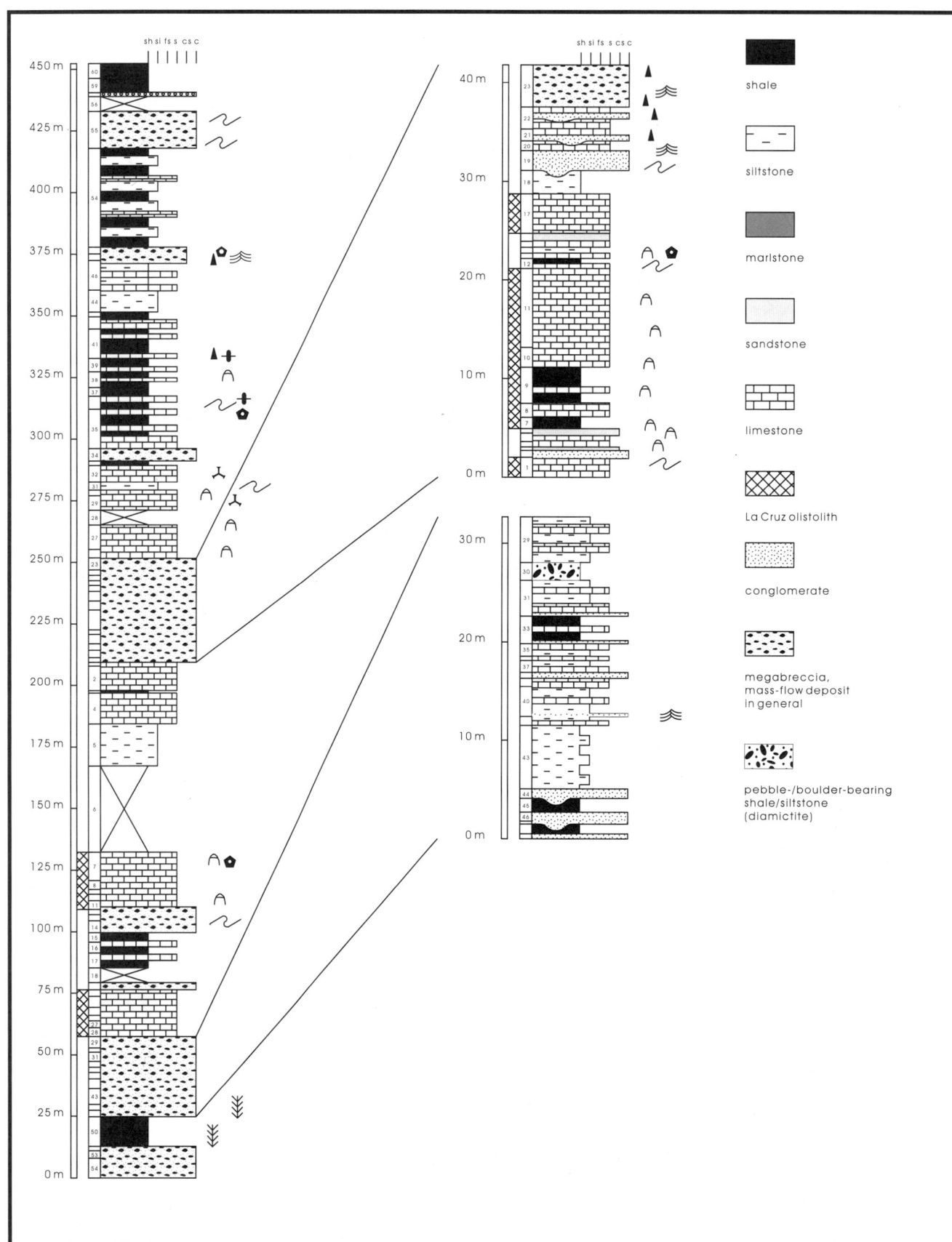

Figure 32 (this and opposite page). Section of Los Sombreros Formation at the type locality.

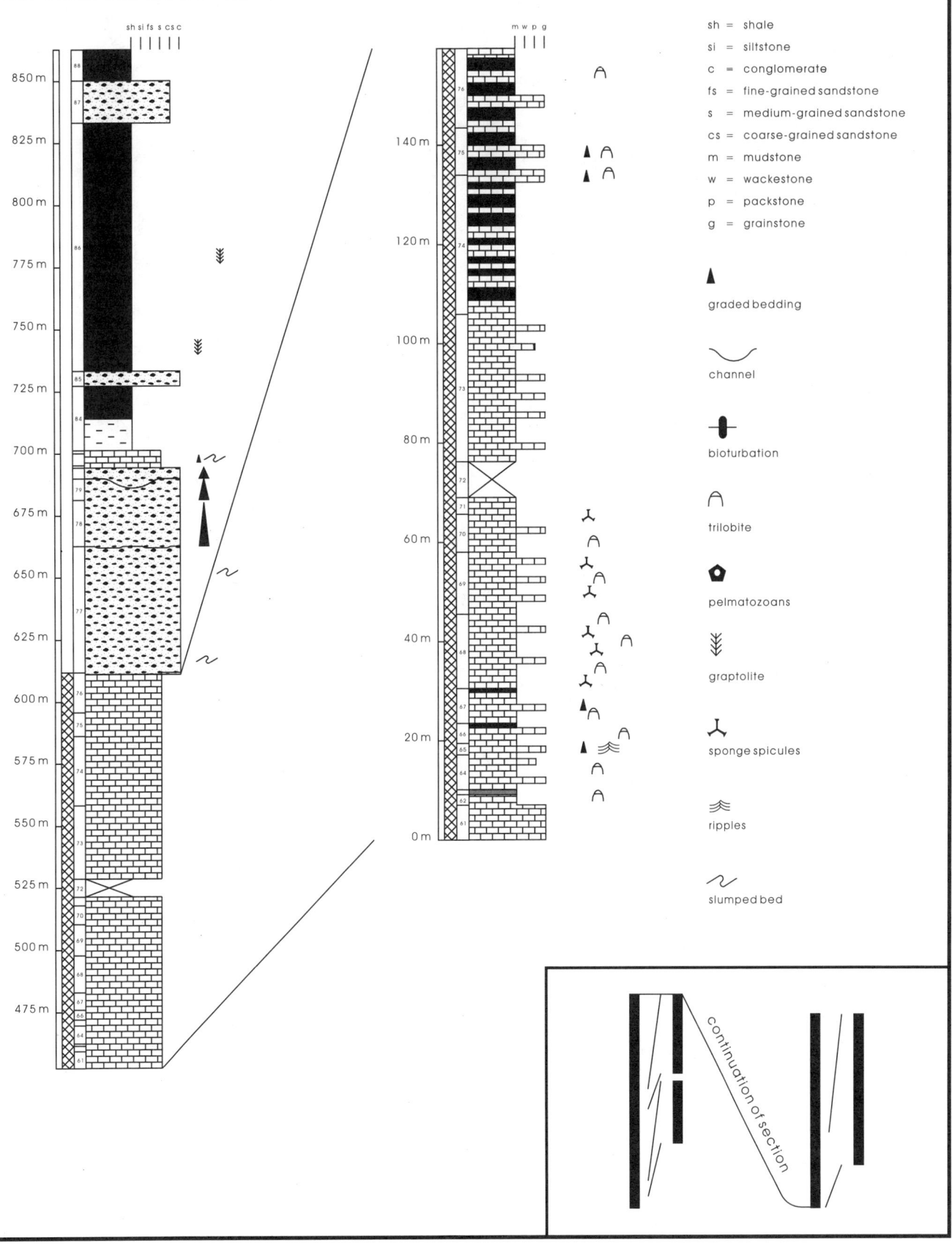

Figure 33. Type section of Los Sombreros Formation. 1: Large olistolith composed of La Cruz Limestones is overlain by thick megabreccia in upper part of Los Sombreros Formation. In places, thin-bedded limestones show large-scale slumping. 2: The upper megabreccia interval consists of two amalgamated thick breccias. Entire succession is >30 m thick. 3: Tabular conglomerate clast in upper megabreccia. These conglomerates are typical facies of turbidite system exposed in Ojos de Agua and km-101 sections of Los Sombreros Formation (hammer tip is 11 cm long). 4: Clast-supported polymict conglomerate with abundant platform-derived limestone clasts (hammer is 32 cm long). 5: Fault-graded bed in upper part of Los Sombreros section. Note small faults that cut through bed. Vertical displacement is large near base of bed, but strongly reduced toward top of bed. These faults probably originated in semilithified sediments during earthquakes. Hammer tip is 11 cm long. 6: Calcareous turbidites in upper part of section contain abundant platform-derived material (e.g., peloids, ooids, algae, etc.) Hammer tip is 5 cm long.

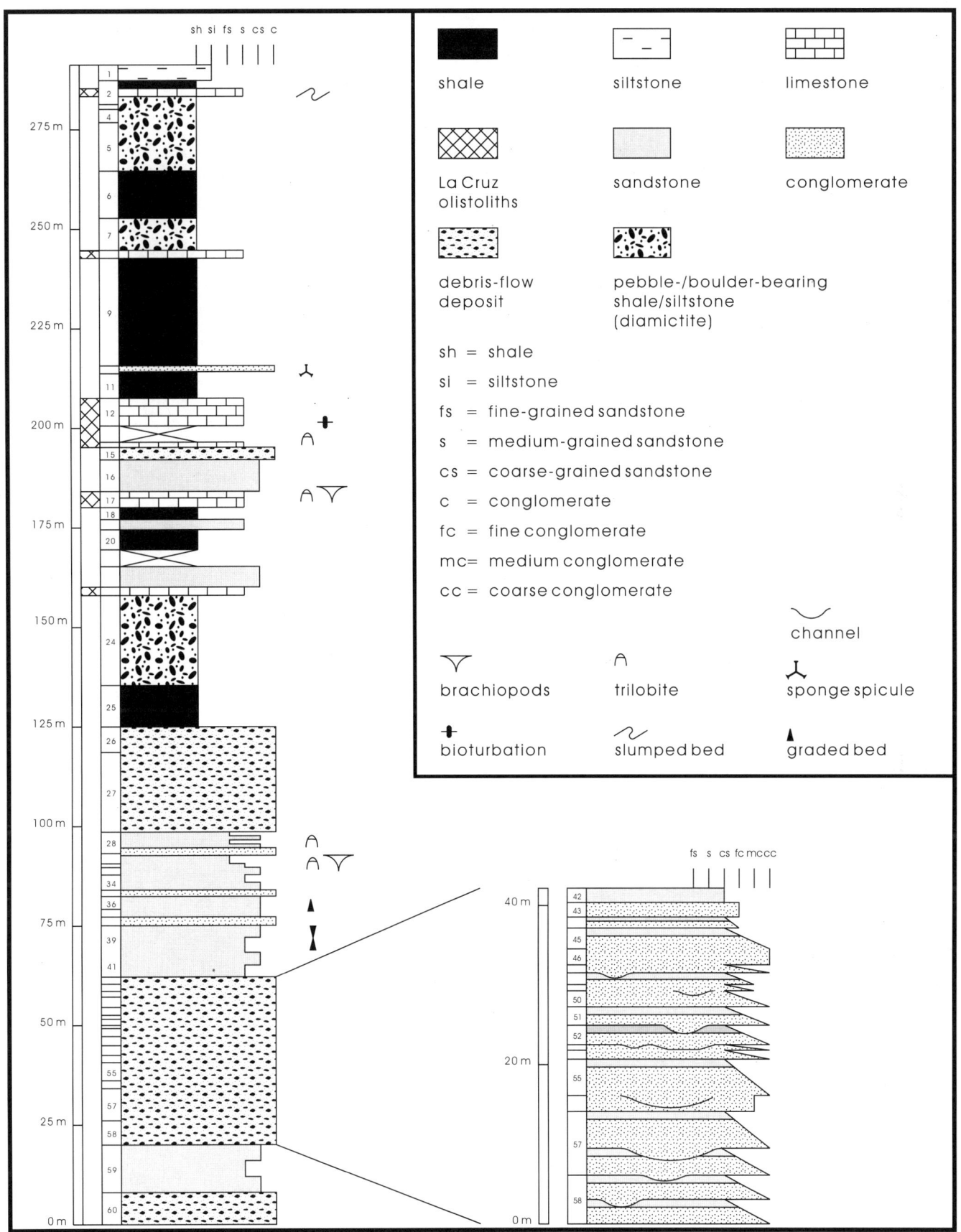

Figure 34. Section of Los Sombreros Formation at Ojos de Agua.

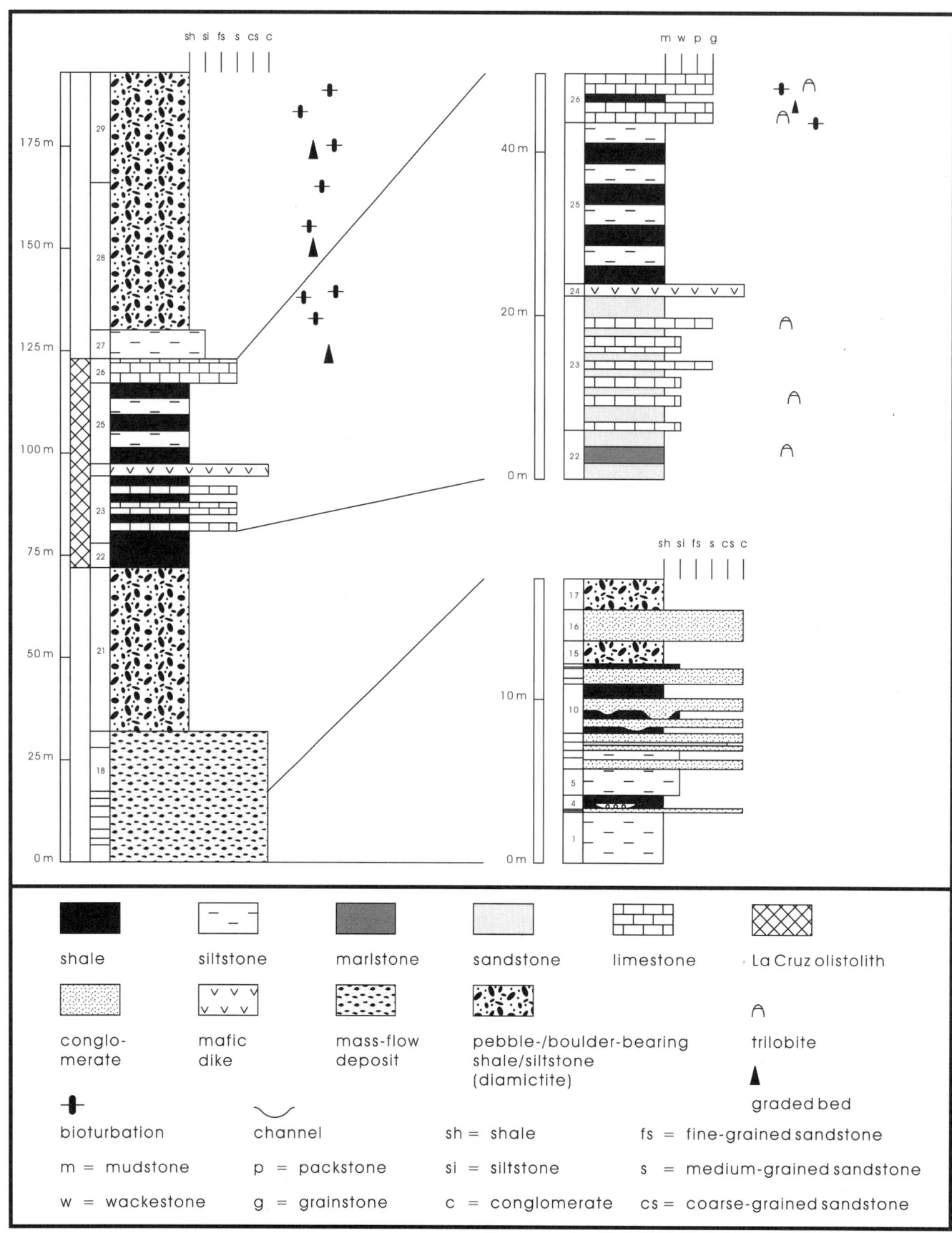

Figure 35. Section of Los Sombreros Formation at Los Ratones, also known as km-101 section

~325 m thick (2 in Fig. 29). The other olistoliths include rocks with a variety of ages, from Early Cambrian to Early Ordovician. In addition, there is a package several meters thick with abundant chert pebbles in a slate matrix. Pebble sizes vary from 1 to 5 cm.

Lower Cambrian olistolith. Only one Lower Cambrian olistolith has been documented (Vaccari, 1987; Benedetto and Vaccari, 1992) in the Los Sombreros Formation. It is present in the Los Túneles section west of Jáchal.

The olistolith shows an alternation of thin-bedded shales and siltstones with frequent intercalations of thin-bedded trilobite grainstones and packstones (Fig. 37). The shales are mostly green, but red has also been observed. Trilobite packstones are the dominant lithology among the carbonates (3 in Fig. 29). Densely packed olenellid fragments form the bulk of the bioclasts in these gray to orange-brown carbonates. Other clasts include pelmatozoans and small intraclasts. Trilobite-intraclast packstones contain trilobites, lithoclasts, and pelmatozoans. Many of the clasts are rounded. Among the intraclasts two different types of microfacies have been observed: dark brown mudstones and brown trilobite wackestones. In addition, millimeter sized oval components are present, which consist of a thin outer layer and a filling of large calcite crystals. The shape of these clasts closely resembles oncoids, which are found in the Middle Cambrian La Laja Formation. In all packstones, the micritic matrix is recrystallized toward microspar and pseudospar and is rich in particles such as hematite and authigenic quartz. A few wackestone horizons are present in the olistolith. They contain trilobites and abundant fragments of the sponge-like *Chancelloria* sp. (1 in Fig. 29).

Sediments of the olistolith were deposited under open marine conditions, probably below fair-weather wave base. This is indicated by the absence of current-induced structures in the siliciclastic sediments that formed the matrix of packstones. It is most likely that these rocks represent distal tempestites, which would explain their thin-bedded nature and the broken and rounded bioclasts. Almost identical sedimentary rocks were described by Palmer and Halley (1979) from Lower Cambrian strata of the western Great Basin (Carrara Formation), and were interpreted to represent sediments of a shelf lagoon, an explanation that is likely for this unique olistolith in the Precordillera which shows a comparable facies.

Los Túneles megaolistolith. The lower part of the megaolistolith (~300 m) consists of Middle Cambrian La Cruz Limestones (Fig. 36). Toward the top, there is an erosional unconformity within the olistolith. Above the contact, a thick breccia with a variety of lithologies is present. The breccia contains clasts of chert to 5 cm × 5 cm, peloidal mudstone, lime mudstone, and deep-water lithologies. In addition, coarse detrital quartz grains, ooids, and bioclasts are present. Many of the ooids are broken. Sorting is poor, the fabric is clast supported, and the clasts are angular to moderately rounded. With better rounding, a transition to conglomerates is present in the overlying horizons. Graded bedding and other sedimentary structures have not been observed.

Upsection, there is a succession of thick-bedded grainstones, conglomerates, few breccias, and thin shale beds. The grainstones are poorly sorted; their main components are ooids. Some of the ooids are broken, and others are well preserved. In addition, there are lithoclasts (peloidal grainstone, intraclast grainstone, black deep-water limestones), bioclasts, and coarse detrital quartz. Some of the grainstones exhibit rippled surfaces, and others fill deep erosional channels. A variety of conodonts was found in the grainstones (Albanesi, cited in Benedetto and Vaccari, 1992), which indicate a late Arenigian or even early Llanvirnian age for the deposits.

Toward the top of the megaolistolith, shales and wackestones become more abundant but grainstones are still present. In the shales, early Llanvirnian graptolites have been found (Ortega et al., 1991).

The lower part of the megaolistolith represents a deep-water environment like that described from the other Cambrian olistoliths of the La Cruz Limestones. However, the Los Túneles section displays a more complete record of sedimentation well into the Ordovician. The breccia, conglomerates, and grainstones are the products of mass movements (Nardin et al., 1979). The three rock types include intraclasts consisting of deep-water limestones, and they contain extraclasts. The almost black intraclasts were clearly derived from the underlying Middle Cambrian strata; they were obviously mistaken by Cabaleri (1989) for "black pebbles" (according to Strasser, 1984) and, together with the oolitic turbidites, were interpreted to represent a peritidal environment of the deposits.

The extraclasts are mainly platform derived, especially the peloidal grainstones and intraclast grainstones, which are typical of the La Silla Formation, as are the ooids. A problem is presented by the quartz clasts and elements of *Teridontus nakamurai*. The latter are badly preserved and belong to a species that is mainly present on the Ordovician La Silla platform (Lehnert, 1997, personal commun.). Hence it seems possible that these conodonts were also transported from the platform to the slope. The provenance of the quartz is even more mysterious. No quartz sands are present on the platform later than the Marjuman, neither as individual sand bodies nor as scattered quartz grains in the carbonates. Their size and shape indicate that they might have a local source. They might have been provided by slope erosion cutting down to the basement or by erosion of quartz veins present close to the continental slope. The breccias are not typical debris-flow deposits because they are clast supported and have only little matrix. However, they lack many features of turbidites (graded bedding, Bouma sequences).

Similar conglomerates and breccias were described by Garrison and Ramirez (1989) from Neogene strata (Monterrey Formation) of central coastal California. Their oligomictic and "petromictic" conglomerates are a good analogue to the sedimentary rocks described here. Transport mechanisms presumed for the Neogene deposits are debris-flow and high- and low-density turbidity currents that gave rise to a variety of geometries in the sediments. Although the clast sizes differ considerably between the California conglomerates and the deposits described here, similar depositional processes are inferred. The relatively

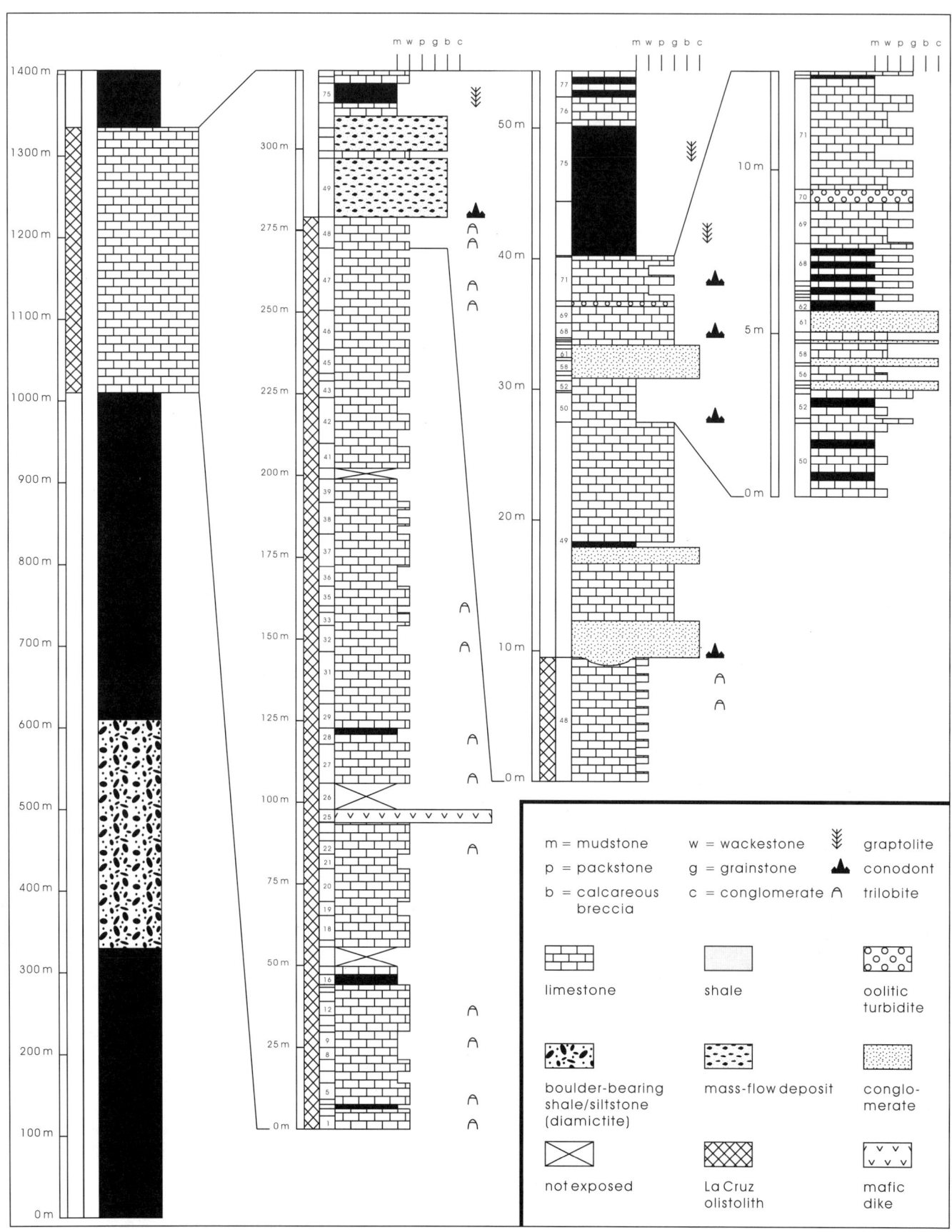

Figure 36. Section of Los Sombreros Formation at Los Túneles locality.

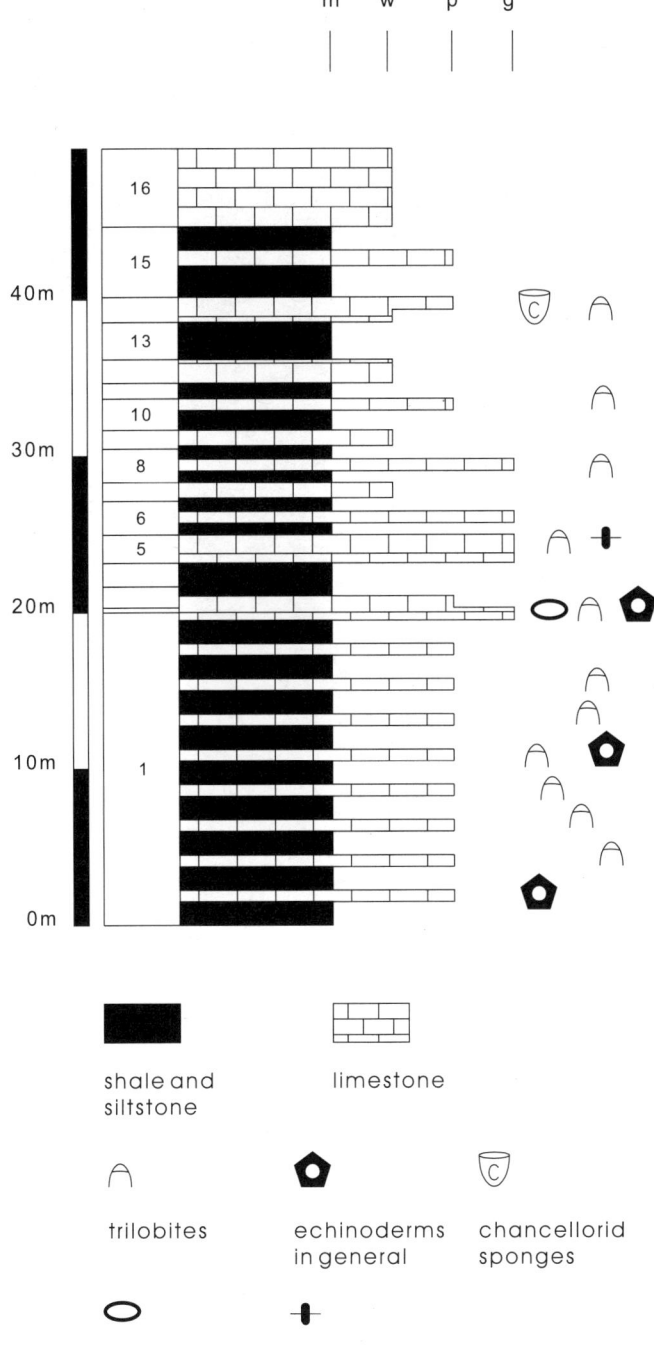

Figure 37. Section measured through Lower Cambrian olistolith at Los Túneles section. Legend as in Figure 16.

fine grained breccias and conglomerates of the Los Túneles section are interpreted to have formed from debris flows and high-density turbidity currents. Both mechanism are operative along slopes and in submarine canyons and fans (Nardin et al., 1979). The restricted outcrops in the olistolith do not permit an assignment of the sediments to either depositional environment. Hence it is more than speculative to infer a slope configuration based on the relatively fine grained sediments, shallow channels, and the sheet-like appearance of the deposits (Nardin et al., 1979).

A major unconformity and a hiatus are present within the olistolith (Keller, 1995). About 2 m beneath the first breccia bed, Middle Cambrian trilobites are present. Conodonts from the base of the breccia (*T. nakamurai*) indicate a Late Cambrian or Early Ordovician age. Therefore almost the entire Late Cambrian section is missing there. The specimens might have been reworked from Ordovician platform strata and it is thus possible that the hiatus present is much greater.

Other olistoliths of the Los Túneles section. One of the most conspicuous olistoliths is composed entirely of quartz pebbles between 1 and 3 cm. Although this olistolith is barely 6 × 7 m across, it still shows that the sediments are thick bedded and cross-bedded. Comparable sediments were deposited around the Laurentian margin during Late Proterozoic rifting and compositionally have a good match in conglomerates of the Appalachian Ocoee Group (R. Hatcher, 1995, personal commun.).

Varieties of small calcareous blocks are also present. The facies of one of them can easily be attributed to the San Juan Formation and this olistolith was dated with conodonts as Arenig (Benedetto and Vaccari, 1992). Many of the other calcareous olistoliths are dolomites similar to the Upper Cambrian Zonda and La Flecha Formations.

Time frame of deposition of the Los Sombreros Formation.
At San Isidro, the oldest autochthonous fossils so far reported (Bordonaro et al., 1993) are present about 250 m above the base of the exposed section, where *Glossograptus hincksii, Climacograptus sp.*, orthograptids, and glyptograptids have been found.

In the Argentine Precordillera, *G. hincksii* appears during the Llanvirn (Cuerda et al., 1986). Although there are another 250 m of sediment beneath the horizon with these graptolites, the presence of olistoliths with Arenigian graptolites (Bordonaro and Peralta, 1987) probably indicates that the base of the Los Sombreros Formation is not older than the Llanvirn.

A multitude of graptolites has been described from the type section (Cuerda et al., 1983, 1986), ranging from the Tremadocian into the Caradoc. Consequently, the authors assumed a Tremadocian through Caradocian age for the deposits. However, from their publication (Cuerda et al., 1986: Fig. 2) it is evident that almost all specimens are found in an interval of about 50 m that is characterized by calcareous breccias and conglomerates. During field work it was demonstrated that among the clasts in the Los Sombreros Formation there are also shale clasts in a shaly matrix, similar to the Arenigian clast in the San Isidro area. Hence it seems likely that the "autochthonous" age is that of the youngest graptolites, i.e., Llandeilo or Caradoc, and that the older graptolites are reworked together with their matrix. This interpretation is supported by the presence of Middle Ordovician graptolites near the base of the measured section (Fig. 32).

In the Los Túneles section, conodonts from a San Juan limestone olistolith indicate an (late?) Arenig age of the "source rock" (Sarmiento *in* Benedetto and Vaccari, 1992). Conodonts

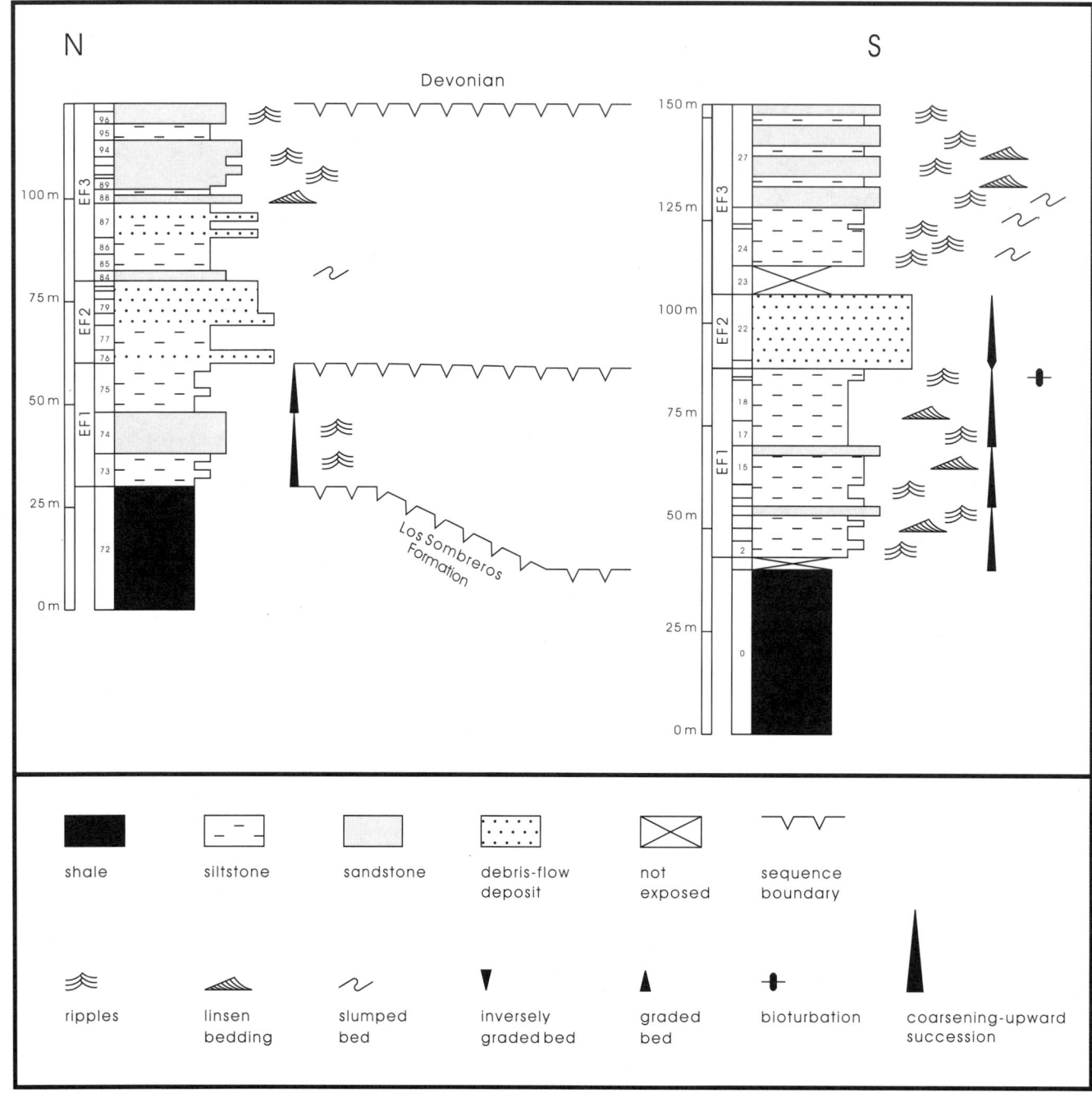

Figure 38. Sections measured in Empozada Formation as defined in this paper. Southern section is proposed type section. EF 1–EF 3 are lithologic units within Empozada Formation. This section corresponds to upper member of the Empozada Formation following Gallardo and Heredia (1995).

and graptolites obtained from the megaolistolith indicate an early Llanvirn age for the top of the allochthonous succession. In contrast, Benedetto and Vaccari (1992) interpreted the upper 25 m of the olistolith to be autochthonous slope sediments, and, consequently, assigned an early Llanvirn age to the deposits of the Los Sombreros Formation. This interpretation would imply that the elements of *T. nakamurai* in this succession (Keller, 1995) are necessarily reworked specimens. Although this possi-

bility cannot be entirely discounted, it is more likely that the Los Sombreros Formation in the Los Túneles section is younger than early Llanvirn. This is in agreement with the data from the other sections and observations by Peralta (1997, personal commun.) that the lower Llanvirnian graptolites found everywhere were in allochthonous blocks. Consequently, the Los Sombreros Formation seems to comprise the late Llanvirn through Caradoc.

Interpretation and depositional environment of the Los Sombreros Formation. Two contrasting sedimentary processes are responsible for the thick deposits of the Los Sombreros Formation. Accumulation of thick, often graptolite-bearing shales or alum shales is the background sedimentation in quiet and deep waters. Together with these feautures, the apparent absence of bioturbation points to stagnant bottom conditions.

The fine-grained sediments form the matrix for mass-movement deposits. Conglomerates and turbidites are major groups and were discussed in detail by Banchig et al. (1990a) for the Ojos de Agua section. They concluded that the depositional environment was proximal and medial to a slope apron. Proximity indicators are the T_{a-c} turbidites, the amalgamation, and the deep erosional features at the base of the turbidites. The turbidites and conglomerates in the Ojos de Agua section clearly represent facies A and B of Mutti and Ricci Lucchi (1972).

The megabreccias give evidence of platform destruction, as Lower Cambrian through Lower Ordovician carbonates are present. In addition, turbiditic sandstone clasts (3 in Fig. 33) demonstrate that the slope or the base of slope, was also affected by submarine erosion. The megabreccias are interpreted as debris-flow deposits, much like those described from the Devonian of Canada (Cook et al., 1972; Mountjoy et al., 1972) or those discussed in Cook and Mullins (1983).

According to Seilacher (1969), fault-graded beds (5 in Fig. 33) are the products of seismic waves (seismites of Seilacher, 1969), disturbing the sediment and documenting increasing consolidation of the sediment with depth. Garrison and Ramirez (1989) related such beds to extensional deformation in a slope and basin setting.

The only possible transport mechanism for the huge olistoliths is rockfall along steep or oversteepened slopes. This implies that the individual olistoliths were deposited not on the slope, but on the adjacent base of slope. In the Los Ratones and the Ojos de Agua sections, these steep slopes affected the deep-water limestones and exposed part of the basement. The dimensions of the fault scarps there are difficult to establish, because of the unknown total original thickness of the La Cruz Limestones.

Another important implication of the presence of the olistoliths is that locally almost the entire carbonate platform succession must have been exposed along escarpments. The main argument for this is the presence of a possibly Precambrian quartz-pebble olistolith in the Los Túneles section and of a proven La Laja Formation olistolith in the San Isidro section. In addition, basement-derived clasts from metamorphic and igneous sources were found in a number of sections. These escarpments must have been on the order of >2100 m, as the maximum thickness of the preserved carbonate succession is about 2100 m. In modern environments, escarpments of this size have been demonstrated bounding the Bahamas platform (Mullins and Newman, 1979; Schlager and Chermak, 1979). There, however, it is not clear whether upwards building of the carbonates (Dietz et al., 1970) or tectonics (Mullins and Lyntz, 1977) is the ruling factor of formation of the steep slope.

In the Precordillera, tectonics seems to be the contolling factor because on the one side, carbonate platform development had essentially ceased by middle Ordovician time and on the other side, during evolution of the platform, no such steep escarpments have been documented. The presence of such an escarpment along much of the eastern border of the western basin, together with the nature of the rocks exposed, is strong evidence that sediment input onto the basin floor occurred along a line source (Schlager and Chermak, 1979). This line source was essentially the carbonate platform, as the majority of the olistoliths are carbonates. Although the carbonate platform was essentially dead by the time the Los Sombreros Formation was deposited, it still behaved like a bypass margin (James and Mountjoy, 1983). Bypass margins of carbonate platforms are characterized by slope and platform edge failure due to oversteepening and the origination of debris flows, slides, and slumps, which are deposited at the toe of slope. The formation of bypass margins is particularly common toward the end of carbonate platform sedimentation (James and Mountjoy, 1983), but in the Los Sombreros Formation it persisted beyond or formed after carbonate platform sedimentation had ceased.

In the description of modern continental margins, essentially three major physiographic areas are recognized (Drake and Burk, 1974): the continental shelf, slope, and rise. A marginal escarpment forms the abrupt and steeply inclined transition from the shelf to the rise where there is no slope. In the Precordillera, the sediments of the Los Sombreros Formation were deposited on the continental rise (or toe of slope) adjacent to a marginal escarpment. This interpretation of the Los Sombreros Formation as representing the continental rise is in contrast to slope interpretations for the Los Sombreros Formation (Banchig et al., 1990a, 1990b; Benedetto and Vaccari, 1992; Keller, 1995); however, a continental-rise interpretation is supported by the scarceness of slumps and slides, which are much more common on the slope (Cook et al., 1982).

Empozada Formation

The Empozada Formation was originally introduced by Harrington (1957) for the Ordovician rocks of the classical San Isidro area (Fig. 30); however, it lacks all formal requirements to satisfy stratigraphical nomenclature. For a long time, these rocks were thought to concordantly overlie Upper Cambrian limestones, the La Cruz Formation of Borrello (1971). However, these Cambrian rocks were interpreted as allochthonous blocks, here within the Ordovician Empozada Formation (Bordonaro et al., 1993).

On the basis of lithologic criteria, the succession above the Cambrian olistoliths was subdivided into two members (Gallardo et al., 1988; Gallardo and Heredia, 1995) separated by a major unconformity. This major unconformity separates a lower succession, in which shales form the matrix for a variety of conglomerates, breccias, and olistostromes, and an upper succession of sandstones and siltstones with a major debris flow in the middle of the unit. Beyond doubt, the lower succession of the Empozada Formation is part of the Los Sombreros Formation as redefined by Banchig and Bordonaro (1994). This poses a nomenclatural problem for the succession at San Isidro, because the name Empozada Formation, introduced by Harrington (1957) for the sedimentary rocks above the Cambrian limestones, clearly has priority. However, in his descriptions (Harrington, 1957; p. 20–21) the formations are not adequately defined. However, the Los Sombreros Formation has been well defined (Cuerda et al., 1983) and redefined only recently (Banchig and Bordonaro, 1994). In order to overcome these problems I propose here to redefine the Empozada Formation and to restrict it to the upper sandy member, following Gallardo et al. (1988) and Gallardo and Heredia (1995). Consequently, the San Isidro area is the only area in which the upper boundary of the Los Sombreros Formation is exposed.

Empozada Formation: Redefinition. The Empozada Formation is a succession of partly calcareous sand- and siltstones with intercalated debris-flow deposits.

Derivatio nominis and type section. The name is derived from the Quebrada de la Empozada in the San Isidro area (Fig. 30), in which the original type section is found. The formation is even better exposed in the Quebrada de San Isidro ~500 m to the south (Fig. 30). This section is proposed as the new type section of the Empozada Formation (Fig. 35).

Boundaries. The lower boundary is found at a level at which sandstones overly black graptolite shales of the Los Sombreros Formation with a sharp contact interpreted to represent a major disconformity.

The upper boundary is drawn where yellow sandstones are abruptly overlain by Silurian-Devonian green shales. Although locally this contact is a fault, the erosional contact is beautifully exposed if the boundary is laterally traced.

The overlying shales are tentatively assigned to the Canota Formation; however, they do not fit the general description of these Devonian sedimentary rocks given by Cuerda et al. (1988a) and Kury (1993).

Age. At San Isidro, high in the black shale succession of the Los Sombreros Formation, graptolites of the *N. gracilis* zone have been described (Alfaro and Fernandez, 1985), and at the top of the black shale unit (below the unconformity) the *Orthograptus quadrimucronatus* zone is present (Alfaro, 1988). No autochthonous fossils were found in the Empozada Formation; however, the conodont *Amorphognathus superbus* has been separated from carbonate clasts within the main debris flow (Gallardo et al., 1988; Heredia et al., 1990). Consequently, the base of the formation is within or is younger than the graptolite zone of *Orthograptus quadrimucronatus*. The presence of *Amorphognathus superbus* in the debris-flow deposit indicates that the overlying sediments are not older than the Ashgillian.

Thickness. Sections of the Empozada Formation as redefined here vary between 120 and 150 m (Fig. 35).

Distribution. At present, outcrops of the Empozada Formation are only known from the greater San Isidro area; however, field observations around Canota to the north indicate that there may be outcrops of the formation in this area.

Lithofacies and the sedimentary succession of the Empozada Formation. Along the southern margin of the Quebrada de San Isidro (Fig. 30), the Empozada Formation is composed of three units (EF1–EF3; Fig. 35). The lower unit (EF1) consists mainly of brown, very thin to thin-bedded fine- to coarse-grained siltstones. Slumping, load casts, and ripple-drift cross-bedding are abundant. In several beds, lenticular bedding with isolated lenses is present. In others, megaripples (Reineck and Singh, 1980) have been observed. The base of individual sandstone beds is often sharp. Three coarsening-upward successions have been observed, which start with fine-grained siltstones and end with fine-grained sandstones (Fig. 35). Within these cycles, there is also a thickening-upward tendency; sand layers that are only millimeters to a few centimeters thick to beds as thick as tens of centimeters. In the northern section only two coarsening-upward successions are preserved.

The second unit (EF2) of the Empozada Formation is a prominent breccia bed, which in its lower 2 m is inversely graded, whereas the upper part (16 m) shows a crude normal gradation. The entire bed is intensely slumped. The breccia is matrix supported; among the clasts are brownish sandstones and siltstones with a lithology similar to that of the underlying beds. Light gray limestones with a variety of textures can be attributed in their majority to the San Juan Formation. In addition, there are dark gray to black deep-water limestone clasts. A big boulder of limestone conglomerate is also present. In the northern section, this major breccia bed splits up into several horizons, which are separated by sandstones and siltstones. Conodonts have been described from clasts of this debris-flow deposit (Gallardo et al., 1988). Only two populations are present, one typical of the middle and upper San Juan Formation and the other, of Caradocian age, similar to the fauna of the Río Sassito carbonates. The breccia can be traced all along the outcrops of the Empozada Formation at San Isidro, and is probably a sheet-like deposit.

The third unit of the Empozada Formation (EF3) is similar to the lower unit; however, grain size varies mainly between coarse silt and fine to medium sand. Current-induced ripples, lenticular bedding, slumping, and load casts (6 in Fig. 29) are very prominent. Indicators of current directions show east-directed as well as west-directed transport. In several horizons in this upper part, small chert pebbles with diameters of 1–2 cm have been observed; the rock is actually a pebbly sandstone or diamictite (5 in Fig. 29).

The upper unit shows four coarsening-upward successions starting with coarse silt and passing into medium-grained sand. The four successions are part of one larger coarsening-upward succession.

Interpretation. In the Empozada Formation, two contrasting styles of sedimentation are observed: sedimentation out of debris flows and sedimentation on a siliciclastic shelf. The units both below and above the debris flow show a variety of structures and a sedimentary succession, which together indicate deposition on a wind- and wave-dominated shelf.

Sand-mud couplets, in which the sands form flat layers or are present as discontinuous lenses (lenticular bedding), are characteristic of the heterolithic facies association of Johnson (1978). The sand-mud ratio varies about between 4:1 and 1:4, reflecting changes in the hydrodynamic conditions or sediment supply. Lenticular, flaser, and wavy bedding are found in subtidal and intertidal settings (Reineck and Singh, 1980). In subtidal environments, sharp-based sandstones, in places graded, with erosional scours are often the result of major storms (Walker, 1984a). However, one characteristic feature of ancient storm deposits, hummocky cross-stratification, (discussion in Dott and Bourgeois, 1982; Walker 1984a), has not been observed in the Empozada Formation. The internal structures of the sand beds in general point to a wave origin or a wave overprint. A similar interpretation of wave domination was given by Gallardo et al. (1988) and Astini (1991) for this part of the Empozada Formation.

In the upper succession there is more evidence of a shallow platform environment, especially the presence of bipolar oriented cross-bedding. Astini (1991) concluded that this part of the Empozada Formation might have been tidally influenced, whereas Gallardo et al. (1988) assumed an environment just above storm wave base. The abundant indications of an oscillatory flow regime, however, are more in agreement with a wave-dominated environment.

An interesting rock type is diamictite. Taking into account their (most likely) Ashgillian age, it is tempting to relate the diamictites to the Late Ordovician Gondwana glaciation. Upper Ordovician glacial sediments are known in the Precordillera from the Don Braulio Formation (Peralta and Carter, 1990a), where striated clasts and dropstones are present. None of these features have been found at San Isidro. However, almost identical diamictites have been described from Portugal and the Prague basin (Brenchley et al., 1991) and were interpreted to be of glacial origin. There, the main arguments against a debris-flow origin are their presence within shallow shelf sandstones and the absence of associated turbidites. The same arguments are true for the sedimentary rocks of the Empozada Formation. Although no conclusive evidence is present that the diamictites in the Empozada Formation are truly of glacial origin, this is a likely possibility and their presence should be kept in mind in future discussions of the Empozada Formation.

If the interpretation of both units is correct, then the entire succession may be regarded as one thickening- and one coarsening-upward succession, separated by a thick debris-flow deposit. In marine environments, they usually form on slopes or the toe of slope (Nardin et al., 1979; Walker, 1984b). However, initiation and sedimentation of debris-flow deposits do not require steep slopes nor do they imply great water depth (Cook and Mullins, 1983; McIlreath and James, 1984). In carbonate depositional systems, two end-member models of slope sedimentation are distinguished, depending on the profile from the platform margin to the slope and basin (McIlreath and James, 1984). Depositional margins have a ramp-like configuration and slopes merge with the basin floor without major break in topography. Bypass margins, in contrast, show a steep slope adjacent to the platform margin, which is often bypassed by platform-derived sediment. On either kind of slope, major intraformational truncation surfaces may develop. The base of the deposit in the Empozada Formation represents one of these truncation surfaces, because at the lower contact sediment is successively cut out. This is seen when the unit is traced to the north (Fig. 35) until only a few meters of the lower unit are left between the black shales of the Los Sombreros Formation and the debris-flow deposit just south of the Quebrada de la Empozada. The sheet-like debris-flow deposit within the Empozada Formation may be taken as a hint to the presence of a slope apron.

Discussion. There are two pecularities of the Empozada Formation: first, there is a debris-flow deposit sandwiched between siliciclastic sediments interpreted to have formed in a shelf environment; and second, most of the clasts within the debris-flow deposit are carbonate platform derived. In addition, the conodonts obtained from the clasts belong to two distinctly different units, the San Juan Formation and the Río Sassito succession. This might indicate that carbonates like those of the Río Sassito succession were also present in the San Isidro area, and second, that a similar hiatus might had been developed between the Caradocian carbonates and the San Juan Formation. The assumption of a situation similar to that at Río Sassito (a horst-like structure with carbonate sedimentation surrounded by siliciclastic grabens) also explains the presence of carbonate clasts in the debris flow. But what about a debris-flow deposit in a succession of platform siliciclastic deposits?

One possibility is that the lower siliciclastic unit was misinterpreted and that it, in fact, represents another depositional environment. Gallardo et al. (1988) assumed a slope environment for the succession based on the presence of slumps and graded beds, which they interpreted as turbidites. In such an environment the presence of debris-flow deposits is nothing unusual. However, they also recognized storm-induced structures in the sandstones and assigned the sediments to the heterolithic facies of Johnson (1978). Following Johnson (1978), this facies is a shallow platform facies. Yet Astini (1991) noted the ambivalence of the environmental indicators and concluded that the sandstones are either a special kind of turbidite or just sands deposited below the depth to which oscillatory flows are effective. However, the presence of lenticular bedding is strong evidence of a shallow platform environment (Johnson, 1978; Elliott, 1978; Reineck and Singh, 1980). As storm surge ebb can produce turbidity flows, it seems likely that the turbiditic nature of many of these beds is a result of storm activity in a deeper platform environment.

If the interpretation of a platform environment for this lower part of the Empozada Formation is correct, then the presence of the debris-flow deposit requires another explanation. The Middle and Upper Ordovician history of the Precordillera has been interpreted as reflecting major crustal extension (Loske, 1992; Von Gosen et al., 1995; Keller, 1995). In addition, the Río Sassito carbonate platform is a platform formed in a horst position (Lehnert, 1995a; Keller and Lehnert, 1998). If it is assumed that block-faulting was still active during deposition of the Empozada Formation, then the debris flow may have originated through oversteepening of the margin of a nearby platform or simply through an earthquake.

ORDOVICIAN OF PONON TREHUE

Since the studies of Wichmann (1928), the presence of Ordovician rocks south of San Rafael, in the area of Ponon Trehue (Fig. 39), has been well established. The limestones, now regarded as the Ponon Trehue Formation, were compared with similar limestones at Cerro de la Cal just north of Mendoza (Fig. 3), the latter beyond doubt belonging to the San Juan Formation. Later, a variety of fossils were described from Ponon Trehue (Baldis and Blasco, 1973; Levy and Nullo, 1975; Heredia, 1982); however, these fossils invariably were younger than the classical San Juan limestones. Only recently was it demonstrated that the Ordovician of Ponon Trehue is composed of two fundamentally different successions (Bordonaro et al., 1996), one entirely calcareous, the other a mixed siliciclastic-carbonate system. Both formations nonconformably overlie crystalline basement, which seems to be typical Grenvillian-type (Ramos, 1995, personal commun.).

Ponon Trehue Formation

Until recently, little was known about the sedimentology of the Ponon Trehue Formation, and its relation to other carbonates of the Argentine Precordillera. A wealth of new data, both sedimentologic and biostratigraphic, now allows a much better discussion of these carbonates and their connection to the Precordillera. The section measured in the Ponon Trehue Formation is shown in Figure 39. According to conodont data, the age of the Ponon Trehue Formation spans the late Tremadoc through middle or late Arenig (discussion in Lehnert et al., 1998).

Lithofacies. Coarse dolosparites are massively bedded, and micritic patches and mound-like structures have been observed only locally. In thin sections, ghost structures are visible which, by their size and shape, are identified as ooids. The rocks are mainly light gray to beige but in places pinkish colors are present.

White and light gray microbial boundstones form massive beds. In many places, no sedimentary or internal structures are visible in the micritic sediments, but there are intercalations of well-developed microbial laminites. In thin sections, these rocks reveal structure grumeleuse, indicating a microbial origin. Fenestral fabrics have only rarely been found in the microbial rocks.

Mudstones and wackestones are dark gray and often have a nodular texture. Bioclasts include crinoids, sponges, nautiloids, gastropods, trilobites, brachiopods, ostracods, and the enigmatic *Monticulipora*. In addition, calcareous algae and *Nuia* are present. Bedding varies from thin to thick bedded. With an increasing amount of debris there is a transition to packstones. The packstone texture, however, was caused by a textural homogenization by bioturbation.

Packstones are thin-bedded accumulations of bioclastic debris. The fauna and flora are essentially the same as in the wackestones; in addition, intraclasts are present.

Nodular crinoidal packstones and grainstones are dark gray and thin bedded. Bioclasts are crinoids and fragments of thick-shelled brachiopods and trilobites. Abundant peloids have been observed in thin sections, giving these rocks a bimodal clast size distribution. Peloidal grainstones are a minor component of the sedimentary succession.

Massive biostromal boundstones are an accumulation of large receptaculitids (>20 cm) and sponges. In places, sponges and algae form small isolated mounds.

Interpretation, facies associations, and the sedimentary succession of the Ponon Trehue Formation. Four units (PT1–PT4) have been distinguished in the field (Fig. 40). Unit 1 is composed of a dolomite association, in which diagenesis has obliterated most of the primary sedimentary features. Some of the rocks near the base were oolitic or oolites; the primary sediment in the other horizons may have been shallow-marine mudstone, as indicated by micritic patches and the conodonts found. In such an environment the obscure mound structures may be remnants of thrombolites.

The microbial-boundstone association characterizes unit 2. The alternation of microbial laminites and pure lime mud is taken as evidence of a supratidal environment, although no desiccation features have been observed. The wackestone association shows an abundant open-marine fauna. Rocks and fauna are identical to the wackestone-intraclast-packstone association of the San Juan Formation, which was deposited on the middle ramp. A similar environment is likely for the rocks of the Ponon Trehue Formation. This is corroborated by the composition of the conodont fauna, which is a temperate-water association. The reef-mound facies association is composed of biostromal boundstones, small isolated sponge-algal mounds, and the intermound sediment. Two horizons are present. In the lower horizon, sponge-algal mounds are associated with mudstone, whereas in the upper interval packstone is the intermound sediment. This facies association is similar to the lower reef-mound horizon at Cerro La Silla; however, in the Ponon Trehue Formation it lacks high-energy grainstones. The sediments were deposited in shallow subtidal areas under changing hydrodynamic conditions. The lower interval reflects a tranquil environment, whereas the upper interval formed under more turbulent conditions; packstones are present and in places flank the mounds.

The wackestone association in combination with the reef-mound horizons are typical of unit 3. The depositional environment is best characterized as subtidal and open marine near normal wave base. Waves or currents only intermittently swept across the platform to stir up the sediment and redeposit it as packstones. In comparison with the San Juan Formation, this environment would correspond to the middle ramp. In the Ponon Trehue Formation, however, it is impossible to tell whether this is really a middle ramp setting or a protected intrashelf basin.

Unit 4 consists of an association of nodular crinoidal grainstones and few packstones. Chert is abundant in this association,

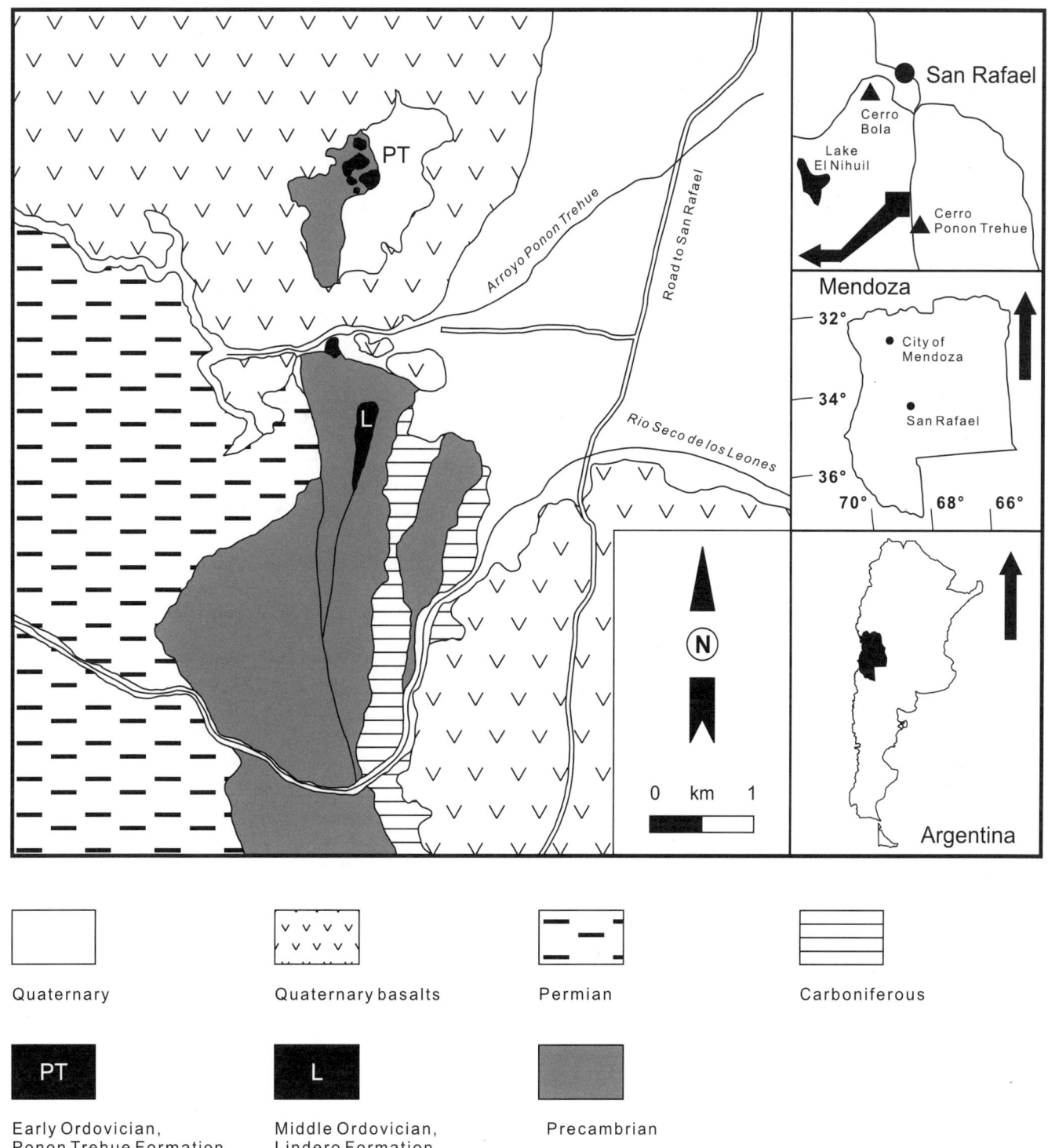

Figure 39. Simplified geologic map of Ponon Trehue area showing classical localities of Ordovician of San Rafael (modified from Bordonaro et al., 1996).

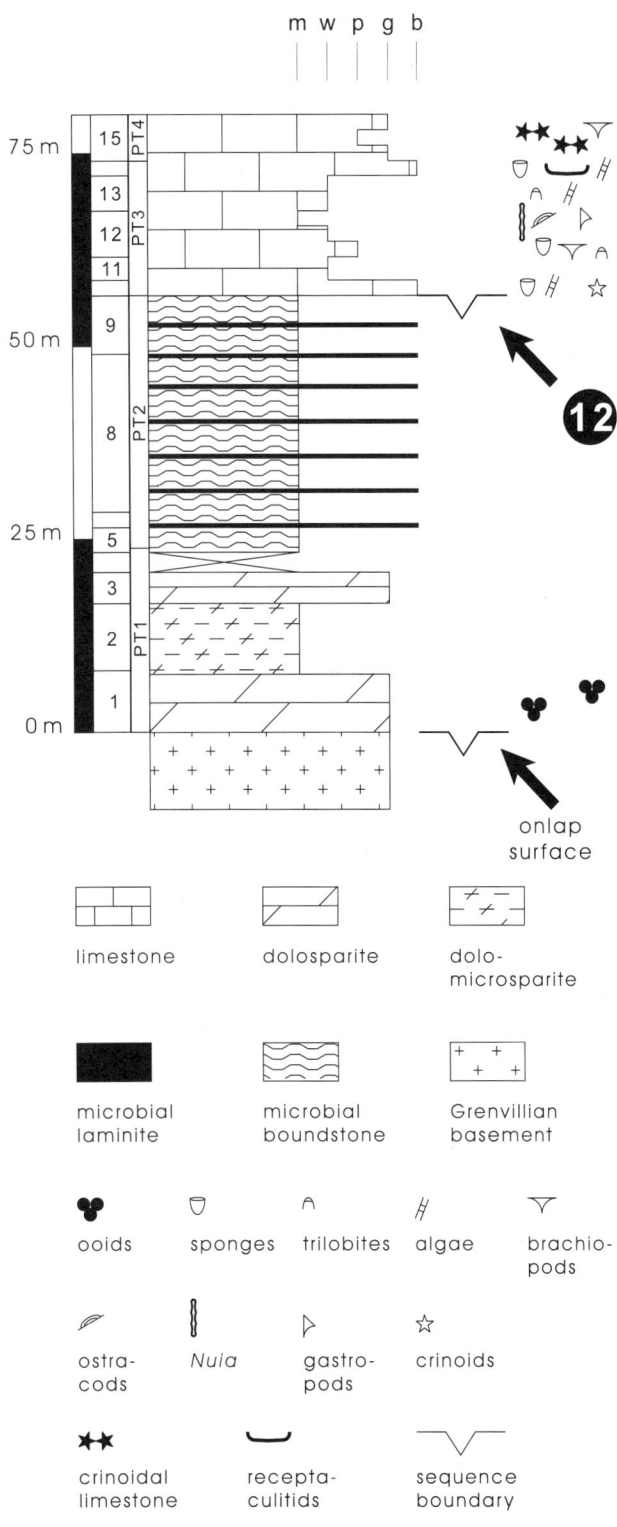

Figure 40. Type section of Ponon Trehue Formation (modified from Bordonaro et al., 1996). 12 = sequence boundary 12 of carbonate platform succession. PT 1–PT 4 are lithologic units distinguished in field. Abbreviations as in Figure 16.

which forms the top of the exposed succession of the Ponon Trehue Formation. The depositional environment is difficult to assess because of the absence of sedimentary structures. The relatively thin beds with intercalations of wackestones do not represent sand bars, but may reflect a platform environment still within the range of waves or currents that periodically affected the platform. The bimodal grain-size distribution (peloids vs. bioclasts) together with remnants of matrix may be taken as evidence of a tempestitic origin of the sediments. Conodonts obtained from these rocks indicate cold-water conditions, which would be in agreement with a deposition in deeper water.

Evolution and sequence stratigraphy of the Ponon Trehue Formation. There are clear limitations to an evolutionary and sequence stratigraphic interpretation of the Ponon Trehue Formation. There is only one 80-m-thick major outcrop of this formation. Fortunately, there is relatively good biostratigraphic control on the upper two units within the formation (Bordonaro et al., 1996; Lehnert et al.,1998). Nevertheless, much of the following is based on the knowledge of the evolution of the La Silla and San Juan Formations in the Argentine Precordillera.

Each of the four units described has sharp boundaries across which there is evidence of a distinct change in environment and depositional style. Unit 1 was deposited immediately above crystalline basement and marks the onlap of the carbonate platform onto the exposed craton during late Tremadocian time. The absence of any siliciclastic material along the lowermost sequence boundary of the Ponon Trehue Formation is remarkable and probably indicates very low relief. However, as is known from central Texas where Cambrian sediments onlap basement rocks, an archipelago of isolated basement islands existed that locally had a topographic relief of >100 m (Donovan et al., 1988; Donovan and Ragland, 1991). Thus, with only such a small outcrop in San Rafael, it is perhaps not possible to assert that the evidence given here really indicates low relief and the absence of a siliciclastic transgressive lag deposit.

Colaptoconus quadraplicatus found in this interval (Lehnert et al., 1998) indicates a latest Tremadocian or early Arenigian age for the deposits of unit 1. As this conodont species is still present at the base of unit 3, it is most likely that the dolomites still belong to the Tremadoc. The transition into unit 2 is equivalent to the transition from a transgressive systems tract to the highstand systems tract. Sediments of the former are shallow subtidal in origin (oolites, mudstones, thrombolites?), and the highstand deposits are intertidal and supratidal (microbial boundstones and laminites). The upper sequence boundary of this interval is located at the transition between units 2 and 3, which marks a pronounced deepening and the onset of submarine deposition. The presence of *C. quadraplicatus* makes the lower interval of the Ponon Trehue Formation roughly equivalent to the lower reef-mound horizon in the northern sections of the Precordillera. The flooding observed between units 2 and 3 at Ponon Trehue is likely to represent the deepening observed at the La Silla–San Juan formational boundary. Consequently, the sequence boundary at the unit 2–3 contact coincides with the type 2 sequence boundary separating the La Silla Formation and the San Juan Formation.

Sediments of unit 3 were deposited during *Oepikodus communis–Prioniodus elegans* time (conodont assemblage zone I of Lehnert, 1993) and record an overall deepening of the environment. This deepening culminates near the base of the *O. evae* zone, where the transition from wackestones with biostromes to a (probably) tempestite-dominated regime is interpreted to be another flooding event. If this interpretation is correct, the flooding is coeval with the flooding at the base of the lower nodular wackestone interval in the San Juan limestones. The deepening interpretation is corroborated by the change from temperate water group conodonts to cold-water forms.

Lindero Formation

The Lindero Formation was defined by Bordonaro et al. (1996) for the Middle Ordovician predominantly siliciclastic succession of San Rafael. Other rocks that contribute to the succession are limestones and volcaniclastic sediments. Conodonts from the limestone intervals indicate a latest Llanvirnian to Llandeilian age for the deposits. Two members were described, which are separated by a hiatus spanning two conodont subzones. The Lindero Formation directly overlies crystalline basement in the type section (Fig. 41).

Lithofacies. At the base, the lower member (Peletay Member; Fig. 41) is composed of dull red or purple arkoses that developed from weathering of the underlying granitoid. The components are mainly angular clasts of feldspar and quartz. Grain size varies from coarse sand to conglomeratic. Higher in the member, clasts are subangular to subrounded, but the most striking features are large slabs (40 × 3 cm) of crinoidal limestones within the arkoses. Locally, limestone slabs are imbricated. These gray grainstones contain a large amount of feldspar clasts and a minor amount of quartz grains.

Similar crinoidal limestones form elongate lenticular bodies near the top of the arkosic interval. The limestones are cross-bedded and thin bedded and contain fragments of trilobites and brachiopods as well as conodonts.

White quartz arenites locally form the top of the member: they are thick bedded, the grain size is medium to coarse, and the grains are subangular to rounded. In places, these sandstones are iron stained, giving them a brown color in outcrop.

The upper member (Los Leones Member; Fig. 41) is mainly composed of fine-grained sandstones and siltstones. The dominant colors of the thin-bedded sediments are brown, green, and gray. Green and brown shales and silty shales are less abundant. In places, trace fossils have been preserved. A few thin beds are composed of arkosic sandstones with a dark gray color.

Diamictites have been observed in two sections. They contain large rounded quartz clasts, and lithic and feldspar fragments in a dark red matrix. Near the base of the horizons, silty intraclasts of the underlying horizons are present.

Thin limestone beds have been found in some of the sections

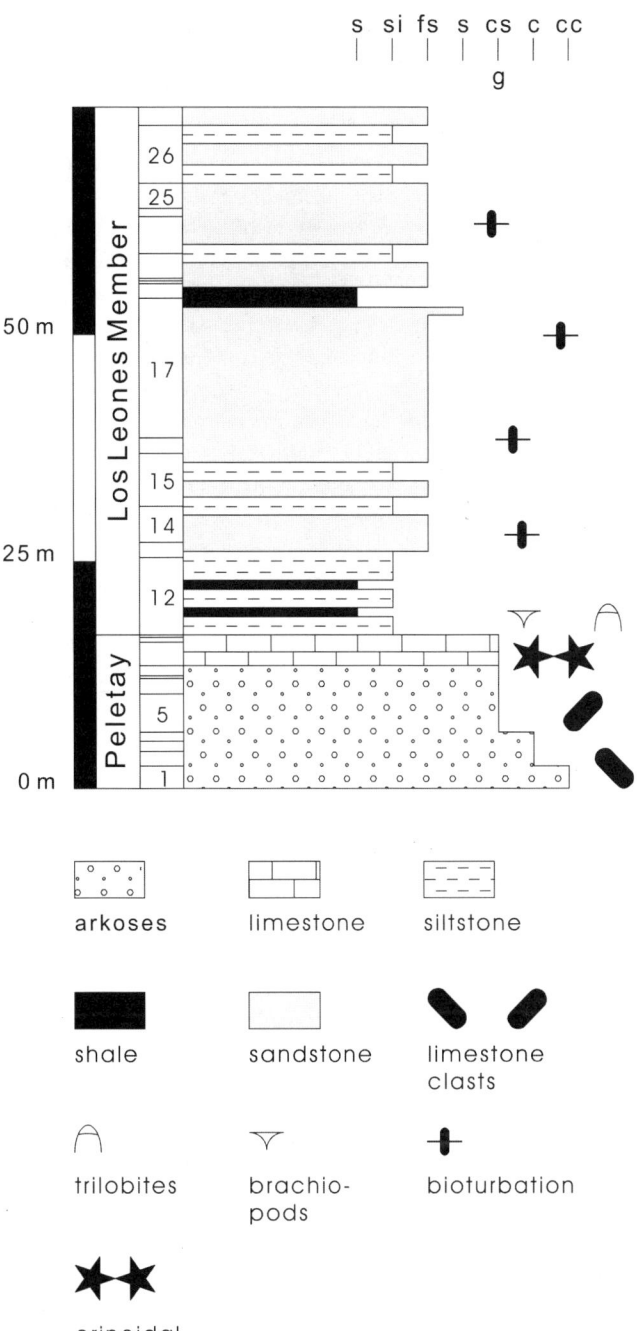

Figure 41. Type section and members of Lindero Formation (modified from Bordonaro et al., 1996). Abbreviations as in Figure 23.

just above the contact with the lower member. These fine-grained grainstones are dark gray and contain conodonts and crinoids as well as fragments of brachiopods and trilobites. Many of the beds are graded and some of them exhibit channel-fill geometries.

Volcanic ashes and breccias are present in one section near the base of the member where they are intercalated into limestone beds. The rocks are green to light gray and thin to medium bedded. Upsection, the ash beds pass into tuffaceous sandstones and siltstones.

Interpretation, facies associations, and the sedimentary succession of the Lindero Formation. The Lindero Formation represents two fundamentally different environments. The arkoses of the lower member document basement uplift, its erosion, and almost in situ redeposition of the detritus. Increasing rounding of the clasts, broad channels, and the limestone slabs indicate increasing transport and reworking of the material. At least in the upper part of the member, deposition took place in a marine environment, as indicated by the autochthonous crinoidal limestones. Better sorting and decreasing grain size toward the top of the member mark the reduction of terrigenous input, either by flooding of the basement or by its erosion. A combination of both processes is most likely. A sudden decrease in sediment input and nondeposition are recorded by the iron-stained sandstones at the top of the member and the absence of two conodont subzones between the two members of the Lindero Formation (Bordonaro et al., 1996).

The upper member is composed of two facies associations, a fine-grained siliciclastic association and a limestone association. The former is composed of sandstones, siltstones, and shales. Trace fossils and the fauna indicate a marine environment; the absence of wave-induced structures and of tempestites points to a setting below storm wave base. The few diamictites are interpreted to be the result of debris flows that transported some coarser material to the depositional environment. Some broad channels with a graded sand fill were formed by turbidity currents; however, sole marks and others indicators of turbiditic deposition have not been observed in the siliciclastic sediments of the Los Leones Member.

The limestone association is mainly composed of allochthonous carbonates, as indicated by abundant graded bedding and the channel-fill geometries, which together point to a distal turbidite origin of the association. The conodonts obtained from these beds are temperate-water forms and commonly are found in deeper water environments.

The entire Lindero Formation represents a rapidly deepening succession, from shallow-marine high-energy environments to low-energy environments well below storm wave base, where much of the sediment was deposited as a result of mass flows. The importance of the outcrops of the Lindero Formation is that

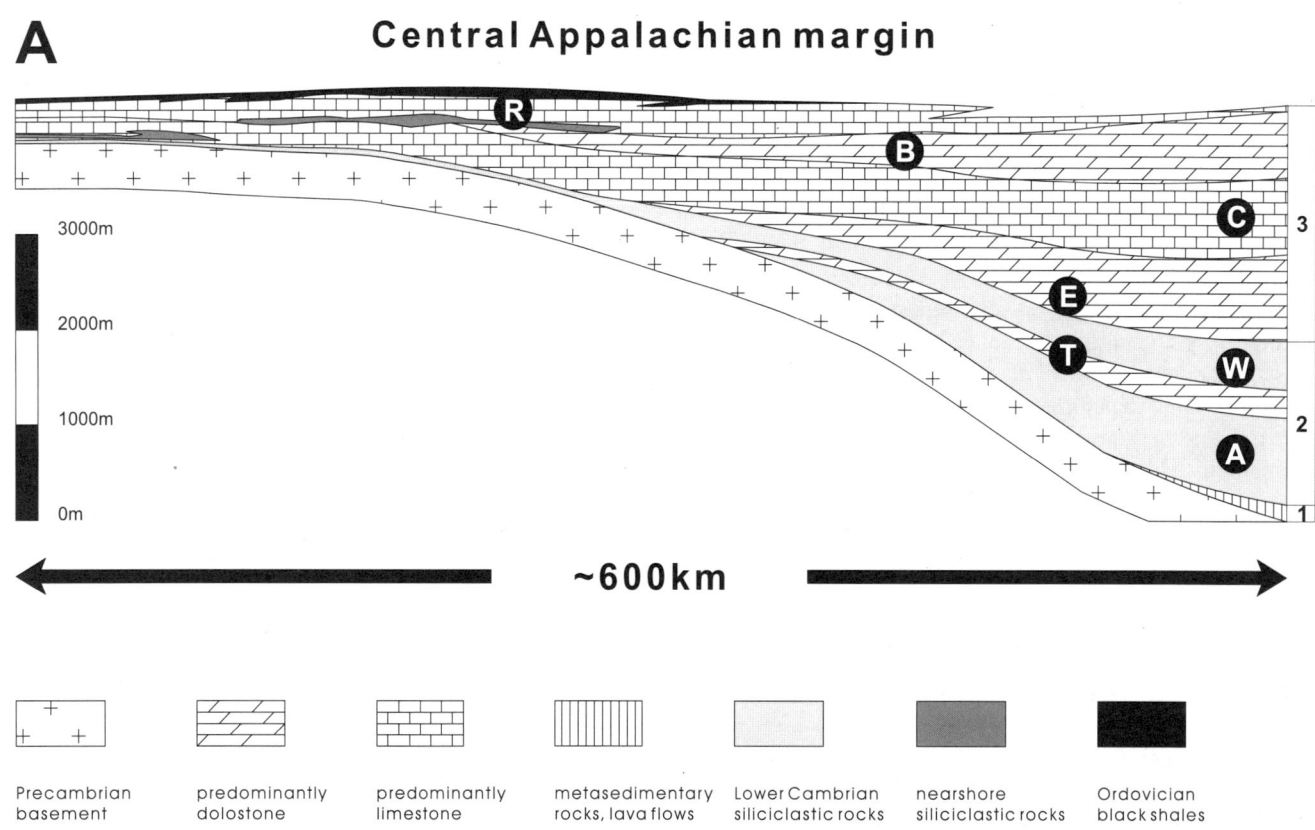

Figure 42 (this and opposite page). Transects through central Appalachians (left side) and Precordillera toward San Rafael showing successive onlap of strata onto craton. In Appalachians, sequence 1 is upper Precambrian stratified sequence; 2 is Lower Cambrian clastic sequence; 3 is Cambrian-Ordovician carbonate platform succession. A = Antietam Quartzite, B = Beekmantown Group, C = Conococheague Limestone, E = Elbrook Dolomite, R = Black River Group and equivalents, T = Tomstown dolomite, W = Waynesboro Formation. Appalachian transect is modified from Skehan (1988). In Precordillera, L = La Laja Formation, Z = Zonda and La Flecha Formations, J = La Silla and San Juan Formations. Note that thickness of carbonate succession is roughly equivalant to Appalachians. Transect through terrane is on the order of 400 km. See text for discussion.

they document uplift and exposure of the basement of the terrane during the Llandeilo; that a local source for much of the detritus found in the Precordillera was present at that time; and that volcanism was active during Llandeilian time.

EARLY PALEOZOIC EVOLUTION AND PALEOGEOGRAPHY OF THE ARGENTINE PRECORDILLERA—GENERAL ASPECTS AND THE CARBONATE PLATFORM

General Aspects

Size of the terrane. As outlined earlier, the Argentine Precordillera is part of a larger terrane, be it Cuyania, Occidentalia, or any other terrane. What evidence do we have about its size?

The isolated Cambrian outcrops in the Precordillera hamper speculations about the size of the terrane. However, during the Early Ordovician more than 600 m of limestones were deposited in the Precordillera (La Silla and San Juan Formations). In the area of Ponon Trehue (Fig. 39), 80 m of upper Tremadocian and Arenigian limestones are present (Fig. 40) that correspond to almost 400 m in the Precordillera proper (Bordonaro et al., 1996). In transects through the Appalachians across strike (Fig. 42), a reduction of about 2000 m of carbonate platform rocks to ~300 m is observed over a distance of ~600 km (Skehan, 1988). Even if we consider that the terrane was smaller and that the Ordovician carbonates onlapped a more steeply inclined depositional surface, a width of about 400 km seems to be the minimum shelf width required for the deposition of the Lower Ordovician limestones (Fig. 42B).

Today, the Argentine Precordillera extends for about 400 km in a north-south direction toward Cerro Cacheuta near Mendoza. The Ordovician of San Rafael is located about 150 km to the southeast (Fig. 2). Still farther south, additional outcrops of limestones are known at Limay Mahuida (Baldis et al., 1985). Although it is rumored that Ordovician fossils have been obtained from these limestones (V. Ramos, 1996, personal commun.), neither conodonts nor other fossils have been found to date in these in places mylonitic rocks (E. J. Llambias, 1996, personal commun.). Hence reliable data give a minimum longitudinal extension of the terrane of about 550–600 km.

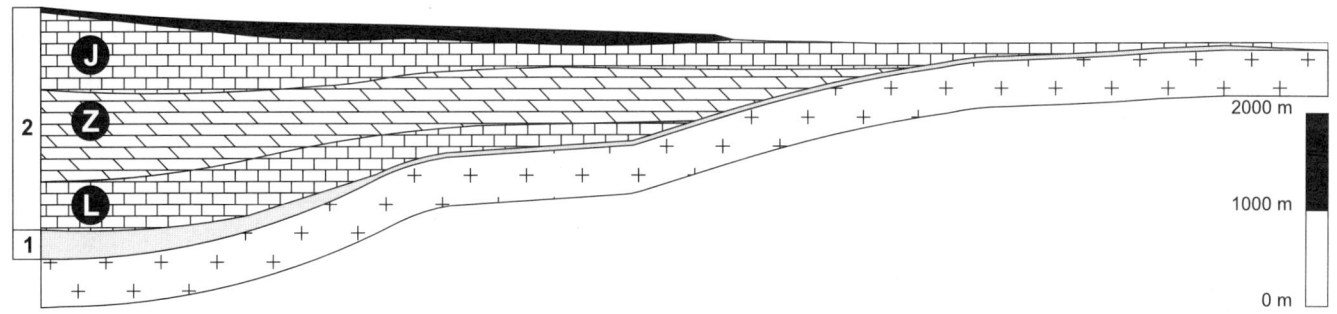

Precambrian basement | predominantly dolostone | predominantly limestone | siliciclastic onlap | black shales, drowning succession

Consequently, a conservative estimate for the size of the terrane is about 550 km by 400 km. If we consider that there was an additional east-facing passive margin on the terrane, the original width might well have been on the order of 1000 km. This is on the order of magnitude envisaged by Thomas (1991) for the microcontinental block that once occupied the Ouachita embayment (Fig. 6) and by Ramos (1995) for the width of the Cuyania terrane.

Biogeographic aspects and faunal provincialism. Biogeographic arguments recently have been used in a number of contributions to the discussion about the provenance of the Precordillera (e.g., Benedetto, 1993, 1998; Keller and Lehnert 1993a, 1998; Astini et al., 1995; Benedetto et al., 1995; Brussa, 1995; Vaccari, 1995) and both benthic and planktonic organisms have been considered.

Among the trilobites, the Cambrian platform fauna shows a high degree of endemism on a species level (100%), but on a generic level there is a good correlation between Precordilleran and Laurentian faunas (Bordonaro, 1990a, 1990b, 1992). About 70% of the genera are Laurentian, 13% are pandemic and 17% are endemic to the Precordillera. Vaccari (1994) reported an even higher degree (>90% on the generic level) of Laurentian Cambrian trilobites. This relation did not change much until well into the Arenig (Vaccari, 1995). From late Arenig time onward, Baltic and cosmopolitan genera became successively more important (Fig. 43). One important observation, however, is the fact that during the entire Ordovician there is a fundamental difference in the composition of trilobite faunas between the Argentine Precordillera and the (today) adjacent terrane of the Sierra de Famatina (Fig. 2), which during the Early Ordovician was the home of Gondwanan trilobites (Vaccari, 1995). Trilobites of the illaenid-cheirurid biofacies and the nileid biofacies, together with bathyurids, point to a low-latitude position of the Precordillera during Early Ordovician time, which is corroborated by epipelagic forms like *Opipeuter* and *Carolinites*. These, according to Cocks and Fortey (1990), have a pan-equatorial distribution.

Brachiopods are well known and described from the San Juan Formation (e.g., Herrera and Benedetto, 1991) and younger strata. The pre-Arenig record, however, is poorly known. Herrera and Benedetto (1991) described five brachiopod assemblages from the San Juan Formation and discussed their biogeographic relationships. Early Arenig brachiopod assemblages show a strong relation to Laurentian faunas. From late Arenig time on, increasing Celtic as well as Baltic elements are observed (Benedetto et al., 1995). Maximum diversity and abundance of brachiopods are recorded from early Llanvirn limestones of the San Juan Formation. Endemic brachiopod genera appear during the mid-Arenig but disappear toward the earliest Llanvirn. A few endemic genera reappear toward the end of the carbonate platform formation. During the early Caradoc, faunal elements of the Scoto-Appalachian and the Anglo-Baltic realm are recognized in addition to cosmopolitan and endemic genera. The early Caradoc, however, is the time of the highest endemism among all benthic faunas (Benedetto et al., 1995; Fig. 43). The first Gondwanan brachiopod, *Tissintia*, appeared during the Caradoc in the Precordillera, which seems to have been isolated from other Gondwanan sedimentary basins, because no faunal exchange is observed (Benedetto et al., 1995; Benedetto, 1998). In contrast, there is still a strong affinity to Anglo-Baltic faunas (Fig. 43). Toward the end of the Ashgillian, the faunas in the Precordillera are characterized by the presence of a typical *Hirnantia* fauna (Benedetto, 1985, 1986), which is widely distributed on the circum-Gondwanan margins. Benedetto et al. (1995: p. 183) stated, that ". . . during the latest Ashgill, the Precordillera became faunistically indistinguishable from other localities characterized by a low-diversity, Kosov *Hirnantia* fauna." As demonstrated by Sheehan and Coorough (1990), the *Hirnantia* fauna is present on almost all continents. In addition, the Kosov faunal province is composed of species that lived in subtropical to temperate environments (Owen et al., 1991).

The advent of *Clarkeia* marks the final and definite change

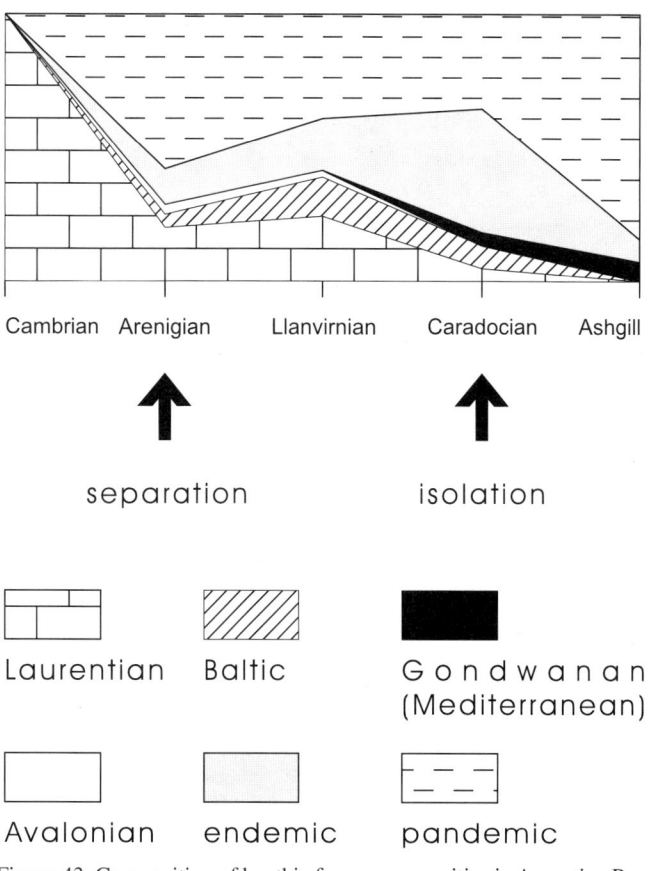

Figure 43. Composition of benthic fauna communities in Argentine Precordillera (AP) (modified from Benedetto et al., 1995). During Late Cambrian, fauna is essentially Laurentian. During onset of extension during late Arenig, first Avalonian and Baltic elements are observed. Note that Gondwanan elements are not present before early Caradoc. In addition, from faunistic point of view Caradoc is time of maximum isolation. This diagram contradicts any model that claims Middle Ordovician collision of Precordillera with Gondwana. For further discussions see Benedetto (1998) and Keller and Lehnert (1998).

toward an entirely Gondwanan faunal signature (Benedetto, 1998; Keller et al., 1998). However, this final change took place not before the Middle Silurian or Late Silurian, because the *Clarkeia* fauna is found mainly in the middle and upper parts of the Los Espejos and Tambolar Formations (Fig. 4) and the *Clarkeia* fauna is a valuable paleogeographic tool predominantly for Wenlockian and Ludlovian time (Cocks and Fortey, 1990).

Graptolites, although planktonic organisms and hence of less value in paleogeographic and biogeographic considerations, have also been considered with respect to the terrane nature of the Precordillera (Brussa, 1995; Maletz and Ortega, 1995; Mitchell, 1997). A biogeographic interpretation of the Precordilleran graptolites (Brussa, 1995) indicates a close relation to Australasia, Texas, northwest Canada, and to a lesser extent to Newfoundland, if mid-Arenigian through late Llanvirnian faunas are considered. Younger graptolites (Llandeilo and early Late Ordovician) show a marked affinity to the eastern margin of Laurentia. According to Maletz and Ortega (1995), all Lower, Middle, and early Upper Ordovician graptolites of the Precordillera belong to the Pacific (warm water) faunal province, whereas the graptolite faunas of all surrounding terranes are typical of the Atlantic or cold-water province. Mitchell (1997) concluded that the composition of the Middle and Upper Ordovician graptolite faunas of the Precordillera contradict a Middle Ordovician collision between Laurentia and Gondwana.

The possibilities and problems of the Precordilleran conodont faunas in paleogeographic studies were discussed by Lehnert and Keller (1993a) and lately extensively by Lehnert (1995a). One important observation is the fact that the occurrence of the different faunas is clearly depth and temperature controlled. Nevertheless, Late Cambrian and Tremadocian elements form associations that typically occur around the Laurentian margins. The incursion of Baltic or North Atlantic elements during the Arenig and early Llanvirn reflects sea-level fluctuations with corresponding temperature changes. Conodonts of Llanvirnian and Llandeilian strata above the carbonate platform reflect a temperate- to cold-water environment and thus are attributed to the North Atlantic or Baltic faunas. This, however, is no surprise because these faunas from the Precordillera were all obtained from deep-ramp to deep-slope environments (Eberlein, 1990; Keller et al., 1993b). A remarkable conodont fauna is present in Caradocian limestones of the Río Sassito section and in allochthonous boulders within the Empozada Formation. *Aphelognathus politus* and *A. rhodesi* are warm-water elements and are well known from eastern Laurentia (e.g., Sweet, 1979). In addition, there are several species of *Panderodus*, *Belodina*, and *Plectodina*, all typical of tropical waters of the Laurentian midcontinent.

Among the organism which in the Precordillera hitherto have been found only in the carbonate platform rocks are sponges, bryozoans (Beresi and Rigby, 1993; Carrera, 1994), and stromatoporoids (Keller and Bordonaro, 1993; Keller and Flügel, 1996). During the Arenig, sponges and bryozoans show clear affinities to Laurentian faunas, whereas during the Llanvirn there are also similarities to Baltica. The only known stromatoporoid from Argentina (*Zondarella communis*, Keller and Flügel, 1996) has been described only from Newfoundland (Pohler and James, 1989).

The sponge-algal facies is important for paleogeographic reconstructions of the position of the Precordillera with respect to Laurentia. Lower-Middle Ordovician sponge-algal mounds are most prominent in tropical, near-equatorial latitudes (Webby, 1992) and are predominantly known from the margins of Laurentia (Pratt and James, 1982; Toomey and Nitecki, 1979). They are concentrated, however, along the northern and eastern margin of the Ouachita embayment (Alberstadt and Repetski, 1989; Fig. 44) where they were studied in detail by Toomey (1970), Toomey and Nitecki (1979), and LeMone (1988a, 1988b). If the Precordillera was originally positioned near the Ouachita embayment (Fig. 6), then the reef-mound facies of the San Juan Formation form a logical continuation of the reef-mound facies along the northern margin of the Ouachita embayment. Although sedimentologically different (Goldhammer et al., 1993) and developed in another paleotectonic setting ("cratonal" sections near El Paso; Fig. 44), the reefs on the Laurentian side of the Ouachita embayment and on the "Precordilleran" side exhibit ecosystems almost identical to each other (Keller and Flügel, 1996; Keller and Dickerson, 1996).

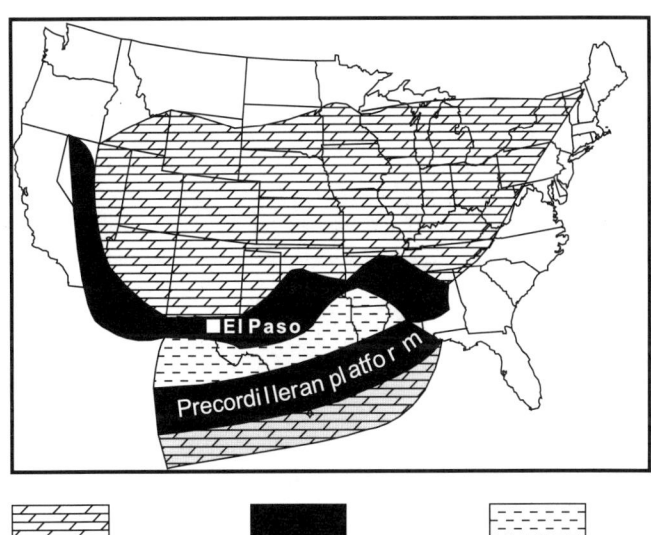

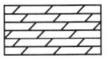  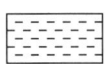

continental interior shelf environment | sponge-algal facies | off-shelf Ouachita facies

hypothetical continuation of AP carbonate platform, eastern passive margin

Figure 44. Distribution of sponge-algal facies along Laurentian margins. If Precordillera was situated to south of Ouachita embayment, these facies have logical continuation in Precordillera (modified from Alberstadt and Repetski, 1989).

K-bentonites. K-bentonites, altered volcanic ashes, may play an important role in testing paleogeographic reconstructions, because in many cases they represent a succession of events over a geologically significant time span. In the Ordovician, their usefulness has been documented in possible intercontinental correlations between Laurentia and Baltoscandia (Huff et al., 1992).

Ordovician volcanic ashes have been described from the Argentine Precordillera (Kolata et al., 1994), where they were found in the upper part of the San Juan Formation and the transition into the overlying Los Azules Formation (Bergström et al., 1994). Although originally described from the Jáchal area (Cerro La Silla and Cerro La Chilca; Bergström et al., 1994), they are also present in the Talacasto and Tambolar sections (Huff et al., 1998), where they occur in the middle and upper part of the preserved succession of the San Juan Formation. Of special importance is their presence in the Tambolar section, the top of which is dated as the Baltoscandic *B. navis/B. triangularis* zone of the latest Arenig (Lehnert, 1995a).

The geochemical and mineralogical characteristics of the Argentinian K-bentonites (Krekeler et al., 1995) point to a parental magma with rhyolitic to trachyandesitic composition, attributed to collision-zone volcanism by Huff et al., (1995, 1998). This is corroborated by granite tectonic discrimination diagrams, where the data from the Argentine Precordillera plot on the boundary between within-plate granites and volcanic-arc granites, which is typical of collision-margin felsic volcanic rocks (Huff et al., 1995). At present (C. Rapela, 1998, personal commun.) it is not clear whether these K-bentonites are related to the intrusion of K-rich monzonites and monzogranites (Rapela et al., 1998) into the basement of the Cuyania terrane at about 481± 6 Ma (Pankhurst and Rapela, 1998). The volume of K-bentonites present in the Precordillera, however, leaves doubt that this magmatic event was the principal source for the ashes.

Volcanic zircons are also present in strata of the San Isidro area (Loske, 1992), where they were described from the Estancia San Martín Formation. The characteristics of these zircons were taken as evidence of an Early Cambrian rhyolitic volcanism in the Precordillera. However, the Estancia San Martín Formation (composed of La Cruz Limestones) has been interpreted to be an olistolith within the Ordovician siliciclastic rocks of the Los Sombreros Formation (Bordonaro et al., 1993). Hence, the zircons described from sandstones (Loske, 1992) in the San Isidro section are also of Ordovician age and are another hint to widespread volcanic deposits in the Precordillera during Ordovician time. Tuffs and volcanic breccias are present around San Rafael. In the absence of geochemical investigations, their relation to the K-bentonites in the Precordillera remains obscure.

Heavy mineral associations. Heavy mineral analysis (Fig. 45) shows that Ordovician through Silurian clastic sedimentary rocks are dominated by a low-diversity–low-quantity association of weathering-resistant minerals (Loske, 1992). The Cambrian samples mentioned by Loske (1992) are taken from the Ordovician Los Sombreros Formation. Characteristic elements of this population are zircons, tourmaline, and rutile. Together they indicate a mature area of provenance with a dominance of non- or only slightly metamorphosed sedimentary rocks.

The other population is a high-diversity–high-quantity association (Fig. 45) composed of the minerals mentioned here and, in addition, garnet, zoisite, apatite, and titanite. This population is present in the Devonian and Carboniferous sedimentary rocks. Hence, from the Devonian onward, higher metamorphic and igneous rocks were exposed in the hinterland. The analysis of individual zircon populations from these heavy mineral associations is further evidence of a major change in the source rocks. In the Devonian sediments, a population is present in which the zircons are angular to subangular, often idiomorphic, and show internal zonations and inclusions (Loske, 1992). This population, absent in the older rocks, indicates the presence of a granitoid area of provenance. The presence of 1.1 Ga zircons (Loske, 1995) in the Punta Negra and the Villavicencio Formations seems to indicate that it was the Grenvillian basement of the terrane that acted as the main sediment source.

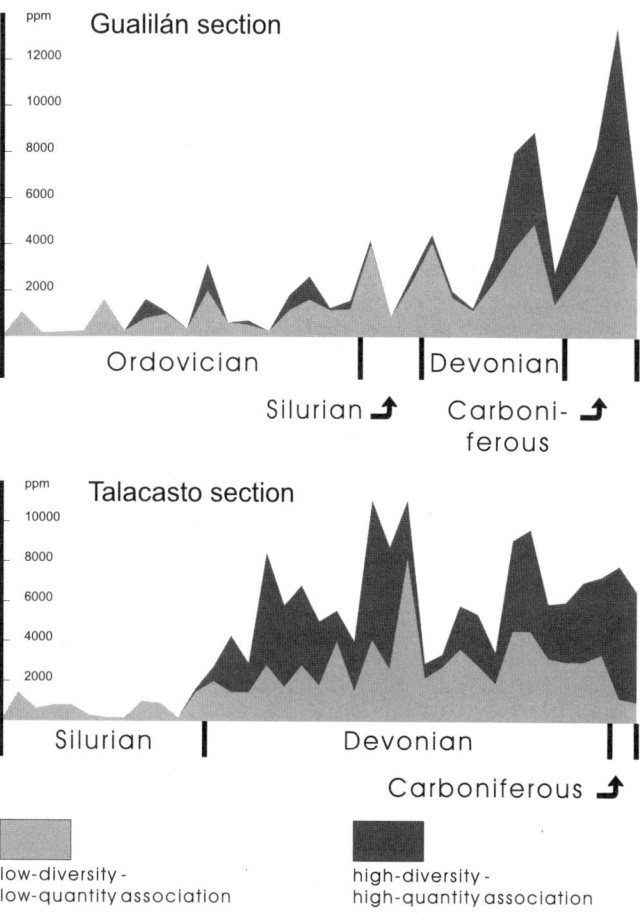

Figure 45. Composition of heavy-mineral spectra in Precordillera through time. Note dramatic change in composition during Early Devonian (modified from Loske, 1992).

Evolution of the Carbonate Platform

Early Cambrian. Lower Cambrian rocks are not widespread in the Argentine Precordillera and are found only in the Guandacol subbasin and the San Juan subbasin. In the Guandacol subbasin, marginal-marine redbeds and evaporites underlie dated uppermost Lower Cambrian limestones and dolostones (Cerro Totora Formation; Astini and Vaccari, 1996). The absence of correlative strata in other parts of the Precordillera makes a regional interpretation difficult; however, the Cerro Totora Formation was interpreted as a synrift deposit by Astini et al. (1995).

The facies of the Lower Cambrian olistolith in the Los Túneles section as well as the facies of the Lower Cambrian outcrops in the San Juan subbasin indicate that by the latest Early Cambrian a passive margin had developed. On this margin, an alternation of siliciclastic sediments and carbonates was deposited (El Estero Member of the La Laja Formation).

Middle Cambrian. The Middle Cambrian was the time of deposition of major siliciclastic-carbonate cycles in the San Juan subbasin. Five third-order sequences are present, the youngest one continuing into the (presumably) Late Cambrian Zonda Formation (Figs. 7 and 46). In the Guandacol subbasin and at Cerro Pelado, partly coeval dolomites are interpreted to have formed in a peritidal environment (Cañas, 1988). These rocks show an intercalation of siliciclastic detritus which, from the description given, is similar to the siltstone association described from the La Laja Formation.

The widespread occurrence of Middle Cambrian deep-water limestones as olistoliths within the Ordovician continental rise deposits implies that a deep-water environment was present from Middle Cambrian time. Mass-flow deposits are absent within the olistoliths, and they contain no siliciclastic material coarser than silt. One conclusion drawn from this observation is that, during Cambrian time, an effective barrier prevented bypass of material from the inner clastic shelf beyond the carbonate belt. This barrier is probably represented by the rocks similar to those at Cerro Pelado and in the Guandacol subbasin, which frequently shoaled up to intertidal and supratidal environments. Early cementation of rocks along the platform margin is well known and regarded as one reason for the absence of mass-flow deposits of Cambrian age in the Precordillera. However, there seems to have been an additional factor operating. The deep-water limestones are present as olistoliths in younger successions. One reason for the absence of mass-flow deposits within the olistoliths may be that blocks with such sediments simply have not been eroded. Another explanation is that the sections, in which Cambrian mass-flow deposits are present, are buried beneath some western thrust sheets (telescoping effect).

However, the outstanding number of these olistoliths in all sections of the Los Sombreros Formation is reason enough to suppose that if megabreccias and turbidites had been deposited they should be represented within the olistoliths. The only expression we find are graded beds interpreted as distal turbidites. In addition, trilobites found in these olistoliths cover the entire Middle and Late Cambrian (Bordonaro and Banchig, 1996), hence there is a continuous record of these deposits without evidence of mass-flow deposits.

The Ordovician continental margin facies testifies to the processes and source area during formation of major relief along a continental slope and rise. In addition to the Cambrian deep-water limestones, platform-derived turbidites and megabreccias are present. The carbonate component of these mass-flow deposits is mainly Lower Ordovician limestone. Considering that the Cambrian deep-water limestones and the Lower Ordovician platform-derived detritus both seem to have come from the same direction (although this is not proven), the debris flows and turbidity currents affected both the platform and much of the adjacent former deep-water depositional environment. In addition, they should have affected the intervening slope, if developed, which is the most likely site of deposition of mass-flow sediments. This situation almost excludes a telescoping effect for the absence of mass-flow deposits in the deep-water olistoliths. The most likely explanation is that during the Cambrian depositional relief between the platform and the deeper environments was not conducive for the formation of mass flows.

The La Laja Formation was deposited in a paleogeographic position at the transition from the inner detrital belt to the carbonate belt (Fig. 9). Most of the terrigenous material was trapped in the inner detrital belt. Consequently, the siltstone units of the La Laja Formation are tongues of this inner belt. The rocks at Cerro Pelado and in the Guandacol subbasin represent the carbonate belt (Fig. 9).

The most likely basin configuration during the Middle Cambrian shows deep-water carbonates in the west, a carbonate belt represented by the shoal complex in the Guandacol subbasin and at Cerro Pelado, and the transition from this carbonate belt into the seaward edge of the inner detrital belt in the sections of the La Laja Formation (San Juan subbasin). The platform configuration inferred here is similar the Early and Middle Cambrian of the southwestern Great Basin, where deep-water limestones actually interfinger with and onlap platform carbonates (Palmer and Halley, 1979), and both rock types occur in the same sections. Depositional relief between both environments was very low and the shifting of the environments was governed by differential rates of relative sea-level rise. This situation resembles the onlap margin of James and Mountjoy (1983), which is characterized by little or no sediment transport toward the slope and basin.

Late Cambrian. The Late Cambrian shows a major change in depositional style. Whereas the Middle Cambrian was dominated by subtidal deposition of limestones, the Late Cambrian is the time of evolution of a vast peritidal platform with the formation of dolostones. In contrast to the underlying sequences, the uppermost succession of the La Laja Formation successively shallows into intertidal and supratidal dolomitic facies of the basal Zonda Formation (Fig. 46). This is the first time that peritidal facies are present in the San Juan subbasin. Three additional third-order sequences formed before the end of the Cambrian (Fig. 46).

In comparison with the San Juan subbasin, rocks of the Guandacol subbasin show a higher proportion of subtidal limestones, some of them fossiliferous. The succession also shows fewer stromatolites and microbial laminites. It is here inferred that this pattern indicates that the Guandacol subbasin was slightly deeper and probably farther offshore than the San Juan subbasin. A reverse trend is reflected in dolomitization; rocks of the San Juan subbasin are almost entirely dolomitized, whereas near Guandacol, there is a higher proportion of limestone. This is interpreted to be the effect of a more frequent exposure of rocks in the San Juan subbasin. This exposure favored fresh-water recharge conducive to mixed-water dolomitization. Frequent exposure is also reflected by the presence of caliche horizons.

The abundance of oolites in the Upper Cambrian platform sediments of the Precordillera is taken as evidence of a platform that may have been rimmed by oolite shoals. However, the actual platform margin is not exposed. These shoals acted as a source area for many of the sheet-like oolites at the base of small-scale shallowing-upward cycles. The vast tidal flats of the Precordillera probably interfingered with oolitic sands in back-barrier areas and the platform achieved a configuration comparable to the aggraded platforms described by Read (1985). Abundant early cementation and a low-relief slope seaward of the oolite rim may account for the lack of mass-flow deposits in the Upper Cambrian deep-water olistoliths. A similar scenario for the Guandacol subbasin was developed by Cañas (1995a).

An interesting succession is present at Cerro Pelado (Fig. 3), where rocks of the Cerro Pelado Formation are unconformably overlain by an upward-deepening succession of deep-water limestones and black shales, the El Relincho Formation of Heredia (1990; Fig. 47). The boundary between both units is a submarine hardground and represents a drowning unconformity (see Appendix 1) caused by a very rapid relative sea-level rise (Schlager, 1981, 1989). The overlying succession, which continues into the basal Ordovician (O. Lehnert, 1998, personal commun.), is free of mass-flow deposits (Fig. 47). Although biostratigraphic resolution on the platform is poor in the Upper Cambrian rocks, nowhere is there an indication of a similar rapid flooding. Consequently, the submergence of the Cerro Pelado platform block well below the photic zone most likely is the result of a local tectonic effect.

It is important to note that during the Late Cambrian the platform must have extended beyond the present outcrops of the Precordillera. Redeposited Upper Cambrian conodonts have been described from the San Juan Formation (Serpagli, 1974; Lehnert, 1995a). However, nowhere within the actual limits of the Argentine Precordillera is there a place where Arenig erosion cut down into the Cambrian strata. Hence these conodonts must have come from a place outside the actual limits of the carbonate platform.

Early Ordovician. During the Early Ordovician, the carbonate platform returned to somewhat more open-marine conditions and limestones were deposited. The Tremadoc shows a very uniform facies development all across the platform. Three third-order sequences are recognized in the La Silla Formation (Figs. 46 and 48). The Tremadoc is an important time in the history of the terrane, because it shows the onlap of the carbonates onto the cratonal basement (Ponon Trehue Formation near San Rafael) and thus a major cratonward shift of the shoreline. In the Precordillera, the Tremadoc documents the incipient but local evolution of a slope with the deposition of breccias and turbidites to the west of the carbonate platform. These slope sediments are present in the megaolistolith of the Los Túneles section and record the input of platform-derived material (ooids, peloids). Platform configuration is extremely difficult to determine because of the uniformity of facies. A platform with a rim of oolite shoals or a distally steepened ramp both seem reasonable. Deeper water environments persisted, and limestones and graptolitic shales are increasingly abundant.

During the Arenig, the platform achieved fully marine conditions. Most or all of the terrane was flooded and relative sea-level excursions can be correlated from the Precordillera into the San Rafael area. Two sequences developed in the San Juan Formation prior to the demise of the carbonate platform during the early Llanvirn (Figs. 46 and 48). Sea-level fluctuations are recorded in the sediments and in the composition of the conodont faunas.

Sedimentation on the incipient slope continued and the platform became a distally steepened ramp. Limestone facies within olistoliths testify to the persistence of deep-water carbonate environments into the Ordovician; in addition, Lower Ordovician deep-water shale facies are preserved in olistoliths.

An important event during the Arenig is the advent of reef-building communities. Two intervals are recognized and both formed during a relative rise in sea level. K-bentonites are present in the upper part of the San Juan Formation and mark the onset of an explosive volcanic history that culminated during the Llanvirn.

The most important event during the Late Arenig, however, was the onset of the demise of the carbonate platform. In the Guandacol subbasin, deep-ramp facies were drowned during the upper *O. evae* zone (conodonts) or *Isograptus victoriae* zone (graptolites) and subsequently, black shales were deposited. In contrast, in both the Talacasto and the San Juan subbasins, this period is marked by the evolution of a highstand systems tract, which is terminated by a type 2 sequence boundary. This is strong evidence that drowning in the Guandacol subbasin was not a eustatic event, but rather controlled by local tectonics. This interpretation is more true for the Cerro Potrerillo section, where drowning is almost coeval with the evolution of the sequence boundary in the Talacasto subbasin. If the interpretation is correct, that since the Middle Cambrian the Guandacol outcrops were located near or at the platform edge, then the disintegration of the platform began along its margin or close to it. The nature of the lower Llanvirnian drowning of the San Juan Formation in the Talacasto and San Juan subbasins is more difficult to establish. First, everywhere within the resolution of biostratigraphy drowning is synchronous (Lehnert, 1995b); second, the early Llanvirn is the time of a major eustatic sea-level rise (e.g., Fortey, 1984). Hence if there was a tectonic factor in the demise of the platform during the early Middle Ordovician, it was most probably masked by eustasy.

Sea-level history of the Precordillera. The following discussion of sea-level history in the Argentine Precordillera is based mainly on data from around Laurentia and from Baltica, two regions with fundamentally different basis for the time scales. Correlations of strata in the Precordillera to both areas are sometimes difficult, and are hampered by the absence of a comparative sea-level curve that shows the similarities and differences between both areas. I am aware that the jump between both time scales is confusing; for better understanding the reader is referred to Figure 49. Figure 48 shows the qualitative sea-level curve for the carbonate platform of the Precordillera and its correlation with the composite sea-level curve for the U.S. Appalachians developed by Read (1989).

The Hawke Bay event. In the Precordillera, the sedimentary record starts in the Early Cambrian. In many places around the Cambrian Iapetus, the Early-Middle Cambrian boundary interval was a period of a major relative sea-level fall, the Hawke Bay regression event of Palmer and James (1980). Depending on the regional geology, an unconformity or deposition of siliciclastic sediments characterizes this interval along the eastern margin of Laurentia.

In the Precordillera, trilobites of the critical interval, which are typical of more complete successions along the western margin of Laurentia, are absent (A. Palmer, 1998, personal commun.). It is likely that the Hawke Bay event is represented by an interval including the upper part of the El Estero Member and the lower part of the Soldano Member (sequences 1 and 2) of the La Laja Formation (Fig. 7). It is surmised that the absence of the *Plagiura-Poliella* and *Albertella* zones reflects this event. The position of the event is indicated by a succession of black shales and quartzitic sandstones at the transition from the El Estero Member to the lower part of the Soldano Member (Fig. 10). An interval of sandstones and calcareous sandstones has also been found in the La Laja and the La Flecha sections at the top of the El Estero Member (Fig. 10).

In the Guandacol subbasin, the top of the Cerro Totora Formation is dated as *Bonnia-Olenellus* zone (Astini and Vaccari, 1996). It is represented by a succession of sandstones and siltstones with intercalated oolites much like the sediments of the El Estero Member (Fig. 10) in the San Juan subbasin. The succession in the Guandacol subbasin is separated from overlying, as yet undated, peritidal carbonates by an erosional unconformity (Cañas, 1988). Hence it seems that the basal part of the Soldano Member correlates to the undated carbonates above the Cerro Totora Formation.

Assuming that the base of the Soldano Member was deposited during the *Glossopleura* chron, the early Middle Cambrian hiatus must be located at the base of the sandstone interval in the upper part of the El Estero Member. It might have a correlative at the base of the unnamed peritidal carbonates in the Guandacol subbasin, as indicated by the erosional unconformity (Cañas, 1988).

In the San Juan subbasin, three sequences (sequences 2–4; Fig. 7) are present within the *Glossopleura* zone of the Middle Cambrian. Sequence 2 begins at the base of the sandstone-grainstone association in the upper part of the El Estero Member. Together with sequence 3, the carbonate succession of this sequence forms the Soldano Member (Fig. 10).

The carbonate succession of sequence 4 is equivalent to the Rivadavia Member. Much of the pre-*Glossopleura* interval is absent in the southern Appalachians as an effect of the Hawke Bay event (Read, 1989; Fig. 48), but a relatively thin succession is present in Newfoundland (Knight et al., 1995).

The top of the Rivadavia Member coincides approximately (Fig. 7) with the boundary between the *Glossopleura* and *Ehmaniella* or *Bathyuriscus-Elrathina* biozones (O. Bordonaro, 1997, personal commun.; see also Bordonaro and Banchig, 1996). Hence, the boundary between the Rivadavia Member and the Juan Pobre Member seems to reflect the Delamaran-Marjuman boundary of Laurentia (Palmer, 1998). The sequence boundary (boundary 5) separating both members has a good match in the sea-level lowstand (Fig. 38) in the Conasauga succession of the Appalachians (cycles 2-2 and 2-3 of Read, 1989) and with the marked lithologic change between the Carrara and the Bonanza King Formations in the southern Great Basin (Palmer and Halley, 1979; Osleger and Montañez, 1996). The top of sequence 5 (which includes the lower part of the Zonda Formation) is younger than the trilobites of the *Bolaspidella* zone, which are found near the top of the Juan Pobre Member of the La Laja Formation and most probably is located in the *Crepicephalus* zone of the Late Cambrian (Fig. 7).

Four sequence boundaries (boundaries 6–9) are present in the Upper Cambrian strata of the San Juan subbasin. The best dated boundary 9 (Fig. 46) separates the La Flecha Formation from the La Silla Formation and corresponds to a global eustatic event.

Sequence 8 was deposited during the *Saukia* chron (sequence 6 of Osleger and Read, 1993) until the Cambrian-Ordovician boundary interval (Figs. 7 and 48). This is deduced from the presence of *Stenopilus convergens* (typical of the *Saukia* zone) about 30 m above the base of the sequence (Fig. 16; Keller et al., 1994; Vaccari, 1994). The sequence boundary (boundary 8; Fig. 7) is surmised to reflect the sea-level lowstand at the base of the *Saukia* zone.

Near the base of sequence 7 (Fig. 7) *Plethopeltis* cf. *P. saratogensis* indicates that the thick basal succession of the La Flecha Formation is not older than the *saratogensis* or *ellipsocephaloides* zone of the Sunwaptan and hence must correspond to the upper part of Laurentian sequence 5.

The ages of sequence 6 (the upper sequence of the Zonda Formation) and of the upper part of sequence 5 in the San Juan subbasin are unclear. However, in the Guandacol subbasin, a fairly well dated sequence boundary is present in the La Angostura section. According to Cañas (1995a) and Vaccari (1994), this boundary is close to the boundary between the *Crepicephalus* and *Aphelaspis* zones (Fig. 7) still within the Marjuman (Palmer,

1998). This sequence boundary reflects the relative sea-level lowstand between depositional sequences 3 and 4 of Osleger and Read (1993).

In the San Juan subbasin, boundary 7, the Zonda–La Flecha sequence boundary, and boundary 6, the intra-Zonda boundary, might reflected the corresponding sea-level event: both correlations seem viable.

If it is assumed that the Marjuman sequence boundary in the Guandacol subbasin corresponds to sequence boundary 7, then the upper part of the Zonda Formation belongs to the Marjuman and is correlative to sequence 3 of Osleger and Read (1993). In this scenario a major hiatus must exist between the Zonda Formation and the La Flecha Formation in the San Juan subbasin. If, however, the alternative is considered to be more likely, then the lower part of the Zonda Formation corresponds to sequence 3 of Osleger and Read (1993), and the upper succession corresponds to sequence 4. As the sequence boundary in the Guandacol subbasin is a type 1 boundary (Cañas, 1995a), the intra-Zonda boundary has to be regarded as a more cryptic type 1 boundary as well. In this scenario, the sequence boundary between the Zonda Formation and the La Flecha Formation corresponds to the 4-5 boundary of Osleger and Read (1993) and the Bonanza King–Dunderberg shale contact in the southern Great Basin (Cooper and Edwards, 1991). A hiatus must be postulated because the basal part of Laurentian sequence 5 is absent in the La Flecha section.

The boundary between sequences 4 and 5 of Osleger and Read (1993) is located close to the *Dunderbergia-Elvinia* biozone boundary (Figs. 7 and 48) which in turn marks the boundary between the Sauk II and Sauk III sequences (Palmer, 1981). Whatever correlation will turn out to be the most likely one, the presumed hiatus between the Zonda Formation and the La Flecha Formation seems to encompass this major unconformity of Laurentia.

In the Precordillera, the Cambrian-Ordovician boundary interval is marked by the transition from the La Flecha Formation to the La Silla Formation (sequence boundary 9). According to the European subdivision (Fig. 49), the first Tremadocian conodonts (basal Early Ordovician according to the Europeans) are found about 172 m above the base of the La Silla Formation (Keller et al., 1994; Lehnert, 1995a) and consequently, the La Flecha–La Silla formational boundary is located in the Late Cambrian. According to North American standard stratigraphy, however, the boundary is located near the Cambrian-Ordovician transition (Fig. 49) as shown by the presence of the trilobite *Plethopeltis obtusus* (Keller et al., 1994; Vaccari, 1994) which is present from the *Saukia serotina* to the *Missisquoia depressa* subzones. This relatively precise assignment is of great importance, because near the Cambrian-Ordovician boundary is the major Lange Ranch eustatic event (Miller, 1984), which is exactly in the time span described here. Consequently, the sequence boundary separating the La Silla Formation from the La Flecha Formation (sequence boundary 9) is here interpreted to represent the eustatic event. At present, it is not possible to

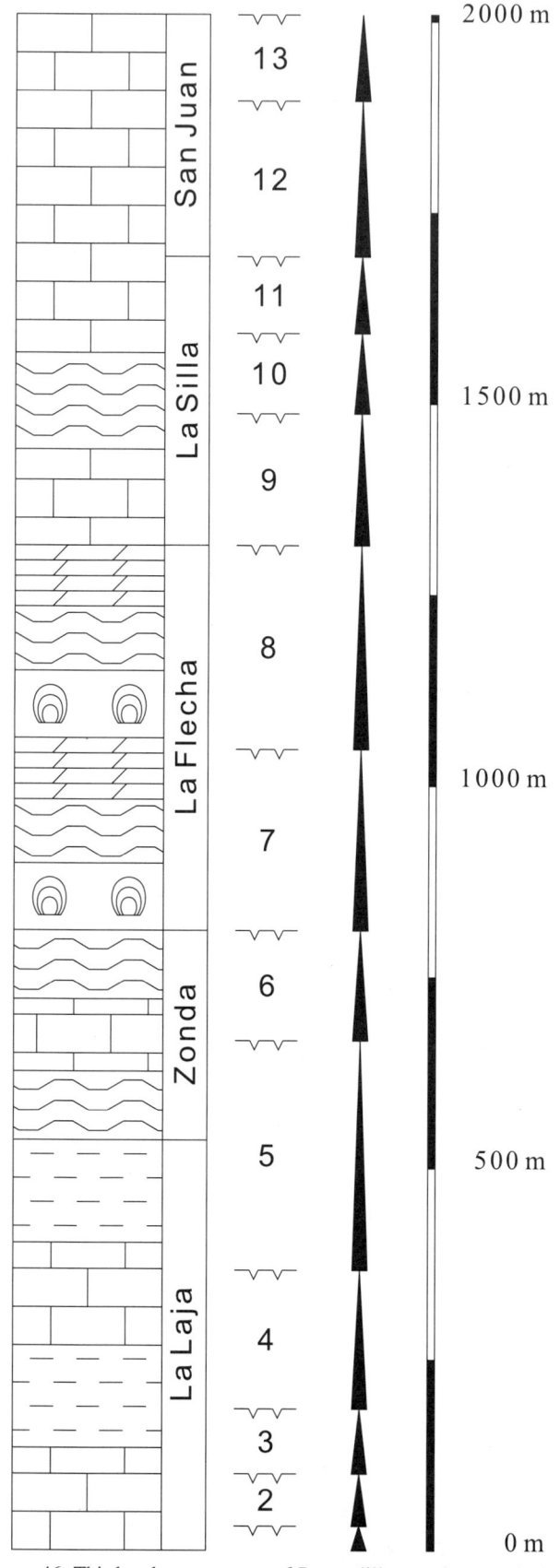

Figure 46. Third-order sequences of Precordillera carbonate platform succession and their thicknesses.

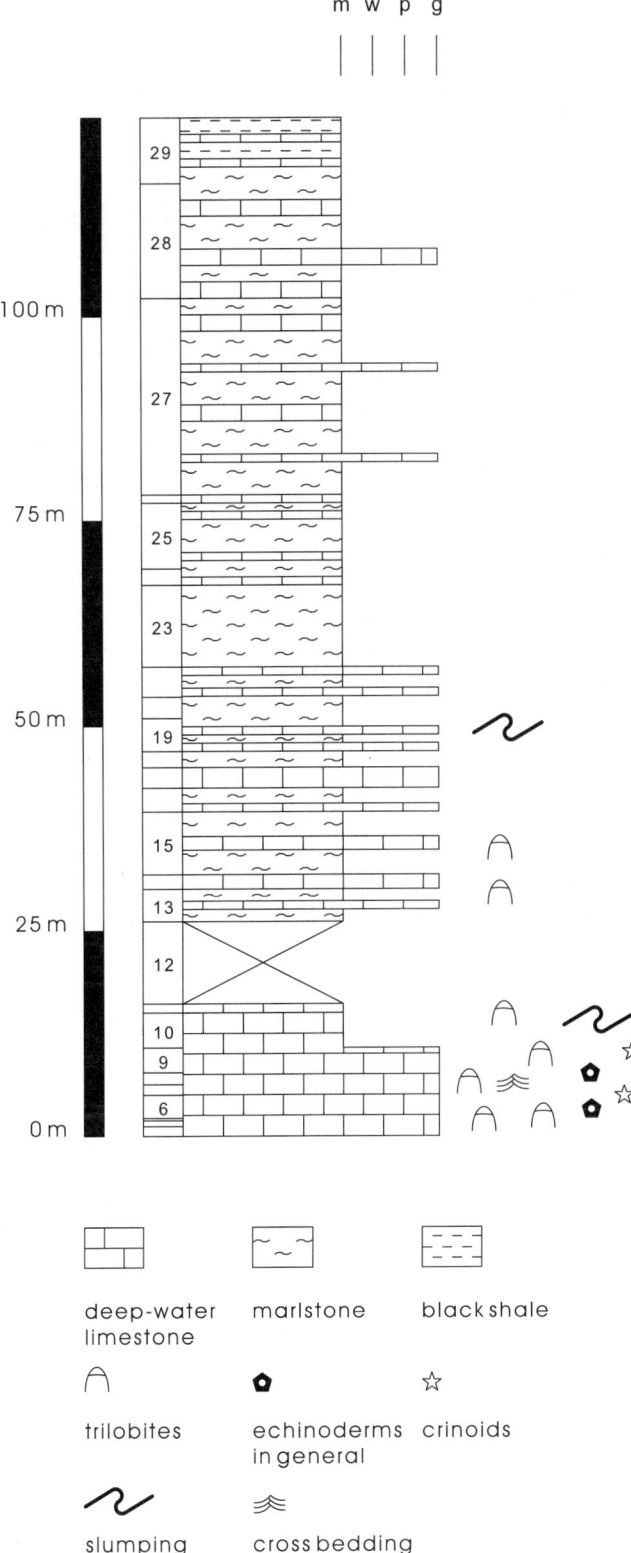

Figure 47. Section of the deep-water limestones, marlstones, and shales of Late Cambrian El Relincho Formation at Cerro Pelado. Succession overlies Middle Cambrian cyclical peritidal rocks of Cerro Pelado Formation. Boundary is marked by hardground and hiatus. See also Figure 7. Legend as in Figure 16.

assign the sea-level fall observed in the Precordillera to one of the two regressive-transgressive couplets that make up the Lange Ranch event (Miller, 1992).

Sequence boundary 10 (Figs. 46 and 48) is located about 25 m above a horizon (point B of Keller et al., 1994) at which conodonts of the *Clavohamulus hintzei* zone (Fig. 49) are present (Lehnert, 1995a). At 75 m above the sequence boundary, conodonts of the *Rossodus manitouensis* zone are found (point C of Keller et al., 1994). Between both conodont zones two major eustatic events, which embrace the base of the Tremadocian (Fig. 49), were recorded by Miller (1992). The global character of a basal Tremadocian transgression was stressed by Fortey (1984). On the North American craton, there is a major flooding event observed, the Stonehenge transgression (Taylor et al., 1992), during the *C. angulatus* zone. This flooding is slightly younger than the two events described by Miller (1992). It will be interesting to see whether biostratigraphy, most promisingly conodonts, will one day be able to obtain a resolution high enough to assign sequence boundary 10 to one of the three events.

The next younger sequence boundary within the La Silla Formation (sequence boundary 11) is reasonably well dated, because conodonts were found about 10 m below the breccia on top of the sequence (point D of Keller et al., 1994). The conodonts indicate a position close to the boundary between the *R. manitouensis* zone and the low-diversity interval or the boundary between the Skullrockian and Stairsian substages of the North American subdivision (Fig. 49). Along the Laurentian margins, the base of the Stairsian is characterized by a major sea-level rise (Ross and Ross, 1995) which is most probably reflected in the rapid flooding at the base of sequence 11 in the Precordillera.

The La Silla Formation and the San Juan Formation are separated by a sequence boundary (boundary 12; Figs. 46 and 48) belonging to the uppermost Tremadocian or mid-Stairsian, respectively (discussion in Keller et al., 1994; Lehnert, 1995a). During the late Stairsian, there is a marked regression around the North American craton (Ross and Ross, 1995). This regression might be reflected by a major eustatic event observed on other continents. Outside North America, it spans the Tremadoc-Arenig boundary (Fortey, 1984). Whether the Laurentian event is identical to this presumably global event has not yet been proven; however, one of these two events is reflected by the sequence boundary separating the La Silla Formation from the San Juan Formation.

Starting with the lower reef-mound horizon, there are several flooding events recognized within the San Juan Formation. Some of them are easily correlated to a sea-level curve worked out for the Arenig of Scandinavia and Australia (Nielsen, 1992). A comparison to North America is somewhat more difficult for various reasons. The San Juan Formation continues well into the Whiterockian (and thus into the Middle Ordovician according to the North American standard; Fig. 49). Along the eastern margin of Laurentia, the Whiterockian sediments record the onset of the Taconic orogeny and the interplay between eustasy and tectonics (James et al., 1989; Read, 1989). The southern and the western

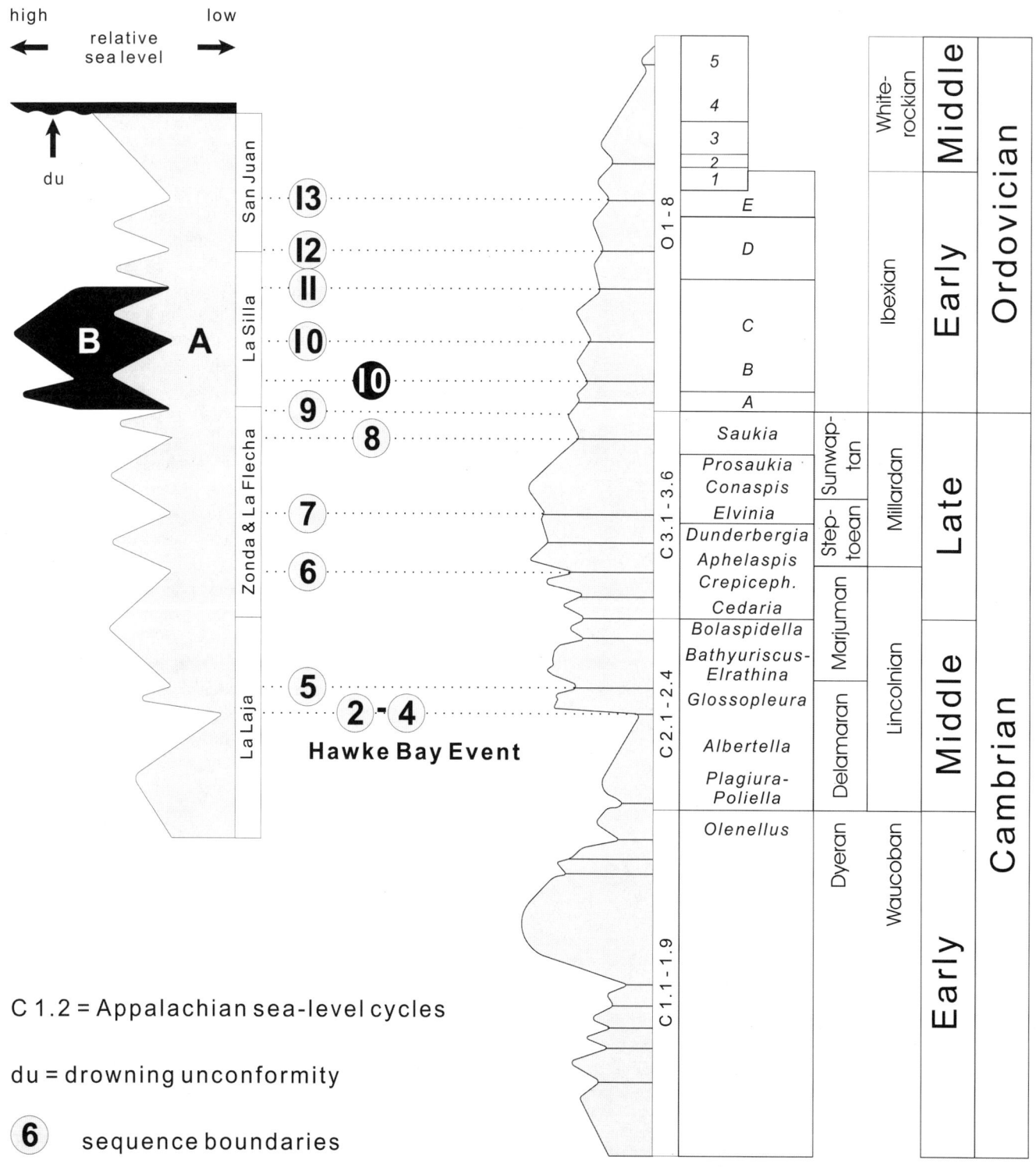

Figure 48. Sea-level curve for Cambrian-Ordovician carbonate platform deposits of Precordillera and its correlation to the southern Appalachians. A and B show alternative correlations for earliest Ordovician sequence boundaries. Appalachian curve is modified from Read (1989).

margins of Laurentia were affected by high-frequency eustatic events in the upper Ibexian and the Whiterockian (Schutter, 1992; Ross and Ross, 1995; Cooper and Keller, 1995; Keller and Cooper, 1995). These sea-level fluctuations are related to the climax of the Sauk sequence deposition. In the Argentine Precordillera, the demise of the carbonate platform was initiated during this climax of sea-level highstand.

Within sequence 12, the basal sequence of the San Juan Formation (Figs. 46 and 48), flooding was most dramatic at the base of the *O. evae* zone, which led to the deposition of the deep-ramp nodular limestones. In some sections, this interval is actually represented by a couplet of nodular limestones with intercalated mid-ramp deposits. Within the resolution of biostratigraphy, this couplet is coeval to similar events in Baltica and Australia. There, they were attributed to major eustatic sea-level changes by Nielsen (1992). It seems plausible to correlate the couplet of deepening events in Baltica and Australia to two of the three sea-level changes in the Blackhillsian (uppermost Ibexian) of North America (Fig. 49); however, at present it is not possible to assign the couplet in the Precordillera to two distinct events around the margins of Laurentia.

The upper reef horizon in the San Juan Formation formed just above the base of the Whiterockian, which is marked by a pronounced regressive event with subsequent flooding (Ross and Ross, 1995). All conodont data from this interval within the San Juan Formation (Lehnert and Keller, 1993b; Lehnert, 1995a) suggest that the sequence boundary below the reefs (sequence boundary 13) is caused by the drop of sea level at the base of the Whiterockian (Fig. 49). This is supported by data from Baltica and Australia (Nielsen, 1992) that also point to a marked sea-level fall in the upper *triangularis-navis* zone (basal late Arenig; Fig. 49). The subsequent history of the San Juan Formation is one of rising sea level that finally led to drowning and the termination of carbonate platform sedimentation. One important factor for this drowning was that global sea level reached a maximum near the Arenig-Llanvirn boundary (Fortey, 1984; Ross and Ross, 1995). However, in the Guandacol subbasin the first signs of tectonic instability are as early as the late Arenig; therefore, rising sea level was one factor in terminating carbonate production, but another was tectonics. At present, it seems impossible to distinguish between the effects of eustasy versus tectonics.

Many of the assignments given here are not, or not well, proven biostratigraphically; however, the interpretations and their correlations may serve as predictions and an encouragement for further biostratigraphic investigations, especially in the Upper Cambrian successions.

Comparison between the carbonate platform of the Precordillera and the eastern margin of Laurentia. Since the 1995 Penrose Conference, it has become widely accepted that the Argentine Precordillera is a Laurentia-derived terrane (Dalziel et al., 1996). Timing of the separation from Laurentia, however, is still highly controversial. Astini et al. (1995, 1996a, 1996b) believe in a separation during the Early Cambrian. In contrast, there is much evidence that the Precordillera was detached during Middle and Late Ordovician time (Dalla Salda et al., 1992a, 1992b; Dalziel et al., 1994; Keller, 1995; Keller and Dickerson, 1996; Dickerson and Keller, 1998). Dalla Salda et al. (1992b) first pointed out that the Ouachita embayment (Fig. 6) is a candidate for the provenance of the Precordillera. Prior to that, much emphasis was placed on stratigraphic and tectonic correlations of the Precordillera to the Appalachians (Fig. 1). In contrast, paleontologists consistently have indicated a close connection to the southern midcontinent and to the western United States, especially the southwestern Great Basin (e.g., Baldis and Bordonaro, 1981b, 1982; Lehnert, 1993, 1995a).

In the following chapter, a comparison is made between the evolution of the carbonate platform in the Precordillera and its counterparts along the eastern and southeastern margin of Laurentia, especially to demonstrate similarities and differences between both areas. This comparison is based mainly on the recent compilations by James et al. (1989) and Knight et al. (1995) for the Canadian Appalachians, and Read (1989) and Read (*in* Rankin et al., 1989) for the U.S. Appalachians. This comparison does not intend to be a detailed stratigraphic comparison like that of Astini et al. (1995), but intends to show general sedimentary and evolutionary patterns.

After a platform-foundation phase during the Late Proterozoic and earliest Cambrian, carbonate platform sedimentation on both the Cordilleran and the Appalachian margins had begun by late Early Cambrian time (Bond et al., 1984, 1988; James et al., 1989; Read, 1989). In many places along the Appalachian margin, the carbonates rest on rifted-margin sediment prisms that terminate with mainly marine, mature quartz sandstones (Fig. 42A). In the Precordillera, there is indirect evidence that most of such a sediment prism is absent. The carbonate platform might directly overlie crystalline basement or is separated from the latter only by a thin siliciclastic veneer (Fig. 42B; Keller, 1995). On both margins of Laurentia, the onset of carbonate platform deposition is broadly correlative with the *Bonnia-Olenellus* trilobite zone (Bond et al., 1989). Similarly, the oldest carbonates in the Precordillera belong to the *Bonnia-Olenellus* zone. On a regional scale, however, the oldest well-developed carbonate platform rocks belong to the Middle Cambrian *Glossopleura* zone (Soldano Member of the La Laja Formation).

Along the Laurentian margins, Middle and Upper Cambrian strata are arranged in grand cycles (Aitken, 1966, 1978; Palmer, 1971a, 1971b; Palmer and Halley, 1979, among others). These cycles show a lower terrigenous half cycle and an upper carbonate half cycle, indicating overall shallowing. In the Precordillera, grand cycles are present although the siliciclastic interval represents the shallowest facies (Fig. 11). The most important difference to Laurentia is that in the Precordillera terrigenous half cycles are only present until the late Middle Cambrian. Younger sedimentary cycles lack siliciclastic units, demonstrating that coastal onlap onto the hinterland had progressed far into the crystalline source area. An alternative explanation for the absence of terrigenous detritus is the development of an intrashelf basin between the actual depositional site of the La Flecha Formation

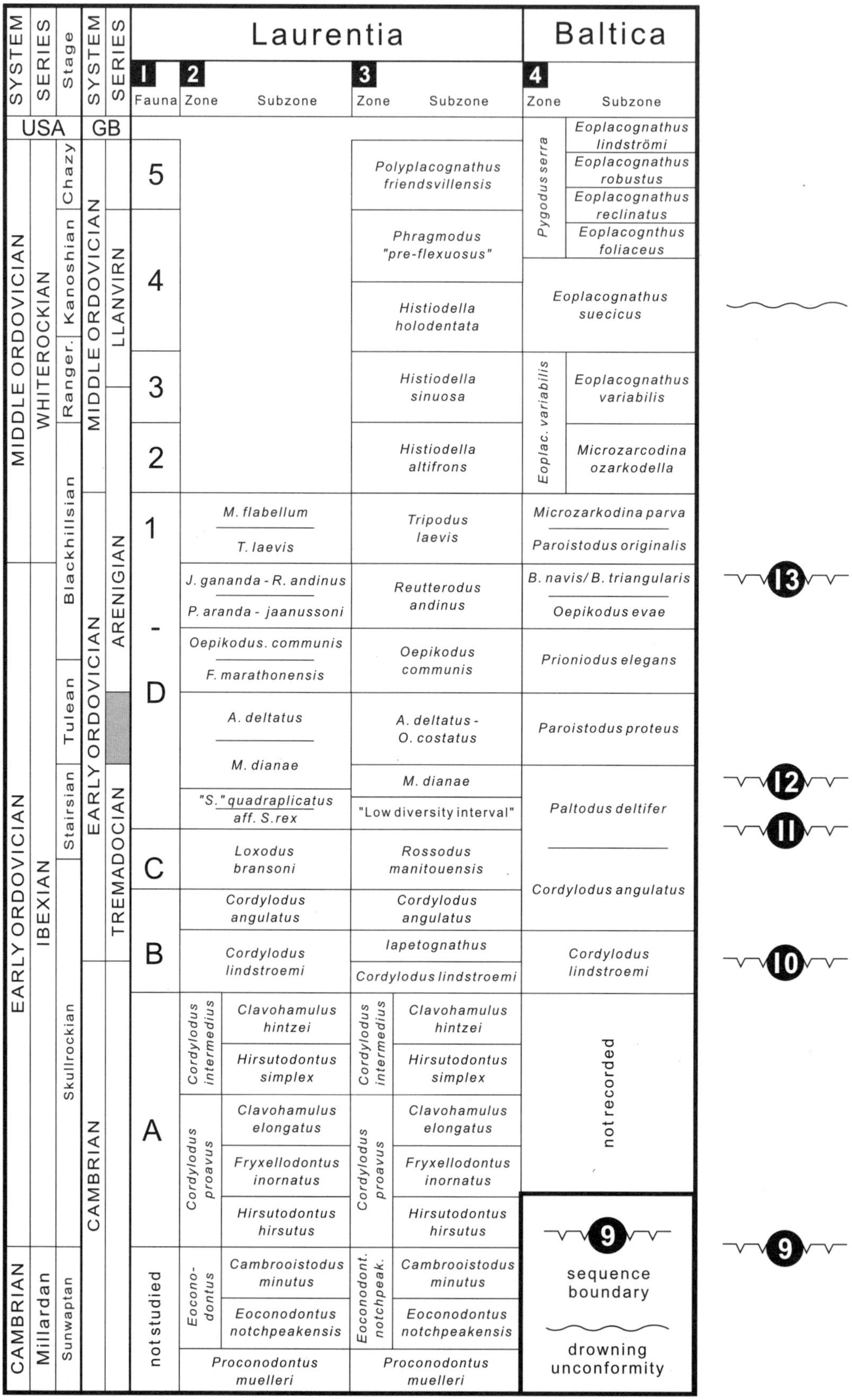

Figure 49. Conodont biostratigraphy of Late Cambrian and Early Ordovician according to North American (USA) and European (GB, Baltic) zonations. Shaded interval marks uncertain interval of Tremadocian-Arenigian boundary. Sources: (1) Ethington and Clarke (1971); (2) Ethington and Clarke (1981); Miller (1988); (3) Ross et al. (1993); (4) Lindström (1971); Löfgren (1978, 1996). Sketch courtesy of O. Lehnert.

and the clastic shoreline. Intrashelf basins are well known from the U.S. Appalachians (Markello and Read, 1982; Rankin et al., 1989), but are called the inner detrital belt along the Cordilleran margin of Laurentia (Aitken, 1978; Palmer, 1971a; see Fig. 9). These basins served as a sediment trap for the siliciclastic materials derived from the adjacent hinterland. Given the present outcrop situation, both scenarios are viable.

In the Canadian Appalachians, there are extremely fossiliferous and bioclastic carbonates of Early Cambrian age, whereas "... middle and upper Cambrian sediments are almost abiotic, composed of peloids, ooids, and intraclasts with microbial buildups and relatively few bioclasts" (James et al., 1989: p. 142). This trend was reversed during the Early Ordovician, when more muddy sediments were deposited and the rocks became progressively more fossiliferous at the expense of ooids and peloids. This evolution, which seems to be recorded all around the Laurentian margin, is also observed in the Precordillera; however, the timing differs slightly. The Middle Cambrian (La Laja Formation) is very fossiliferous, containing abundant trilobites, hyolithids, and eocrinoids. The Upper Cambrian Zonda and La Flecha Formations mirror the Canadian Middle and Upper Cambrian deposits in being oolitic and intraclastic with abundant microbial structures. At the Cambrian-Ordovician boundary, the reverse trend was initiated with the flooding above the La Flecha–La Silla formational boundary (sequence boundary 9). Muddy sedimentary rocks prevail in the La Silla Formation, and the main components are peloids and, to a minor extent, ooids. At the base of the San Juan Formation, a fully marine fauna started to flourish; in contrast, ooids and peloids are subordinate constituents.

A striking similarity between the Appalachians and the Precordillera is the tripartite subdivision of the Ibexian. In Canada, this interval is represented by a succession of subtidal limestones sandwiched between peritidal carbonates (Knight et al., 1995). Almost identical strata are present farther south (Chepultepec interval, Bova and Read, 1987) and a similar succession is recognized in the Precordillera. Three sequences are present in the La Silla Formation (sequences 9–11; Figs. 46 and 49); the middle sequence is the thickest and has the largest proportion of subtidal facies.

During the Early Ordovician, the morphology of the carbonate platform in the Precordillera changed from a Late Cambrian platform with a high-energy rim toward a carbonate ramp during the Arenig. A similar evolution is observed in the U.S. Appalachians (Read, 1989) and is there related to flexure of the continental crust during incipient collision. In the southern Appalachians, the Knox-Beekmantown unconformity (Bird and Dewey, 1970) marks the transition from passive-margin sedimentation to convergent-margin deposition. Mussman and Read (1986) demonstrated that karstification and erosion are most prominent in the south and gradually decrease toward the north. Rocks beneath and above the unconformity demonstrate that the main events took place during the middle Whiterockian, corresponding to the Llanvirn and Llandeilo in the Precordillera. Demise of the carbonate platform in the northern Appalachians started with block faulting and local erosion of the platform strata, followed by rapid subsidence and the final burial under the prograding clastic wedge (Knight et al., 1995). In the northern Appalachians, the main tectonic events are dated as Middle Ordovician.

In the Precordillera, the carbonate platform was entirely drowned by the early Llanvirn. Although sea level seems to have played the leading role, sequence stratigraphy and the sedimentology of the pre-drowning succession (Fig. 50) indicate that tectonics might have been an important factor. Figure 50 shows a relatively uniform carbonate platform evolution and sequence development during much of the Early Ordovician. No major differences in platform configuration or sediment thickness are recorded in the Tremadocian La Silla Formation. The La Silla Formation is by far the most uniform succession present in the platform deposits, which probably indicates that the entire terrane was flooded and that eustasy was the only ruling factor.

The basal sequence of the San Juan Formation (sequence 12; Fig. 46) demonstrates that accommodation was greatest in the Guandacol subbasin but it was almost matched by the southern and western sections of the Talacasto subbasin (Fig. 50). The decreasing thickness of this sequence toward the sections around Jáchal and finally toward the San Juan subbasin reflects successive onlap in an eastern and southeastern direction, a pattern not observed in the La Silla Formation. In addition, the intermediate position of the northern sections of the Talacasto subbasin is documented by the presence of a shelf-margin systems tract at the base of the San Juan Formation, which is present neither farther west nor in the San Juan subbasin. The onlap pattern of the sediments, together with the fact that Cambrian conodonts are found in the San Juan Formation, indicates that sea level was lower during the time of deposition of the San Juan Formation than during La Silla Formation deposition or that tectonic uplift affected the terrane in some place. Global sea level was lower in the Arenig than in the Tremadoc (Fortey, 1984; Read, 1989), so global sea-level fall is the most likely interpretation.

In the upper sequence of the San Juan Formation (sequence 13; Fig. 46) the uniform facies and thickness evolution is no longer observed (Fig. 50). In contrast to the underlying sequences, the sections in the San Juan subbasin show the thickest successions of all subbasins, whereas the sections of the Talacasto subbasin reveal highly variable thicknesses (Fig. 50) between the sequence boundary and the drowning unconformity. This heterogeneous evolution of the carbonate platform toward its end is interpreted to reflect differential subsidence caused by the onset of crustal extension. Consequently, the Talacasto and the San Juan subbasins record the continuation of block faulting, which started to affect the platform during the early late Arenig in the Guandacol subbasin.

In the Precordillera, incipient block faulting in combination with an important sea-level rise at the beginning of the Middle Ordovician led to the drowning of the carbonate platform, which subsequently underwent major block faulting, uplift, and erosion. In contrast, in the Canadian Appalachians block faulting led to local uplift and erosion of the carbonate platform prior to subse-

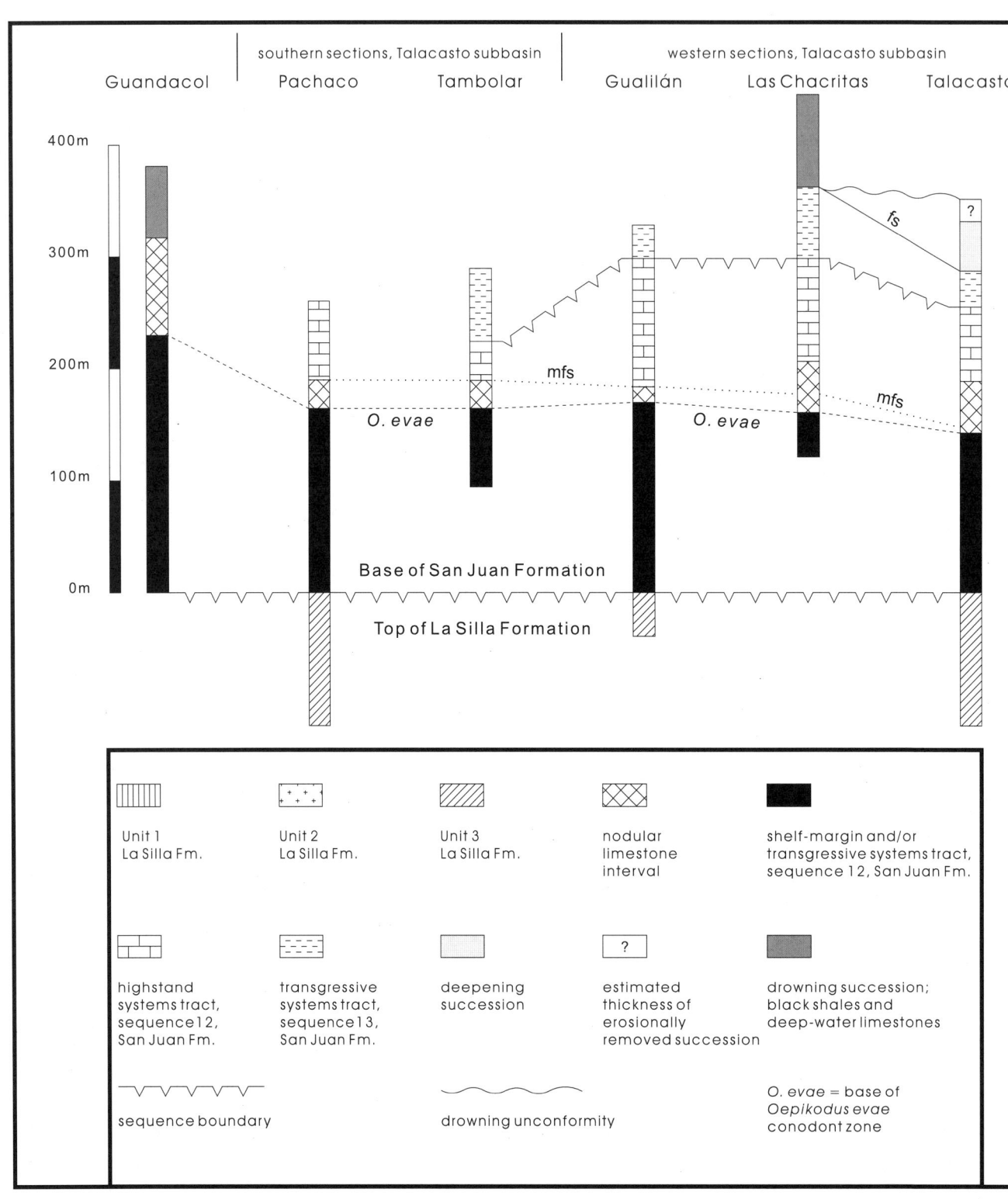

Figure 50 (on this and opposite page). Facies associations and sequence stratigraphy of Early Ordovician carbonate platform deposits show uniform development during deposition of La Silla Formation and basal San Juan Formation. Upper sequence within San Juan Formation, with its succession of flooding surfaces and its irregularly distributed facies associations, is interpreted to reflect onset of crustal extension. Sections are arranged in their presumed relative position across ancient shelf. The Guandacol section is in most offshore position, whereas sections at Los Berros and Las Lajas are most cratonward sections.

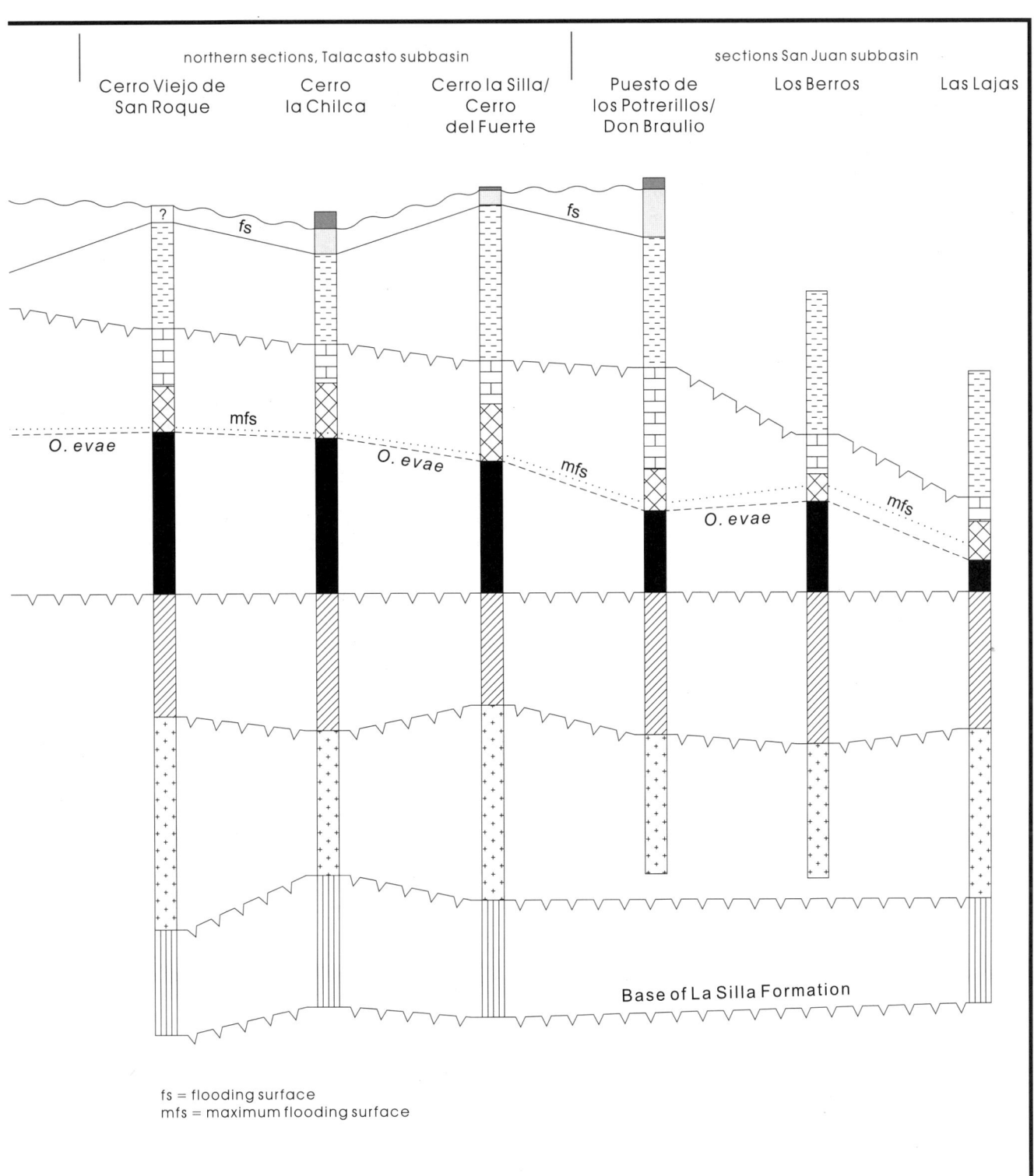

quent flooding, drowning, and the deposition of a clastic wedge.

These differences in the evolution of both areas, the Argentine Precordillera and the Appalachians, and in the demise of the carbonate platform seem to be related directly to the geotectonic environment in which the sediments were deposited. In the Appalachians, block faulting is related to crustal extension on a subducting plate, i.e., it is related to the Taconic orogeny. In the Precordillera, incipient block faulting seems to be related to the onset of crustal extension, which finally led to continental breakup and the separation of the Precordillera from Laurentia (see following discussion).

EARLY PALEOZOIC EVOLUTION AND PALEOGEOGRAPHY OF THE ARGENTINE PRECORDILLERA—THE MIDDLE ORDOVICIAN THROUGH DEVONIAN SILICICLASTIC SUCCESSIONS

Middle Ordovician

After the demise of the carbonate platform, sedimentation continued during much of the Llanvirn; however, major facies differentiations are recognized. In the San Juan subbasin, Middle and Upper Ordovician strata are exposed in the Don Braulio section (Figs. 3 and 28). There, the San Juan Formation was drowned during the *E. suecicus* zone, deposition of the Gualcamayo Formation followed. Sedimentation of the Gualcamayo Formation continued probably into the earliest Llandeilo. The hiatus present between the Gualcamayo Formation and the La Cantera Formation is caused by the erosional event that accompanied deposition of the basal conglomerate of the La Cantera Formation.

There are indications that the Gualcamayo Formation covered a much larger part of the San Juan subbasin. Sarmiento et al. (1986) described early Llanvirnian conodonts from an olistolith within the Rinconada Formation and mentioned overlying graptolite-bearing shales. In most sections of the San Juan subbasin, however, Middle Ordovician strata have been eroded and the extent of erosion cuts down to different levels within the San Juan Formation (Lehnert and Keller, 1994; Lehnert, 1995b; Keller and Lehnert, 1998).

A similar evolution to that of the San Juan subbasin is recorded in the Talacasto subbasin. Black shales of the Gualcamayo Formation are preserved in several sections around Jáchal, but there are others (Las Aguaditas, Las Chacritas; Fig. 3) where carbonate sedimentation continued. The Las Aguaditas section shows the early Llanvirnian drowning event and the evolution of calcareous slope facies with megabreccia sedimentation (Keller et al., 1993b). Although eustatic sea-level fluctuations may have been a factor during deposition, the formation of a slope apron, which acted as a source for the mass-flow deposits in the Las Aguaditas Formation, is taken as evidence of block faulting. An important statement of Keller et al. (1993b) was that within the Argentine Precordillera there is no evidence of a carbonate platform that might have acted as a source for the hemipelagic limestones.

At Las Chacritas, sedimentation continued with nodular, probably deep-ramp limestones above a marked flooding surface on top of the San Juan Formation (Figs. 21 and 50). This succession, the age of which is not clear, is capped by a latest Ordovician chert-pebble conglomerate. In the sections toward the west and south, Ordovician erosion removed the strata on top of the platform and much of the platform. One exception and a key point for the interpretation of the geodynamic history is the Río Sassito section, where Caradocian limestones are present (Lehnert, 1995a; Keller, 1995). These overlie deeply eroded San Juan limestones, demonstrating that the main period of Ordovician erosion took place prior to the Caradoc (Keller and Lehnert, 1998).

In the Guandacol subbasin, deposition of black shales continued across the Early-Middle Ordovician boundary. As in the Don Braulio section, a succession of turbidites, conglomerates, and olistostromes (Las Vacas Formation; Fig. 4) overlies the black shales with an erosional unconformity (Astini, 1991). As in the Don Braulio section, sedimentation of the chaotic and catastrophic deposits started around the Llanvirnian-Llandeilian boundary with erosion of the underlying strata. Limestone blocks of the San Juan Formation within the Las Vacas Formation, more than 70 m long, attest to rapid uplift and increasing relief. The composition of the clast population within the conglomerates is similar to that of the La Cantera Formation; however, mafic and ultramafic magmatic clasts compose >20% of the population (Loske, 1992). Overlying the Las Vacas Formation is the Las Plantas Formation (Fig. 4) which was deposited in a slope-apron environment (Astini, 1991). Typical sedimentary rocks are autochthonous hemipelagites and allochthonous debris-flow deposits which are as young as the early Caradoc. The entire succession of Gualcamayo Formation–Las Vacas Formation–Las Plantas Formation shows the same general evolutionary pattern as the Don Braulio section; however, sediment thicknesses are clearly different. In both sections, the Llandeilian through lower Caradocian rocks display an overall fining-upward trend. The successions mirror several events during which conglomerates were delivered to the basin and resedimented. Both successions reflect sedimentation in slope-apron environments. Hence the most likely depositional scenario for both units includes grabens or half grabens with bounding faults along which material was locally eroded (limestone slabs, shale slabs), but that also received material from the hinterland. This coarse detritus was resedimented to form thick conglomerates. In the Guandacol area, block rotations are reflected in the succession by several angular unconformities (W. Von Gosen, 1997, personal commun.), especially between the Las Vacas conglomerates and the Las Plantas Formation. Locally, the angular unconformity between the two formations is as great as 30°.

Near San Rafael, Middle Ordovician rocks (Lindero Formation) are also present (Nuñez, 1979; Baldis and Blasco, 1973;

Bordonaro et al., 1996). The lower member of the formation is composed of arkoses and fine conglomerates (Fig. 41) which are the products of insitu weathering of the granitoid basement. Above a marked hardground, sandstones, siltstones, volcanic breccias, and locally limestones form the upper member. Carbonates, contained both in the lower and the upper member, yielded conodonts (Bordonaro et al., 1996) that indicate that sedimentation of the lower succession continued toward the Llanvirn-Llandeilo boundary, and that the upper member was deposited during the late early Llandeilian and early Caradocian. The importance of the outcrops is that they demonstrate that basement was exposed during early Middle Ordovician time and that the composition of this basement makes it a logical source for the coarse detritus found in many of the Middle Ordovician sandstones and arkoses in the Precordillera. This does not necessarily imply that the area of San Rafael was the source area, but that similar outcrops in other parts of the terrane might have acted as a source.

Although the outcrops of the Lindero Formation do not permit a regional interpretation, the facies succession from arkoses to deep-water shales and limestones indicates that the Ponon Trehue area was rapidly submerged during the Middle Ordovician. Hence the area might represent a local fault-bounded block that underwent major downwarping.

The most important event during the Middle Ordovician is the evolution of a continental margin facies and the western basin. The onset of basin formation is still a matter of debate (Keller, 1995; Astini et al., 1996a); however, the presence of Llanvirnian faunas within olistoliths demonstrates that the main events, especially the emplacement of the major olistoliths, took place during the latest Llanvirn but predominantly during the Llandeilo. The stratigraphic and paleogeographic relationships between the Middle and Upper Ordovician units in the western basin are not clear. There seems to be a westward fining tendency: rock-fall deposits of the Los Sombreros Formation to conglomerates of the Portezuelo del Tontal Formation to turbidites of the Don Polo Formation to shales of the Alcaparossa Formation (Figs. 4 and 51). West-directed transport is observed in turbidites and megabreccias in the Los Sombreros Formation as well as in the conglomerates and turbidites of the Portezuelo del Tontal Formation (Spalletti et al., 1989). In the Don Polo Formation, however, there are indications that sediment transport was directed toward the southeast and south (Astini, 1991). This pattern is present in the area along the Río San Juan and from there toward the south. Whether there are similar tendencies farther north has to be resolved once there is more understanding of the stratigraphy in those areas.

Late Ordovician

In the western basin, Middle Ordovician sedimentation continued into the Late Ordovician without a major change in sedimentation patterns. There is a general tendency toward finer grained siliciclastics, especially black shales (Alcaparossa Formation), but the most important feature is Caradocian pillow basalts (Fig. 51). The only evidence of shallow-water deposition is preserved in the San Isidro area, where (glacially influenced?) platform sandstones overlie the continental margin facies.

In the San Juan subbasin, Upper Ordovician sediments are present only in the Don Braulio section. In addition to the uppermost part of the La Cantera Formation, the Late Ordovician is mainly represented by the Don Braulio Formation, from which the first evidence of the Late Ordovician Gondwana glaciation was described (Peralta and Carter, 1990a). Although Astini and Buggisch (1993) favored a terrestrial origin of the diamictites, the presence of coeval marine sediments (Sánchez et al., 1991) seems to indicate that the lower member of the Don Braulio Formation is of marine origin as well. At least two unconformities have been recognized in the succession at Don Braulio, one at the base of the diamictites and the one at the base of the conglomerate, which marks the base of the upper member of the Don Braulio Formation (Fig. 28). Locally, both unconformities merge to one surface. Both unconformities have been described (Baldis et al., 1982; Sánchez et al., 1991) and differently interpreted. Baldis et al. (1982) favored a tectonic origin ("Villicum phase"), whereas Sánchez (1991) interpreted both surfaces to be related to sea-level changes in connection with the Late Ordovician glaciation.

In the Talacasto subbasin, there are only isolated outcrops of Upper Ordovician sediments. The importance of the temperate-water Sassito limestones has already been mentioned. East of Jáchal, near Cerro del Fuerte (Fig. 3), Ashgillian shales are exposed (Sánchez et al., 1991) and interpreted to have been deposited below storm-wave base. A conglomeratic interval near the base of the exposed succession may be a correlative to the Don Braulio conglomerate. In most sections of the Talacasto subbasin, a hiatus is present between the upper part of the San Juan Formation and the latest Ordovician transgressive sediments of the La Chilca Formation. The latter belong to a Silurian sedimentary cycle and will be discussed in the next chapter. The main erosional event took place prior to the deposition of the Sassito limestones. The effects and the importance of the post-Caradocian erosion, i.e., the post-Sassito limestone erosion, are difficult to establish.

In the Guandacol subbasin, sedimentation rates and subsidence increased toward the Caradoc-Ashgill boundary, as reflected by more than 1500 m of predominantly shales and turbidites (Astini, 1991). Calcareous megabreccias attest to a tectonically unstable environment and the local erosion of strata down to the San Juan Formation. Concomitantly, another source area provided the detritus deposited as quartz arenitic turbidites. Paleocurrent indicators (Astini, 1991) demonstrate that the megabreccias originated along the margin of the depocenter whereas the turbidites were transported longitudinally through the basin.

Upper Ordovician sedimentary rocks are also present in the San Rafael area (Cingolani and Cuerda, 1996). These outcrops at Cerro Bola (Fig. 39) are not directly connected to those at Ponon Trehue. At Cerro Bola, a thick succession of Caradocian tur-

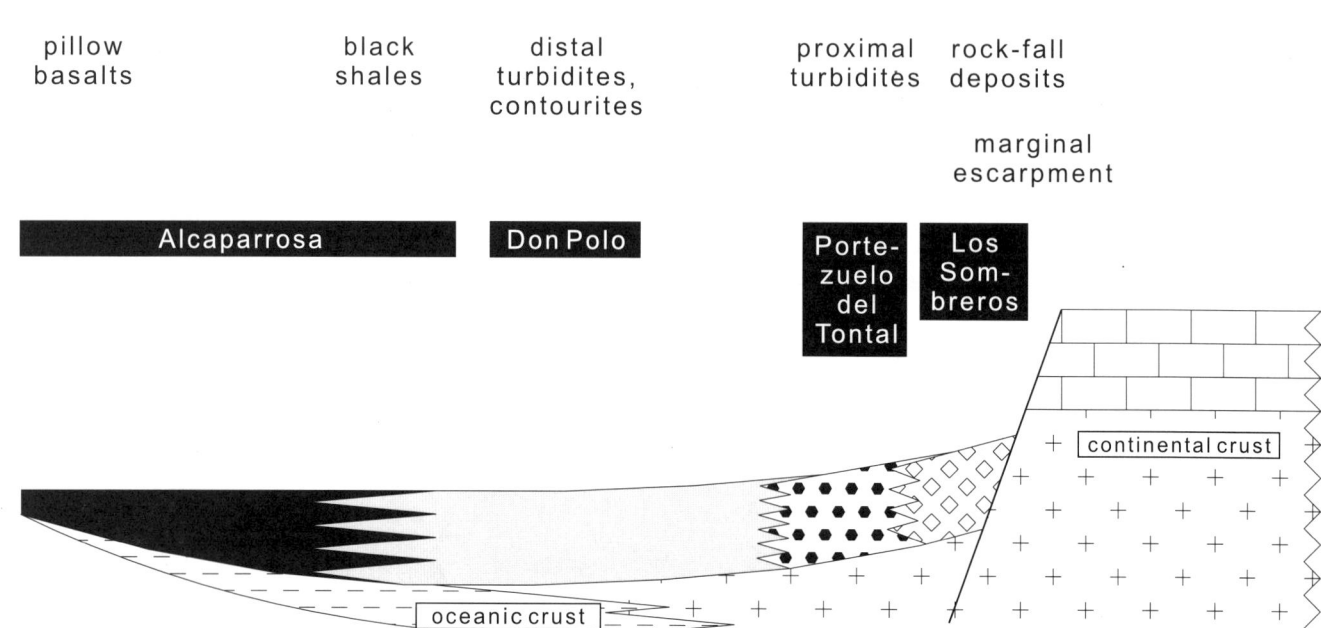

Figure 51. Depositional environments and sedimentary rocks of Ordovician of western basin. Rocks reflect formation of Atlantic-type continental margin during Middle and Late Ordovician.

bidites and shales crops out, which might be a correlative to some of the units in the western basin of the Precordillera. The presence of pillow lavas (Wichmann, 1928) similar to those of the western basin in the Precordillera (Davicino and Sabalúa, 1990) suggests a correlation to the Alcaparrosa Formation.

Discussion of the Middle and Upper Ordovician evolution of the Argentine Precordillera

In the Precordillera, the Middle Ordovician was the time of major change in depositional style. Carbonate platform sedimentation was terminated by drowning during the early Llanvirn, and only locally did carbonate deposition continue with deep-water limestones and slope hemipelagites. Increasing tectonic instability postdating the late Arenig is indicated by the diachronous drowning of the platform; drowning in the Guandacol subbasin was coeval with shallowing in the Talacasto and San Juan subbasins. In the eastern basin, the Llanvirn was dominated by black shale sedimentation (Gualcamayo and Los Azules Formations), but locally carbonates were still produced. Toward the end of the Llanvirn, the basement of the terrane was locally exposed (San Rafael) and subject to intensive weathering. Near the base of the Llandeilo, the entire eastern basin shows the effects of the Guandacol tectonic phase.

Guandacol event. The Guandacol tectonic phase was introduced by Furque (1972) to explain the presence of thick conglomerate successions in the Precordillera. This concept was later followed by Furque (1979), Furque and Cuerda (1979), and Baldis et al. (1982). As already stated by Furque (1979), the conglomerates document uplift and erosion of the central and eastern Precordillera (sensu Ortiz and Zambrano, 1981). The effects of the Guandacol event are recognized everywhere in the eastern basin: conglomerates and turbidites were deposited coevally in the Don Braulio section and in the Guandacol area. The first and most important megabreccia in the Las Aguaditas Formation is of early Llandeilian age (Keller et al., 1993b). The pre-Caradocian erosion that affected much of the Talacasto subbasin is here interpreted to reflect the Guandacol event. The Guandacol event is also recognized in the San Rafael area, where the hiatus between the Peletay and Los Leones Members of the Lindero Formation represents the early Llandeilo. Deposition of the Llandeilian rocks everywhere seems to be related to crustal extension and the formation of graben structures (Furque, 1979; Loske, 1992; Von Gosen, 1992; Keller et al., 1993b); for most of the formations a slope-apron configuration of the sediment source has been reconstructed.

The Guandacol event is also recognized in the western basin. There, formation of the continental slope was initiated during the late Arenig or early Llanvirn (Astini et al., 1995; Keller, 1995); however, most of the sediments assigned to a slope or basin environment were deposited during the Llanvirn and especially during the Llandeilo (Spalletti et al., 1989; Cingolani et al., 1989). The Los Sombreros Formation contains blocks as young as early Llanvirn (Keller, 1995) and it is not difficult to imagine that the emplacement of the olistoliths is related to the Llandeilian Guandacol event. It must be stressed here that during the Guandacol event a steep escarpment formed, locally more than 2100 m high, as indicated by the olistolith content, which comprises the entire carbonate platform stratigraphy, as well as clasts of a crystalline basement. None of the stratigraphic units in the western basin yielded (autochthonous) fossils older than Llanvirn, and most of them were deposited during the Llandeilo and early Caradoc. Hence, the Guandacol event marks not only an extensional regime in the eastern basin, but also the formation and the initial major subsidence of the western basin. This basin formation was related to major crustal extension leading to continental margin and basin plain environments. Sediments were supplied from line sources (Spalletti et al., 1989; Keller, 1995).

Toward the Caradoc, almost all areas show an overall fining of the sediments and the return to black shale deposition, especially widespread during the *N. gracilis* zone. Important features of Late Ordovician sedimentation are pillow basalts that extruded into the shales of the Alcaparrosa Formation. Upper Caradocian and lower Ashgillian sediments are volumetrically important (Trapiche Formation) but are areally restricted to the Guandacol subbasin and to the San Isidro area (Empozada Formation as defined herein) of the western basin. In the other areas a hiatus is present that spans the late Caradoc through most of the Ashgillian. Although there are local indications of erosion associated with this hiatus (Río Sassito section), the dimensions of erosion are not clear. In most places, the Upper Ordovician unconformity merges with the Middle Ordovician unconformity caused by the Guandacol event. Above the Upper Ordovician surface latest Ashgillian sediments of the La Chilca Formation are found. The only exceptions from the distribution pattern of Upper Ordovician sediments are a few sections where glaciomarine diamictites (Peralta and Carter, 1990a; Sánchez et al., 1991) are present, indicating that the effects of the Gondwana glaciation, to some extent, affected the Precordillera (Buggisch and Astini, 1993).

Silurian. The age and sedimentology of Silurian sedimentary rocks in the western basin are insufficiently documented to allow their discussion and comparison with other parts of the Precordillera. In the Guandacol subbasin, neither Silurian nor Devonian rocks are preserved, whereas they are well developed in the Talacasto subbasin. There, sedimentation started with a chert-pebble conglomerate above the Upper Ordovician unconformity. The lowermost horizons of the La Chilca Formation (Salto Macho Member of Baldis et al., 1984b) still belong to the latest Ashgill (Cuerda et al., 1988b, 1988c) and these beds are traceable from the area east of Jáchal (Cerro del Fuerte) to the Talacasto area (Cuerda 1985; Benedetto et al., 1986; Cuerda et al., 1988b; Peralta, 1990; Sánchez et al., 1991). However, this lower member of the La Chilca Formation seems to be absent immediately south of Jáchal (Peralta et al., 1989; Lehnert, 1990). In general, the Silurian sedimentary rocks of the Talacasto subbasin were deposited on a marine shelf with environments from the high-energy shoreface to muds deposited well below storm wave base (Aceñolaza and Peralta, 1986; Peralta, 1990; Peralta and Carter, 1990b; Astini and Maretto, 1996, among others). There is a consistent pattern of north-south polarity in the sediments of the Silurian. Paleocurrent data indicate predominantly a north-south to north-northeast–south-southwest transport (Astini and Maretto, 1996).

In the Jáchal area, sedimentary rocks are thicker, they are in general more coarse grained, and the sections are more complete than toward the Río San Juan. For example, the thickness of the La Chilca Formation decreases from 90 m in the north (Lehnert, 1990) to about 30–40 m in the Talacasto area (Baldis et al., 1984; Fig. 52). The La Chilca Formation is absent in the sections along the Río San Juan. The Los Espejos Formation, which overlies the La Chilca Formation, displays a similar thickness trend (Fig. 52): from more than 450 m around Jáchal (Peralta et al., 1989; Lehnert, 1990) to 250–300 m around Talacasto (Peralta, 1990). The Tambolar Formation (Fig. 4), only present in the Río San Juan sections (Heim, 1952), is coeval with the Los Espejos Formation (Fig. 4) and is regarded as a distal facies of the latter (Peralta, 1993). In comparison to its proximal counterpart, the thickness is strongly reduced and varies between 120 and 30 m in different sections. There is not only a north-south polarity but also an east-west trend. The Tambolar Formation thins from 120 m at Río Sassito to 71 m at Tambolar and 28 m at Pachaco (Fig. 52). Similar values were described by Astini and Maretto (1996) in a compilation of thicknesses of the Los Espejos Formation. This compilation also demonstrates that there is a north-south–trending ridge connecting the sections of Las Aguaditas with Las Chacritas. Along this ridge, the Los Espejos Formation shows thicknesses between 260 and 300 m. In contrast, sections southwest and east of Las Aguaditas are much thicker (e.g., 450 m at Cerro del Fuerte; Fig. 52).

In the San Juan subbasin, the Silurian shows a completely different development. The upper member of the Don Braulio Formation starts with the main conglomerate which is overlain by shales of varying thicknesses. The shales contain the Ashgillian *Hirnantia* fauna. Important features toward the top of the formation and yet within the Silurian are iron oolites and associated chert pebbles. This part of the succession is absent in the Rinconada area; however, it is likely that it had been present as is shown by reworked glacial clasts in the overlying Rinconada Formation (Von Gosen et al., 1995). The Rinconada Formation is a thick succession of shales and siltstones containing abundant olistoliths and mass-flow deposits. Deposition of this chaotic succession continued at least into the basal Ludlow (Peralta and Medina, 1986; Peralta, 1990). In the Don Braulio section, the olistoliths within the Rinconada Formation mirror the entire underlying succession down to the San Juan Formation.

Discussion of the Silurian evolution of the Argentine Precordillera

Two principal depositional environments are recognized in the Silurian of the Precordillera: a vast siliciclastic platform in most of the Talacasto subbasin and an olistostrome along the

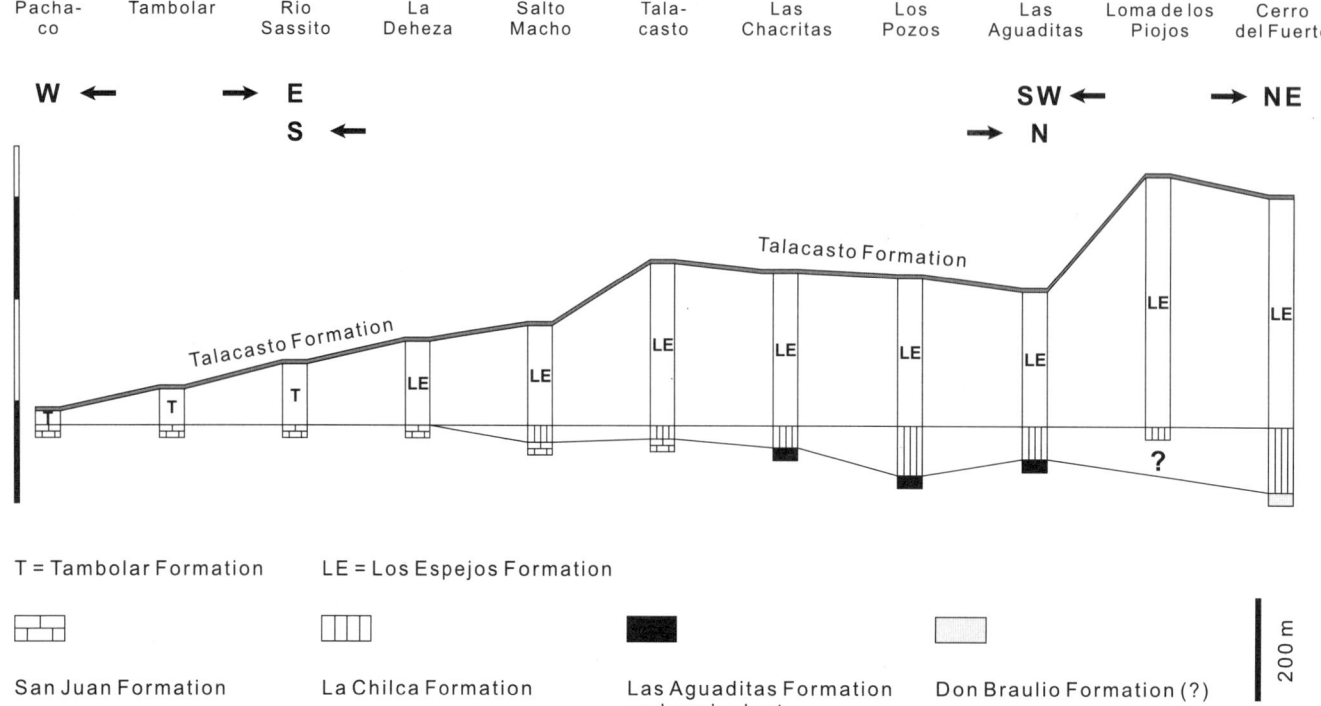

Figure 52. Sediment thicknesses of Silurian deposits in Talacasto subbasin. Note Jáchal depocenter and strongly reduced thicknesses along Río San Juan transect. Also note onlap character of La Chilca Formation and the dramatic cut-out of section beneath Devonian Talacasto Formation.

eastern margin of the San Juan subbasin. The locally derived olistoliths and their inverse stratigraphy point to the formation of a rapidly subsiding, elongate depocenter and pronounced uplift along the flanks. The western flank separated this graben from the platform area to the west. However, it is important to note that glacially influenced sediments are present in the Talacasto subbasin and that the basal Silurian deposits in the Don Braulio section (above the glacial diamictites and beneath the sedimentary melange) are platform facies much like those of the basal Salto Macho Member of the La Chilca Formation. Hence it seems that for a short period during the latest Ordovician and the earliest Silurian a shallow-marine platform extended over much of the Talacasto subbasin and the adjacent San Juan subbasin and that there was no facies differentiation at that time.

On the basis of thickness distribution of the Silurian succession, Furque and Caballé (1990) interpreted the platform of the Talacasto subbasin to consist of north-south–trending basins and ridges. A depocenter existed in the Jáchal area, although there are strong local differences in thickness (Fig. 52). However, the reduction of thickness toward the Río San Juan is much more dramatic and is accompanied by an increase in the number of hiatuses (Benedetto et al., 1992; Herrera, 1993; Peralta, 1990). Concomitantly, the time span involved in each of these hiatuses increases. These hiatuses include the one at the base of the overlying Devonian Talacasto Formation; the hiatus also increases in magnitude toward the south (Fig. 53). Thinning of the sections toward the south seems to be mainly due to cut-out of section downward from the top (Fig. 53). This is true for the top of the La Chilca Formation as well as for the top of the Tambolar Formation (Fig. 52) and is indicated by basal, almost synchronous, conglomerates above the unconformities (Astini and Maretto, 1996). In contrast, the top of the successions beneath the unconformities locally shows strong variations in age (Franciosi, cited in Astini and Maretto, 1996), pointing to considerable local erosion (Figs. 52 and 53).

The formation of the depocenter along the eastern margin of the San Juan subbasin (Villicum graben of Loske, 1992) has been attributed to crustal extension (Loske, 1992; Von Gosen et al., 1995). Strong uplift along the western margin of the Villicum graben is thought to be responsible for the asymmetric geometry of the depocenter and the gravitational transport of the carbonate megaolistoliths (Amos, 1954; Cuerda, 1985; Von Gosen et al., 1995). Strong crustal extension with the formation of ridges (Zonda arc: Padula et al., 1967; Gonzalez Bonorino, 1975b) and basins (Villicum graben) affected the eastern margin of the San Juan subbasin during the Silurian. Hence it is plausible to assume that the same extensional regime also affected other parts of the Precordillera. The basins and ridges described by Furque and Caballé (1990) are here interpreted to be the result of (rotational?) block movements. However, these

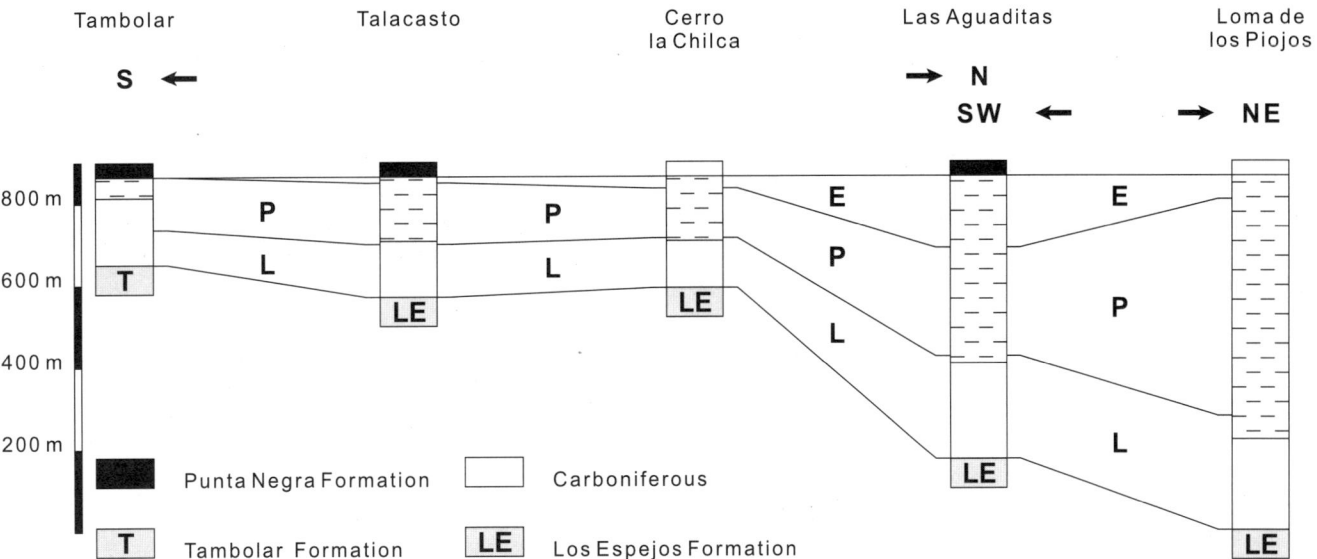

Figure 53. Thicknesses of Early Devonian Talacasto Formation show dramatic loss of section from north to south. This loss is caused mainly by loss of section beneath overlying Punta Negra Formation. L = Lochkovian; P = Pragian; E = Emsian (modified from Herrera, 1993).

movements were less effective and, in addition, had a south-north tilting component. This tilting toward the north created the Jáchal depocenter, in which most of the coarse detritus was trapped. Basin-parallel transport distributed the finer grained siliciclastic material toward the south. An apparent problem is that the depositional environment was deeper in the south than in the north. This, however, is explained by a sedimentation rate in the north capable of keeping-up with high accommodation, and insufficient sediment supply toward the south to fill inreasing accommodation there. A detailed model and a sequence stratigraphic interpretation were given by Astini and Maretto, (1996).

Devonian

In the western basin, east of Calingasta, several units are present that are assigned a Devonian age (Quartino et al., 1971; Selles Martinez, 1986). Their stratigraphy and sedimentology are insufficiently known to allow a comparison with the better known strata of the Talacasto subbasin. West of Mendoza, a thick succession of turbiditic sandstones is present, which is commonly referred to as Villavicencio Group (Harrington, 1957; Padula et al., 1967). This succession was assigned an Ordovician and Devonian age with probable Silurian parts (Cuerda, 1988). As demonstrated and discussed by Kury (1993), the Ordovician parts (Caradocian black shales) of the Villavicencio Group belong to the Empozada Formation and hence to the Los Sombreros Formation as defined in this paper. The turbidite succession of Devonian age constitutes the Villavicencio Formation in the sense of Kury (1993), which is equivalent to the Canota Formation of Cuerda (1988). Vascular land plant fossils indicate an Early Devonian age for these rocks (Cuerda, 1988). The turbidites were derived from the east; their composition indicates a recycled-orogenic and cratonal source area (Kury, 1993). The tectono-sedimentary environment of the Villavicencio Formation (sensu Kury, 1993) is interpreted as a passive continental margin or the continental side of a backarc system.

In the Talacasto subbasin, the Silurian-Devonian boundary coincides approximately with the formational boundary between the Los Espejos and Talacasto Formations (Herrera, 1993; Fig. 53). The Talacasto Formation was deposited on a marine shelf. The sedimentary succession within the formation is characterized by the transition from basal shale-dominated deposits toward sandstones near the top of the formation. The Talacasto Formation is overlain by the flysch-like deposits (Gonzalez Bonorino, 1975a) of the Punta Negra Formation. Biostratigraphic data from the base of the Punta Negra are somewhat sparse; however, the top of the Talacasto Formation is well dated (Herrera, 1993) and clearly shows that the topmost beds get older to the south, toward the Tambolar area. This trend is mirrored by dramatic decrease in thickness in the same direction. It is not clear whether this reflects a major erosional event prior to the deposition of the Punta Negra Formation or a strongly diachronous origin of the boundary.

The Punta Negra Formation consists of a thick succession of turbidites that was described by Gonzalez Bonorino (1975a), who reconstructed a fan-delta environment for the Punta Negra Formation. The composition of the graywackes and paleocurrent data indicate two source areas, one to the northeast of Jáchal that provided detritus from a mixed igneous and metamorphic source, and a metamorphic source area to the southeast of San Juan.

Everywhere in the Precordillera, a hiatus separates the Carboniferous sedimentary rocks from the Devonian or older strata. In many places this hiatus is represented by an angular unconformity.

Discussion of the Devonian evolution of the Argentine Precordillera

The Devonian strata record the change from crustal extension and continental drift to a compressional regime. This change is indicated in the geohistory plot curve for the Precordillera (Fig. 54) which shows a passive-margin evolution until the end of the Silurian and the beginning of the Devonian (Gonzalez Bonorino and Gonzalez Bonorino, 1991). During the Early Devonian, the shape of the curve changes and attains the shape typical of foreland basins (Kominz and Bond, 1986). This change is also reflected in the style of deposition and sediment composition. During the Early Devonian, there was the demise of the siliciclastic platform with fine-grained sediments. Subsequently, coarser grained sediments were deposited and, at the base of the Punta Negra Formation, sedimentation from turbidity currents became dominant. An accompanying evolution of sediment composition is observed with the change from quartzose sands during the Silurian (Loske, 1992; Astini and Maretto, 1996) toward arkosic sands in the Punta Negra Formation and the change in composition in the heavy-mineral spectra.

All sedimentologic data document the uplift of a cratonal source to the east of the Precordillera. West-directed thrusts and imbricates (Von Gosen, 1992; Von Gosen et al., 1995) affected the eastern margin of the Precordillera in post-Silurian but pre-Late Carboniferous time. Similar structures are observed in seismic lines, where they are interpreted to be located in the crystalline basement of the western Sierras Pampeanas (Cominguez and Ramos, 1990). Considering only the eastern basin of the Precordillera, the entire set of data presented here can be interpreted as the accretion of the Precordillera to Gondwana.

Concomitantly, however, the western margin was also affected by compression, documented by a metamorphic event between 425 and 410 Ma (Buggisch et al., 1994b) that affected the western margin of the Precordillera. The effects of this metamorphic overprint rapidly decrease toward the east (Keller et al., 1993c). The corresponding compressional deformation affected strata as young as the Devonian Talacasto Formation (Von Gosen, 1997). Hence, deformation along the western margin of the Precordillera is roughly coeval to compression along its eastern margin. Compression along the western margin of the Precordillera is interpreted to reflect the approach of Chilenia (Ramos et al., 1984, 1986).

GEODYNAMIC EVOLUTION OF THE ARGENTINE PRECORDILLERA AND ITS PARENTAL TERRANE

History of the Precordillera

Precordillera as part of Laurentia. Grenvillian-type rocks that are present beneath the Precordillera and in the western Sierras Pampeanas seem to be closely related to basement rocks of the southeastern Laurentian Grenville belt (Kay et al., 1996),

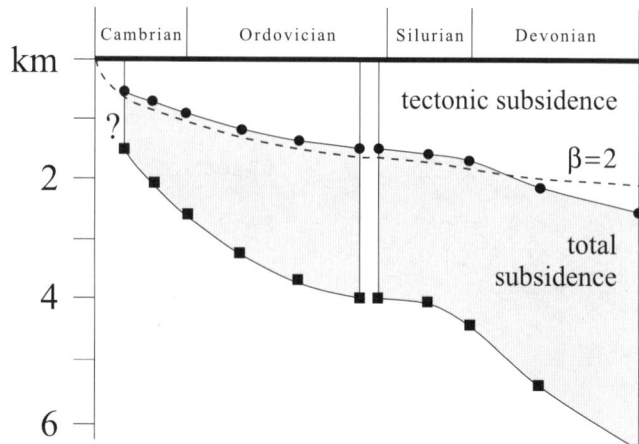

Figure 54. Thermal subsidence curve for lower Paleozoic sediments of Precordillera (modified from Gonzalez Bonorino and Gonzalez Bonorino, 1991). Note that not before latest Silurian are effects of a possible foreland evolution visible. Late Ordovician hiatus, interpreted to have been caused by continental breakup, does not affect shape of curve. Sketch courtesy of W. Buggisch.

especially the Llano uplift area (Fig. 6). Ramos et al. (1993) and Ramos (1995) took this evidence to postulate that the Precordillera and the western Sierras Pampeanas might have been part of southeastern Laurentia within the supercontinent of Rodinia (Fig. 5), which resulted from the Grenvillian orogeny.

The breakup of Rodinia started at about 725 Ma (Hoffman, 1991; Powell et al., 1993) along the western margin of Laurentia. For the Appalachian margin, rifting and continental breakup are inferred for the late Neoproterozoic and earliest Cambrian (Bond et al., 1984; Hoffman, 1991). Within the alternative models this would have been breakup of Pannotia (Fig. 5). The basic configuration of the early Paleozoic Appalachian-Ouachita passive margin of Laurentia was outlined by Thomas (1991, 1993). Along the Appalachian margin, the rift-drift transition occurred near the Precambrian-Cambrian boundary (Bond et al., 1984, 1989; Read, 1989; Thomas, 1996). Carbonate platform sedimentation started during the Early Cambrian, interrupted only by a short break reflecting the Hawke Bay regression event (Read, 1989; Fig. 48). The Ouachita margin displays a somewhat different history. The rift-drift transition seems to be younger than along the Appalachian margin (Thomas, 1996) and passive-margin sedimentation did not start before the Middle Cambrian. By Sunwaptan time, the entire Ouachita margin was the site of carbonate platform deposition. The oldest off-shelf sediments are also of Late Cambrian age.

Important features along the entire Appalachian-Ouachita margin are intracratonic fault systems that indicat incipient rifting. The most important fault system is known as the Mississippi Valley–Rough Creek–Rome graben system and is interpreted as resulting from crustal extension paralleling movements along the Alabama-Oklahoma transform (Thomas 1991, 1993). The graben fill of this system and that of the Birmingham graben (Thomas 1991) consists of Lower Cambrian fine-grained redbeds and local

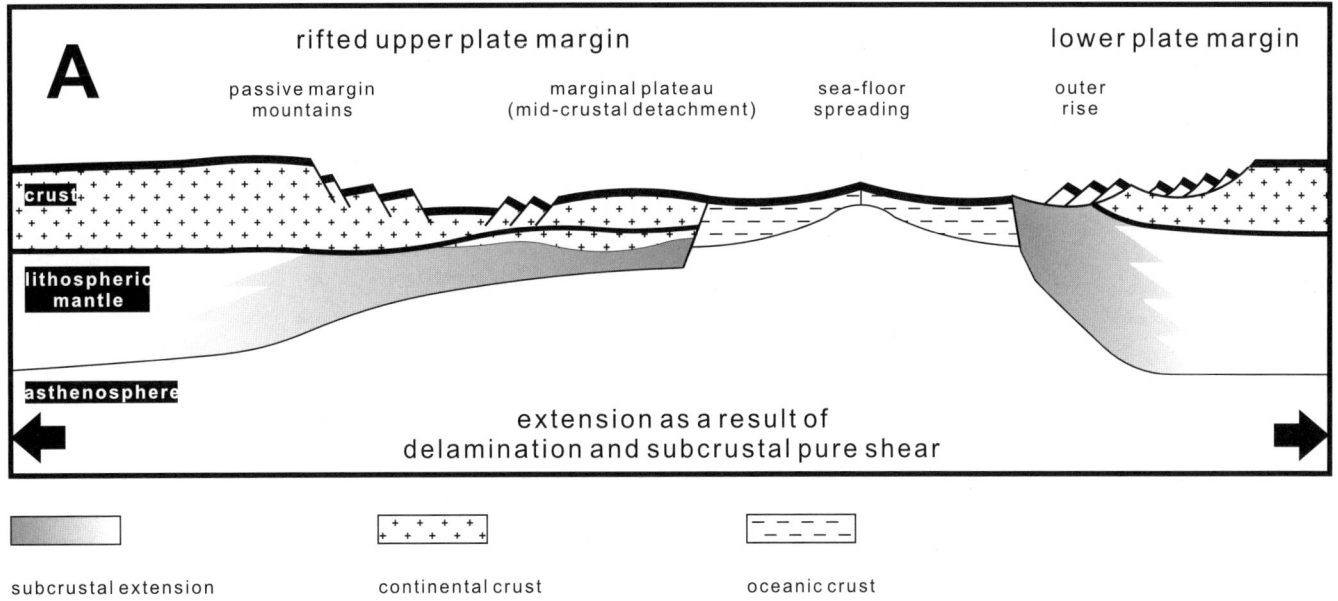

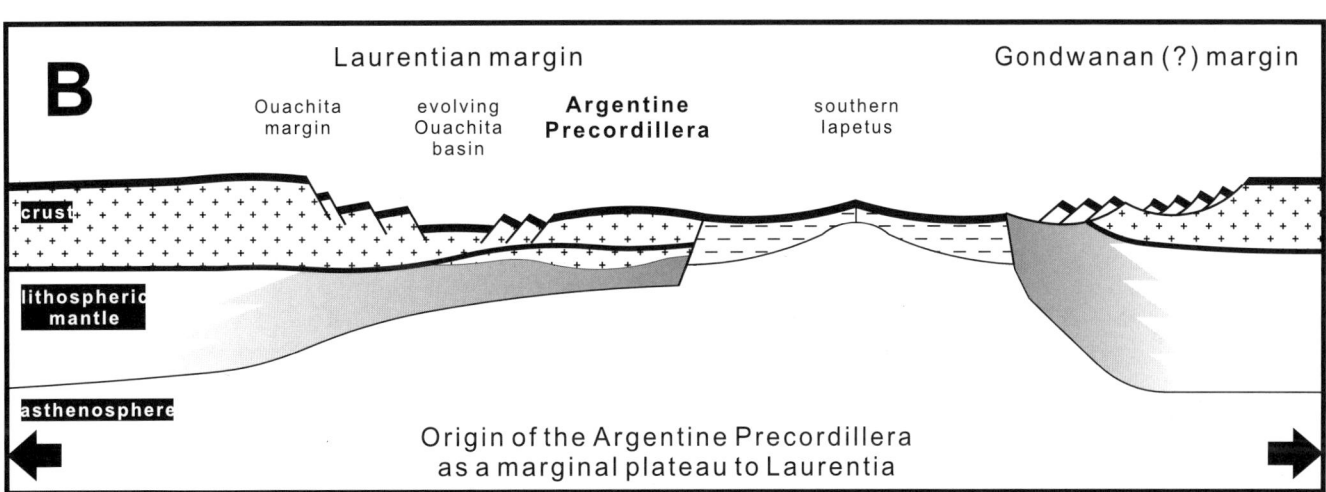

Figure 55. A: Principal and main components of low-angle detachment model (Lister et al., 1991) in case of extension as result of delamination and subcrustal pure shear. B: Application of model to Precordillera, which, during Cambrian-Ordovician time, is interpreted to have been marginal plateau to Laurentia. See text for further explanations.

evaporites overlain by marine siliciclastic material deposited until the Late Cambrian (Read, 1989; Thomas, 1991). Crustal extension documented by these successions, however, was not accompanied by the formation of oceanic crust. This is an important fact to remember for the discussion of the separation of the Argentine Precordillera from Laurentia.

Thomas (1993) applied the low-angle detachment model for continental rifting (Wernicke, 1985; Lister et al., 1991) to the eastern margin of Laurentia and concluded that the Ouachita margin represents an upper plate configuration. This interpretation is based on the general lack of synrift deposits (Viele and Thomas, 1989) and relatively thin passive-margin deposits (<1000 m). Subsequently, the discussion of the rifting geometry was amplified to include the Precordillera as part of the model, and Thomas and Astini (1996) concluded that the Precordillera and the Ouachita rifted margin constitute a conjugate rift pair, the Ouachita margin being the upper plate and the Precordillera being the lower plate.

Discussion. In low-angle detachment models of rifting of the continental lithosphere (Lister et al., 1991; Fig. 55A), lower plate margins are characterized by thick synrift sediments that are deposited in isolated, fault-bounded graben structures. Late synrift and early passive-margin strata unconformably overlie older deposits in the grabens and associated horsts (breakup unconformity of Falvey, 1974). Passive-margin successions are thick, much thicker than on the conjugate upper plate. The latter usually lacks thick synrift deposits; if they are present, they are laterally and temporally very restricted.

In the Precordillera, the Lower Cambrian redbed-evaporite succession of the Cerro Totora Formation (~300 m thick) together with the quartz-pebble olistolith in the Los Túneles section have been used as arguments for major rifting and a lower plate setting of the Precordillera (Astini et al., 1995; Thomas and Astini, 1996). However, there is strong evidence that, on a regional scale, the carbonate platform was established on the crystalline basement without an intervening succession of thick siliciclastic sediments (Keller, 1995). One of the main arguments is the abundance of locally derived angular basement blocks together with carbonate-platform olistoliths in the Ordovician slope facies. In contrast, in most sections, siliciclastic olistoliths that might be attributed to a siliciclastic cover sandwiched between the crystalline basement and the carbonates are absent. Considering all evidence, thick synrift sediments have not yet been proven to be present in the Precordillera; instead, redbeds and evaporites are localized facies. Hence the Precordillera rift is much like the Ouachita margin and, during its earliest history, seems to correspond more to an upper plate setting than to a lower plate setting. The thickness of the carbonate-platform succession, in contrast, matches that of mature passive margins in the Appalachians corresponding to lower plate settings. Hence the rift-drift history of the Precordillera reflects a composite history.

One approach to the tectono-sedimentary history of the Precordillera and, in part, to some aspects of the Ouachita history is the assumption that during Cambrian through Middle Ordovician time the Precordillera was a marginal plateau to Laurentia (Fig. 55B). Marginal plateaus (Lister et al., 1991) are large, relatively unstructured crustal fragments, which owe their characteristics to a mid-crustal detachment and the concomitant pull-out of middle and lower crust and, in places, of mantle material from beneath the future marginal plateau (Fig. 55A). Modeling of rifting and of continental breakup demonstrates that marginal plateaus are to be expected on both the upper and the lower plate, depending on the specific nature of the detachment. Marginal plateaus are separated from their main plate by a ramp syncline or hanging-wall basin, which may be relatively deep (Lister et al., 1991). Extension faults are present in the inboard basin; however, these faults are not shallowly dipping rotational faults.

The Texas plateau model of Dalziel (1997) is also based on the hypothesis that the Precordillera may have been a plateau marginal to Laurentia.

Argentine Precordillera as a marginal plateau to Laurentia. Marginal plateaus are more commonly developed on upper plate margins (Etheridge et al., 1989) than on lower plate margins. Upper plate margins are characterized by relatively thin synrift deposits (Lister et al., 1991) and hence it is no surprise that only areally restricted rift-related sediments are present in the Precordillera. The rift-drift transition along the Appalachian margin occurred during the late Neoproterozoic and earliest Cambrian (Bond et al., 1984; Hoffman, 1991) and might have affected the southern margin of Laurentia, the Ouachita area (Fig. 6), as well. However, thermal subsidence data (Bond et al., 1984) indicate that the rift-drift transition in the Ouachita embayment proper is younger than elsewhere. If the Precordillera was indeed a marginal plateau to Laurentia, then the main rifting took place *outboard* of the Precordillera and a passive margin must have developed there. This conclusion was also drawn by Dalla Salda et al. (1992b) in their reconstruction of the Occidentalia terrane. The same conclusion was independently drawn by Thomas (1991) in his discussion of the Ouachita rifted margin, before it became clear that the Precordillera might represent his microcontinental plate (see also Thomas, 1991, Figs. 6 c and d). The timing of the rift-drift transition outboard of the Precordillera might coincide with the main rifting events along the southern and eastern margin of Laurentia. Not much is known about this passive margin, which might be represented by the Caucete Group (Borrello, 1969) or Caucete metamorphics of Dalla Salda and Varela (1984). These sedimentary rocks consist of metaquartzites and metacarbonates that underwent their main metamorphic overprint in Ordovician through Devonian time. Upper Precambrian trace fossils have been described from these rocks (Bordonaro et al., 1992), which seems to support the considerations about the age of these poorly dated rocks. However, it seems surprising that trace fossils are preserved at all in these marbles, so that this datum should be taken with utmost caution.

In a later stage of rifting, crustal thinning led to the incipient separation of the Precordillera from Laurentia to form a marginal plateau (Fig. 55B). This stage correponds to the ridge-jump event in the model of Thomas (1991). Sedimentologic evidence from the Precordillera records the following steps (Fig. 56). Crustal thinning during the Early Cambrian is indicated by the redbeds and evaporites of the Cerro Totora Formation. From the late *Bonnia-Olenellus* chron onward, carbonate sedimentation prevails. The presence of a widespread hiatus indicates the presence of the Hawke Bay regression event in the Precordillera. During the *Glossopleura* chron of the Middle Cambrian, a fully developed carbonate factory became established. The subsequent history of the Precordillera marginal plateau mirrors that of the normal passive margins of Laurentia; however, as it consists mainly of unextended lower crust with rift-related uplift playing only a minor role (Lister et al., 1991), the plateau was rapidly flooded. During the latest Marjuman, much earlier than in other parts of Laurentia, siliciclastic input from a crystalline source area had ceased, testifying to the flooding of the plateau. From the Steptoean to the early Arenig, sedimentation of the carbonate platform proper was almost exclusively controlled by eustasy. The marginal-plateau model helps to explain the absence of margin-derived megabreccias and other mass-flow deposits in the Cambrian deep-water limestones. The gradient between the platform and the slowly subsiding basin to the west was not big enough to permit gravity-driven processes to operate on a large scale (Fig. 55B).

The basement of the Precordillera had been flooded by the Late Cambrian. However, the sequences deposited in cratonal

Argentine Precordillera: Sedimentary and plate tectonic history, Laurentian crustal fragment, South America 105

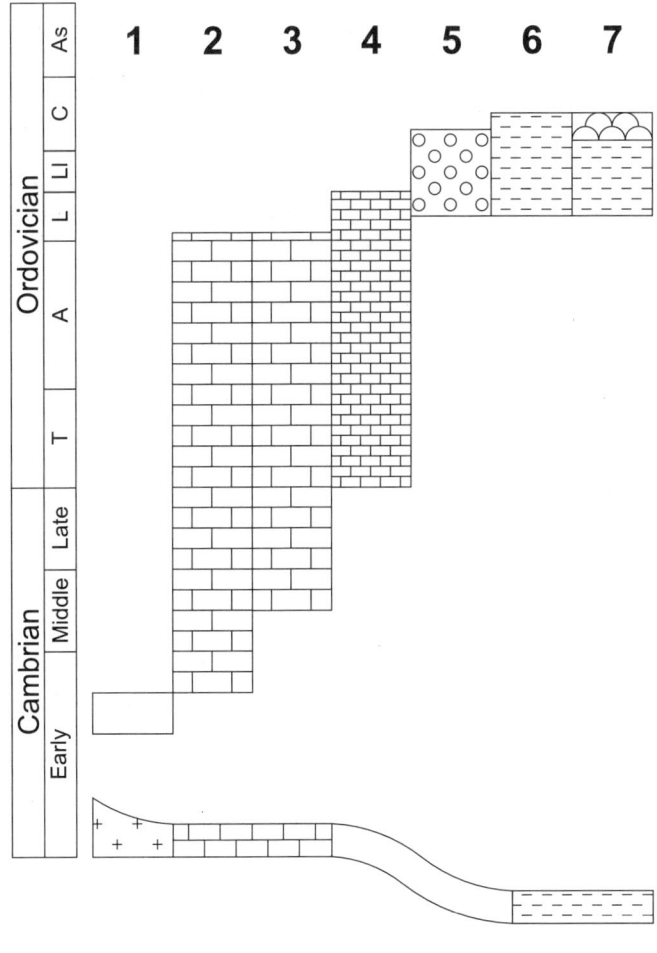

Figure 56. Evolution of sedimentary environments in Precordillera through time. 1 = Redbeds and evaporite in Guandacol subbasin as result of crustal stretching; 2 = carbonate platform deposits with passive-margin history; 3 = deep-water limestones; 4 = carbonate slope deposits as preserved in Los Túneles olistolith. 5 = continental rise deposits with rock-fall and turbidite sediments along nascent continental margin; 6 = distal turbidites and basinal shales. 7 = basinal shales and intercalated pillow basalts.

positions must have been thin and repeatedly were subject to erosion, dominantly during eustatic sea-level events. This is shown by Cambrian conodonts in the San Juan Formation, which can only have been derived from a cratonal section. This is also shown by the Tremadocian onlap of carbonates onto crystalline basement near San Rafael (Fig. 42B).

The low relief on the marginal plateau is responsible for the absence of mass-flow deposits in the Cambrian deep-water limestones. These limestones seem to have been deposited on a very gently inclined ramp situated between the Precordillera platform and the incipient Ouachita embayment, which represents a hanging-wall basin (Lister et al., 1991). The main dip was toward the cratonward Ouachita basin.

This basin, which was established during the *Glossopleura* chron, separated the Precordillera from the Ouachita margin; it is probably underlain by attenuated continental crust (Fig. 55B), a necessary assumption for the model of a marginal plateau (Lister et al., 1991). However, as documented by Dalziel (1997), the Falkland-Malvinas Plateau, a marginal plateau to South America, is separated from the main continent by a deep graben floored by oceanic crust. Oceanic crust with an approximate age of 565 Ma has been described from the suture between Chilenia and the Precordillera near Uspallata (Fig. 2; Davis, 1997). It is not clear to me whether this crust was part of the Precordillera or whether it represents obducted ocean floor of the Chilenia terrane. In any case, its presence does not necessarily contradict the argumentation followed here. For the time being it is assumed that there is no definite proof of Cambrian oceanic crust in the Precordillera.

During Late Cambrian through Ordovician time the basin accommodated the off-shelf facies of the Ouachita embayment (Fig. 57, A and B). These facies were described by Lowe (1985, 1989) and McBride (1989), who pointed out the divergent interpretations of the Ouachita sediments as being shallow-water deposits or deep-water deposits, respectively. As discussed by Keller and Dickerson (1996), the assumption that the Precordillera was situated outboard of the Ouachita embayment during Cambrian-Ordovician time and that the Ouachita embayment developed above a rift-related fault-block topography with horsts and grabens (Fig. 57, A and B) gives a good explanation for the opposing interpretations.

Geophysical evidence suggests the presence of highly attenuated continental crust or oceanic crust to the southeast of the Alabama-Oklahoma transform (G. Keller et al., 1989). These data have been taken as evidence of the opening of a Cambrian ocean in the Ouachita embayment (Thomas 1991, 1993). They have been taken also as evidence of a Cambrian separation of the Precordillera from Laurentia by Astini et al. (1995, 1996b). The positive Bouguer anomalies characteristic of this attenuated or oceanic crust delineate the interior zone maximum, which extends all along the interior side of the Ouachita orogenic belt (Fig. 6). This Bouguer anomaly is taken as an indicator of the early Paleozoic margin of Laurentia in this part of the North American craton (Viele and Thomas, 1989). To my knowledge, there is no evidence as to the age of the deeply buried crust described by G. Keller et al. (1989). Lowe (1985) argued that the Ouachita off-shelf rocks were deposited in a failed rift that was a two-sided basin. The southern continental block contributed quartz-rich sediments and magmatic pebbles to the siliciclastic facies of the Ouachita trough. Consequently, Lowe (1985, 1989) argued that the original continental margin of Laurentia was to the south of this continental block. The marginal-plateau model seems to be a viable solution to the contrasting views of the nature of the southern Laurentian margin (Fig. 57). A passive margin must have bordered the Precordillera to the present-day east while it was attached to Laurentia as a marginal plateau. This passive margin is the equivalent of the Late Proterozoic–Early Cambrian passive margin as present in the Appalachians. The inboard basin of the Precordillera plateau is represented by the Ouachita trough (Fig. 57A) in the sense of Lowe (1985), who concluded that this basin might have been a narrow structure.

The Precordillera and Laurentia shared a common history as part of the same plate until the Middle Ordovician, when renewed rifting led to the final separation of the Precordillera from Laurentia (see later discussion). This rifting produced oceanic crust, part of which is preserved in the Precordillera (Haller and Ramos, 1984). A major proportion of this crust, however, remained with Laurentia and today may be represented by the interior zone maximum. This zone marks the "... gravity signature of the early Paleozoic margin of the North American craton from the Ouachita Mountains to the Marathon region of Texas" (Viele and Thomas, 1989: p.699). The absence of age data on this crust facilitates the interpretation that the crust described in the subsurface might be of Ordovician age.

The model outlined here is based on sedimentologic and regional evidence, mainly from the Precordillera but also from the Ouachita embayment. On the basis of paleomagnetic data and global plate reconstructions, Dalziel (1997) also developed a model showing a promontory at the southeastern side of Laurentia during Late Proterozoic and earliest Paleozoic time. The model presented here and the Texas plateau model of Dalziel (1997) agree in the timing of the final separation of the Precordillera from Laurentia during the Caradoc, as originally proposed by Dalla Salda et al. (1992a, 1992b; 1993) but differ in the timing of collision with or accretion to Gondwana.

Continental breakup and the rift-drift transition. The carbonate platform of the Precordillera was drowned during the early Llanvirn; however, in the Guandacol subbasin this event is older and is dated as late Arenig. Drowning of the carbonates in the Guandacol area is considered mainly as tectonically triggered (see Astini et al., 1995) because it is coeval with shallowing in other parts of the basin. The drowning event in the Guandacol subbasin is here interpreted to be the first expression of the onset of crustal extension and rifting that finally led to the separation of the Precordillera from Laurentia.

A comprehensive model for rifting, continental breakup, and subsequent drifting of continental margins was proposed by Falvey (1974), who postulated two important unconformities on extensional margins (Fig. 58): a rift-onset unconformity and a breakup unconformity. The rift-onset unconformity is the result of erosion due to thermal expansion of the crust. Thermal expansion is followed by the formation of grabens and horsts characteristic of the rift-valley stage. The breakup unconformity separates the rift stage from the passive-margin stage and is interpreted to be approximately coeval with the onset of ocean-floor formation. This unconformity is mainly recognized in the continental-margin sediments and on the shelf adjacent to the newly opened ocean. Embry and Dixon (1990) developed several criteria for the recognition of the breakup unconformity, e.g., an abrupt change in subsidence rates below and above the breakup unconformity. Pre-breakup strata are mainly deposited in actively subsiding half grabens that are bound by steep faults cutting into the basement. Adjacent horst structures are subject to considerable erosion; however, they may also be covered with thin sediments. Volcanic strata related to synrift extension are present predominantly below the unconformity.

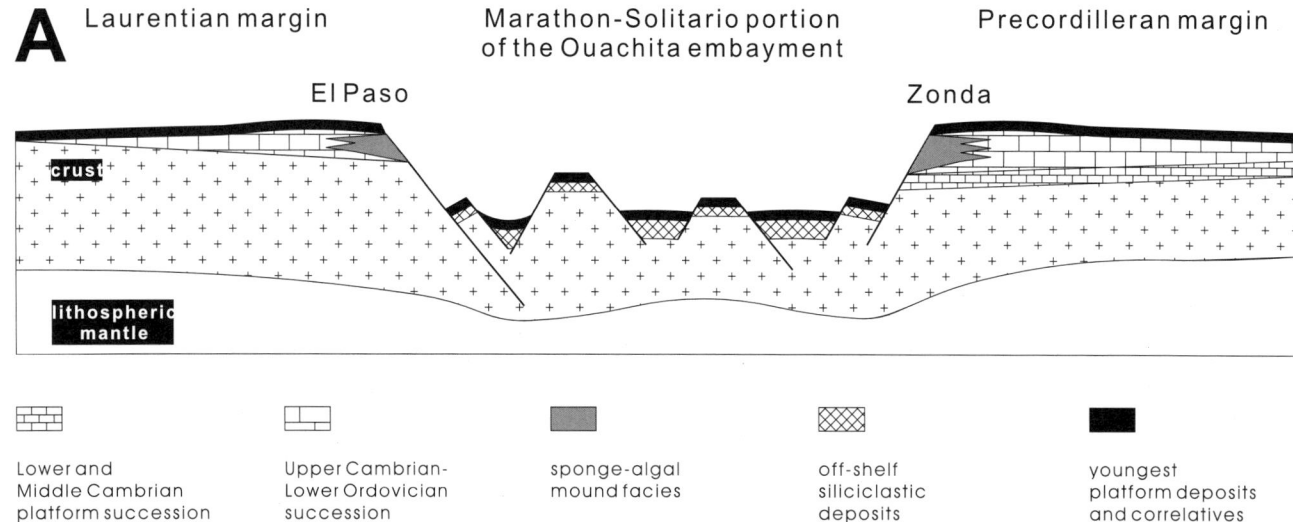

Figure 57 (this and opposite page). A: Cross section from Laurentian craton across Marathon-Solitario portion of Ouachita embayment to carbonate platform of Precordillera during Late Cambrian and Early Ordovician time. Pinpoint for correlations is sponge-algal facies as exposed in Scenic Drive section near El Paso and in Sierra Chica de Zonda of Precordillera. Fault-block geometry in Ouachita embayment is responsible for highly varied sedimentary successions, in part interpreted to be of shallow-water origin, in part interpreted to be of deep-water origin. B: Same cross section during Middle and Late Ordovician time shows effects of continental breakup. Precordillera is drifting away, and oceanic crust is formed between Precordillera and El Paso. Ouachita margin of Laurentia and eastern basin of Precordillera both show crustal extension and concomitant erosion of fault blocks. Marathon-Solitario basin underwent major deepening; large Late Ordovician hiatus is interpreted to reflect breakup unconformity. For details see Keller and Dickerson (1996) and Dickerson and Keller (1998).

The Middle and Upper Ordovician rocks of the Precordillera were deposited under a regional extensional regime (Spalletti et al., 1989; Loske, 1992; Von Gosen, 1992; Keller, 1995). Horst and graben structures are interpreted to be responsible for the highly varied sedimentary succession in the eastern basin. As outlined earlier, the Guandacol tectonic event marks uplift and erosion of individual blocks and the onset of deposition in grabens. Consequently, in the context of the Falvey (1974) model, the surface separating the black shales (Gualcamayo Formation and equivalents) from the overlying conglomerates is a likely candidate for the rift-onset unconformity (Fig. 58). Deposition of the Las Vacas and La Cantera conglomerates represents the rift-valley stage of sedimentation, which continued into the Late Ordovician. Concomitantly, the carbonate platform was eroded to highly varied levels, but in a later stage, the Sassito limestones were deposited on one of the horst structures (Fig. 58). Boulders in the Los Sombreros Formation prove the exposure of the basement and of deep-seated faults affecting the entire platform succession and cutting down into the basement (Fig. 58).

In the eastern basin, sedimentation in grabens and half grabens continued into the Caradoc. In most sections of the eastern basin, sediments younger than the *N. gracilis* zone are conspicuously absent. The timing of the onset of sea-floor spreading in the western basin is not well established. Pillow lavas extruded into the Caradocian Alcaparrosa Formation (Fig. 56) and are interpreted to represent part of an ophiolite complex (Haller and Ramos 1984). Coeval basic magmas that are geochemically comparable to those of the Precordillera extruded in the San Rafael area (Davicino and Sabalúa, 1990). Because pillow basalts are typical components of ophiolite suites, which testify to the former existence of oceanic crust, the pillows basalts of the western basin might represent part of such an ophiolite suite, although their chemical signature is not unambigious (Kay et al., 1984). Following Falvey (1974), the extrusion of the pillow basalts as part of developing oceanic crust should be accompanied by lateral heat flow causing uplift and erosion on the adjacent continental margin. The widespread late Caradocian–Ashgillian hiatus (Fig. 4) separating the older Ordovician rocks from the uppermost Ordovician and Silurian La Chilca Formation is here interpreted to reflect the erosional event during continental breakup. The breakup unconformity (Fig. 58) is marked by the chert-pebble conglomerate and associated facies at the base of the classical Silurian succession in the Talacasto subbasin. Identical rocks, although slightly younger, are also found in the San Juan subbasin (Don Braulio section), demonstrating that the localized facies development during the Middle and Late Ordovician had given way to uniform sedimentary environments just above the breakup unconformity.

In the Don Braulio section, two unconformities are present, one between the La Cantera Formation and the diamictites of the Don Braulio Formation and one separating these diamictites from the overlying upper member of the Don Braulio Formation (Fig. 28). The absence of much of the Caradoc and a part of the Ashgill below the diamictites indicates that the unconformity marking the La Cantera–Don Braulio contact corresponds to the breakup unconformity recognized elsewhere (Fig. 58). The upper unconformity is correlative to the global sea-level fall accompanying the Late Ordovician glaciation. It is the subse-

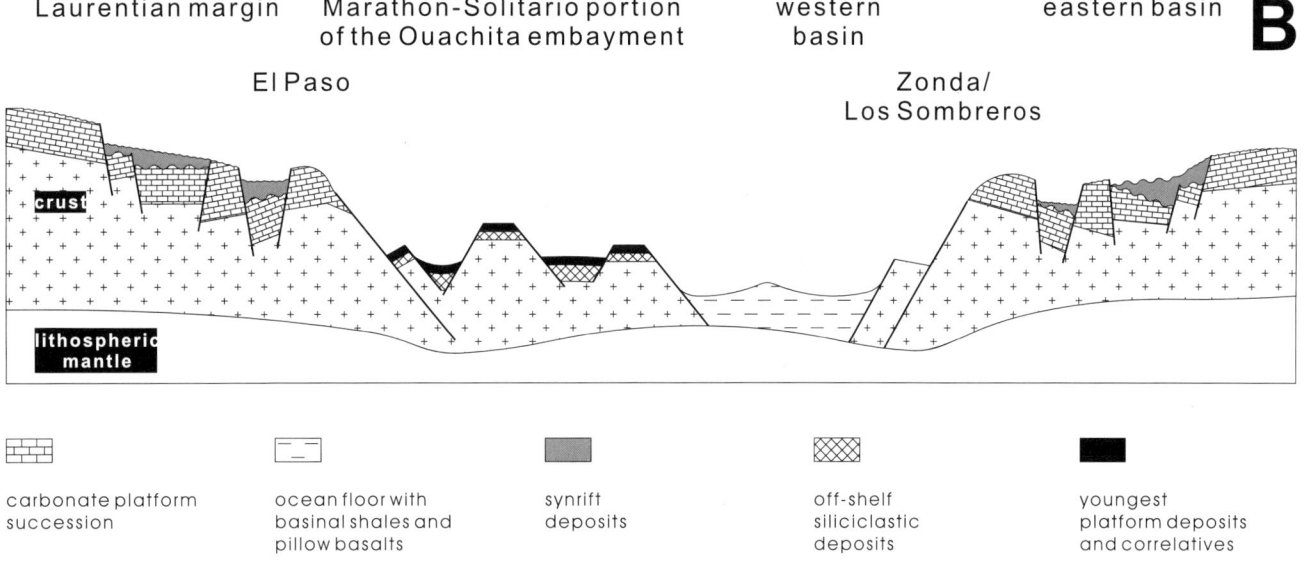

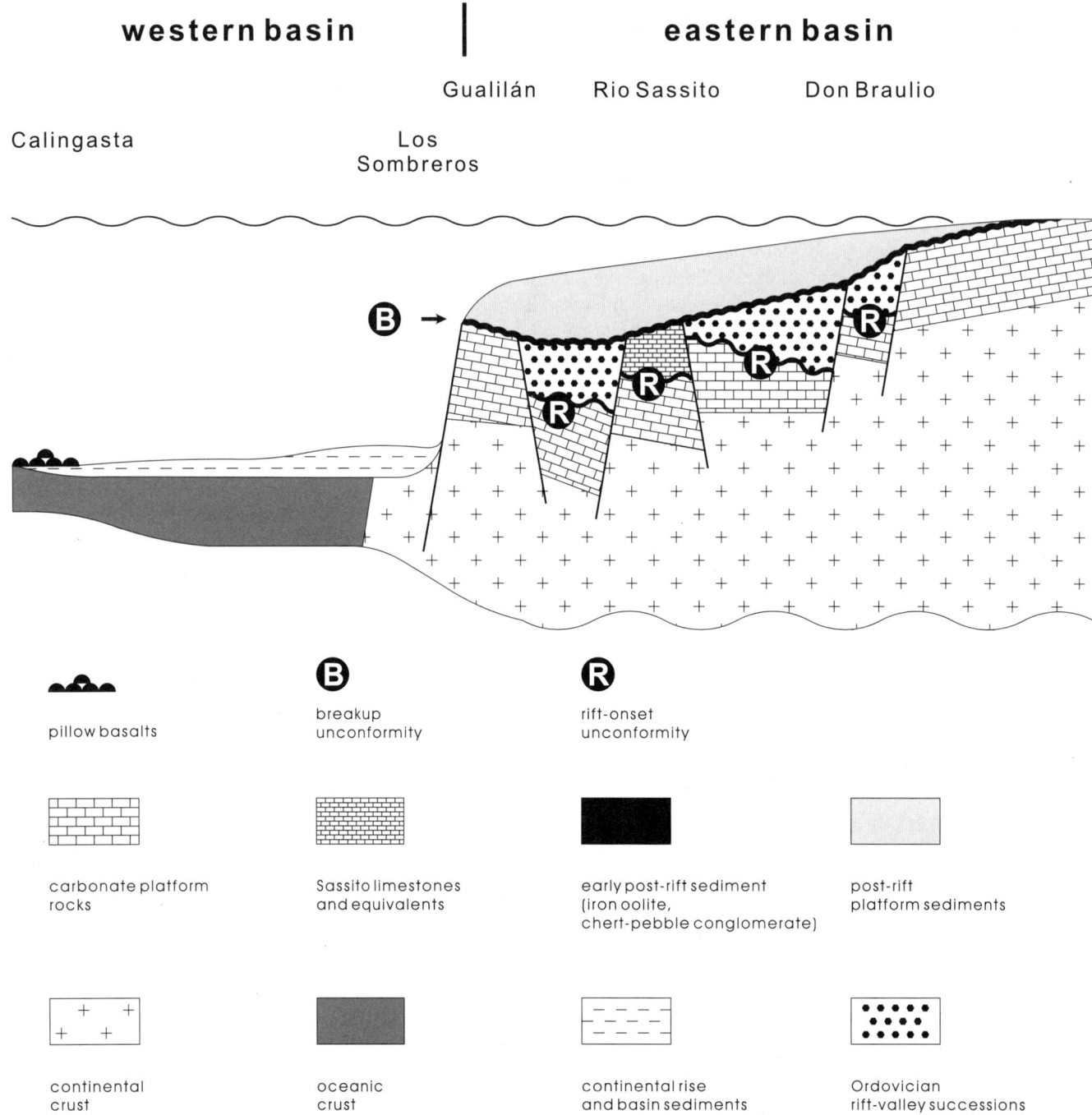

Figure 58. Falvey (1974) model of continental breakup, applied to Precordillera. Note that there are places where rift onset and breakup unconformities merge. See text for further discussion.

quent sea-level rise that was responsible for the flooding of the entire terrane. Looking at the Don Braulio section alone, the sedimentary succession deposited during this sea-level rise can be interpreted as reflecting a glacio-eustatic rise. However, if the sections in the Talacasto subbasin are incorporated, this sedimentary succession of latest Ordovician age represents the post-breakup succession. This shows that the Late Ordovician history of the Precordillera was controlled by the interplay between eustatic events and tectonic events.

Discussion. The Middle and Upper Ordovician rocks of the Precordillera have been interpreted as part of a clastic wedge related to the Middle Ordovician collision of the Precordillera with Gondwana (Astini, 1992; Astini et al., 1995, 1996b). However, there are no compressional structures older than the Devonian in the Precordillera; in contrast, the Ordovician was dominated by crustal extension. It might be argued that these phenomena are caused by extension on a subducting plate and especially the passage of the peripheral bulge. The passage of a peripheral bulge might cause uplift of as much as some 300–500 m (Jacobi, 1981). In the Precordillera, this is contrasted by the formation of an almost linear fault scarp locally more than 2100 m high, which seems to be incompatible with the passing of a peripheral bulge. In addition, the thickness of typical clastic wedges is several thousand meters. This generalization is true for, e.g., the late Paleozoic Marathon, Valverde, and Arkoma basins (Fig. 6) in the provenance area of the Precordillera, the Ouachita embayment (Houseknecht, 1986, Wuellner et al., 1986). Hiscott (1995) described 4000 m of Taconic flysch in the northern Appalachians; the flysch is broadly coeval with the presumed collision of the Precordillera with Gondwana and its timing as proposed by Dalla Salda et al. (1992a, 1992b) and Dalziel et al. (1994). These thicknesses are contrasted by a mere 150 m of the La Cantera Formation. Other objections to a Middle Ordovician collision include the composition of the heavy-mineral spectra of the Ordovician rocks (Alcaparossa Formation, Los Sombreros Formation, Don Braulio Formation, among others; Fig. 45). Most important *all* faunal data from benthic and even from pelagic organisms directly contradict this Middle Ordovician collision (Fig. 43). The geohistory plots for the Precordillera of Bond et al. (1984), Gonzalez Bonorino and Gonzalez Bonorino (1991; Fig. 43), and Williams (1997) do not show any excursion from passive-margin evolution until the Silurian. Following Williams (1997), it is unlikely that the shape of the subsidence curve of the Precordillera reflects an Ordovician collision.

The observed phenomena in the Middle and Upper Ordovician rocks are much better explained by rifting than by continent-continent collision. This rifting is responsible for the separation of the Precordillera from Laurentia (Fig. 56) and for the concomitant formation of a continental margin in the Precordillera. This continental margin is marked predominantly by the deposits of the Los Sombreros Formation and by the Portezuelo del Tontal Formation (Spalletti et al., 1989). The history of the marginal plateau of the Precordillera and the evolution of depositional environments during rifting become very evident if we focus on the oldest sediments of each environment. Figure 54 reflects this history of incipient crustal stretching during the Early Cambrian, carbonate platform evolution until the Llanvirn, and the formation of the western basin as a rift structure.

The fault scarp, which separates the platform from the nascent basin (Fig. 58), resembles a marginal escarpment (Drake and Burk, 1974) that bordered the carbonate platform rocks toward the evolving western basin. This basin is characterized by attenuated continental crust, which seems to pass into oceanic crust (Fig. 58). As argued earlier, there are no indications of an Ordovician carbonate platform margin or a platform slope at the present western limits of the carbonate outcrops. The platform must have extended farther west; hence it seems reasonable to assume that continental breakup during the Ordovician occurred within the platform and that a part of the carbonate platform was left behind on the conjugate margin. Consequently, this block might have been a source for the enigmatic southward-derived Middle Cambrian boulders in the Pennsylvanian Haymond Formation of the Ouachita orogen, which were described by Palmer et al. (1984). This remnant might have been also the source for the southward-derived boulders in the Ordovician deposits described by Young (1970).

In the Precordillera, a succession of turbidites and shales was deposited in the rapidly deepening western basin (Fig. 51). Close to its eastern margin, east to west-directed transport prevailed (Spalletti et al., 1989) and by far the coarsest sediments are observed along this margin. There is a general fining westward of the sediments accompanied by a change in transport directions into a more northeast to southwest direction. Astini (1991) reports north to south transport in the Don Polo Formation, implying margin-parallel (and consequently basin-parallel) transport. According to their lithology and sedimentology, many of the thin sandstone beds bear characteristics of contourites, as discussed by Reineck and Singh (1980). A transect across the Ordovician passive margin of the Precordillera (Fig. 51) almost mirrors the present-day North Atlantic margin, e.g., as discussed by Heezen (1974). Along the North Atlantic margin, the continental shelf drops down to the continental rise along the slope. Relief is on the order of several hundred meters to more than 1000 m as in the Precordillera, where a marginal escarpment was formed. Turbidites are the major source for continent-derived detritus, which is deposited on the continental rise. Farther offshore, longshore currents become increasingly more abundant and redistribute the sediment to form contourites. In areas not affected by contour currents, thick shales accumulate on the continental rise (Heezen, 1974). In the Precordillera, much more research on the western clastic basin is needed to include all units present there into a model of sedimentation along a passive continental margin during Ordovician time.

Keller and Dickerson (1996), in their discussion of the joint history of the Precordillera and the Marathon-Solitario area of the Ouachita embayment, concluded that the Middle and Upper Ordovician sediments as well as the important hiatus observed in the Ouachita strata are related to rifting and the subsequent opening of an ocean between the Precordillera and Laurentia (Fig. 57A). Here the idea is put forward that the Caradocian hiatus in the Marathon-Solitario area corresponds to the breakup unconformity observed in and described from the Precordillera.

The early history of the Precordillera is here interpreted as a marginal plateau to the evolving Ouachita margin (Figs. 55 and 57). This margin was interpreted as an upper plate margin (Thomas, 1993). Remnants of the upper plate, such as marginal plateaus, are commonly bounded by cratonward-dipping major faults (Etheridge et al., 1989; Lister et al., 1991), which typically develop in late stages of crustal extension. Hence these faults might be preserved as the west-dipping master faults (Von Gosen, 1992, 1997) of the Middle and Late Ordovician rifted margin (Figs. 57B and 58). These faults probably originated from the

Cambrian rifting event, implying that these faults are old features that were reactivated during the separation of the Precordillera from Laurentia.

Approach of the Argentine Precordillera toward Gondwana. The following chapter does not intend to solve all the problems and uncertainties with respect to the accretion of the Precordillera to Gondwana and its relations to the Sierra de Famatina and the Sierras Pampeanas. This would be far beyond the scope of this paper. Instead, some of these problems are outlined and a few facts are combined to give a general idea about the possible timing and mechanisms of the accretion of the Precordillera to Gondwana.

The current models for western South American geology argue for a Middle Ordovician collision between the Precordillera and the Sierras Pampeanas. Dalla Salda et al. (1992a, 1992b, 1993) include this collision in a much larger framework of an overall Laurentia-Gondwana collision. Astini et al. (1995, 1996a) favor an accretionary event between their Precordillera terrane, which during the Early Cambrian had become an independent microplate, and the Sierra de Famatina during the mid-Llanvirn to Llandeilo.

The most recent geochronological and geochemical data (Rapela et al., 1998) from the Famatinian mobile belt, however, suggest that this belt formed as a response to east-directed subduction beneath western Gondwana, starting at about 490 Ma and lasting until 450 Ma. The Famatinian magmatic arc was rooted on continental crust, its plutonic rocks show almost identical Sm-Nd crustal residence ages to those of the Cambrian plutons of the eastern Sierras Pampeanas. It seems that there was no "contamination" from another plate with continental crust, be it Laurentia or a Laurentia-derived terrane. Consequently, Rapela et al. (1998) argued in favor of a Siluro-Devonian accretion of the Argentine Precordillera to Gondwana.

In the Precordillera, neither plutonic nor deformational events are preserved that might be related to the Ordovician events observed both in the Sierras Pampeanas and the Sierra de Famatina. The only vestiges of magmatic activity are the intrusion of quartz monzonites and monzogranites at 481 ± 6 Ma (Rapela et al., 1998; Pankhurst and Rapela, 1998), the K-bentonites in the San Juan Formation, and the volcanic breccias in the San Rafael area. The oldest deformational structures hitherto described affect strata as young as the Early Devonian Talacasto Formation in the Jáchal area (Von Gosen, 1997).

Along the eastern margin of the Precordillera, steeply dipping carbonate platform rocks and the Silurian rocks are unconformably covered by Carboniferous sediments. Deformation is estimated to have taken place in post-Late Silurian (Rinconada Formation) but in pre-Late Carboniferous times (Von Gosen, 1992; Von Gosen et al., 1995). Deformation was east-west directed and produced west-vergent structures. Hence this deformation is mostly coeval to the deformation in the west and produced similar structures. Corresponding effects of this deformation are unknown in the Talacasto and Guandacol subbasins, a problem already addressed by Baldis and Chebli (1969).

Discussion. In the Precordillera, there are no compressional structures that can be interpreted as being related to an Ordovician collision of the Precordillera with Gondwana. If we accept the evidence that the Middle and Upper Ordovician sedimentary succession of the Precordillera is related to crustal extension rather than compression, then a Middle Ordovician collision of the Precordillera with Gondwana is highly unlikely.

Larvae of modern brachiopods and arthropods are able to migrate about 1000 km and if the actualistic approach to fossil faunas is valid, the faunal evidence from within the Precordillera is a critical test to the model discussed here and the alternative paleogeographic scenario. Paleomagnetic data allow a maximum separation of Laurentia and Gondwana at the latitude of the eastern Sierras Pampeanas of ~2500 km (Dalziel, 1997) between 475 and 465 Ma, i.e., prior to the Caradocian rifting. If we consider maximum separation of 2500 km (Fig. 59C) between Laurentia and Gondwana during the Middle Ordovician and an intervening ocean 1500 km wide, then there remain 1000 km (probably the maximum terrane width) to accommodate the Precordillera in its position attached to the Ouachita margin and still prevent faunal exchange. If Laurentia and western Gondwana were juxtaposed during the onset of rifting during the late Middle Ordovician (Fig. 59, D and E), a new ocean opened to the west but the ocean to the east was successively closed. In an intermediate position within the oceans on either side, faunal exchange with both Laurentia and Gondwana was possible. With increasing distance to Laurentia, Laurentian faunas rapidly disappear.

More realistically, the majority of the larvae may travel a distance between 1000 and 2000 km, some as much as 3000 km (discussion in McKerrow and Cocks, 1986). Maintaining maximum separation between the Laurentian and Gondwanan plate but increasing the possible range of larval migration to about 1500 km, then the distance between Gondwana and the Precordillera was not insurmountable for the larvae. Why then was there no faunal exchange between at least the Precordillera and Gondwana? In contrast, faunal exchange with Avalonia and Baltica is observed, two terranes which were (following the reconstructions of Dalziel, 1991, 1997) in a far more distant position. In addition, if the Precordillera was trapped between Gondwana and Laurentia, the Precordillera might have served as a stepping-stone for the further migration of the larvae toward Laurentia; but hitherto, no typical Gondwanan faunal elements have been reported from the Appalachian margin. Was there some climatic or ocean-current control that prevented larval migration from one terrane to the other? Furthermore, if the Precordillera had been accreted to Gondwana during the Middle Ordovician, why are the faunas composed of warm- to temperate-water elements, whereas the Gondwanan faunas are temperate- to cold-water dominated?

Vaccari (1995), on the basis of trilobite studies, explicitly stated that during the Arenig, the Precordillera and the Sierra de

Famatina must have been separated by an ocean wider than 2000 km. Although we don't know the effects of oceanic circulation during the Ordovician, if a drift rate of 15 cm / yr is assumed which is within the speed limit even of large plates (Meert et al., 1993), then 10 m.y. are needed to bring the two areas within a range of 500 km and some 13 m.y. until the ocean basin is entirely closed. Even if 20 cm / yr are taken as the drift rate, 10 m.y. are needed to close the ocean. If we accept the terminal Arenigian distance of 2000 km as a starting point and the drift rates discussed here, then a Middle Ordovician collision is, at the very least, difficult to explain. These drift rates, in contrast, show that by Late Ordovician time the Precordillera may have moved to high latitudes, as documented in the sediments by the Ashgillian glacial diamictites (Peralta and Carter, 1990a).

A Middle Ordovician collision between the Precordillera and Gondwana seems highly unlikely. Consequently, the Precordillera must have been accreted to Gondwana during a later event. The following facts and observations from within the Precordillera and from outside, compiled by Keller et al. (1998) and Rapela et al. (1998), may help in elucidating the timing of the accretion to Gondwana.

1. In the Precordillera, compression is documented for the Late Silurian, Devonian, and pre-Late Carboniferous interval on both sides of the Precordillera (Von Gosen, 1997). The western structures, which are found in strata as young as the Lower Devonian Talacasto Formation, are supposedly related to the collision of Chilenia (Ramos et al., 1984, 1986). This collision apparently closed the western Ordovician ocean, leaving behind the rocks of the Caradocian ophiolite complex (Haller and Ramos, 1984) as a marker of the suture between the Precordillera and Chilenia. The corresponding metamorphism was dated as Late Silurian to Early Devonian (Buggisch et al., 1994b).

In the southern segment of the Sierras Pampeanas, compressional deformation and a corresponding metamorphism were described by Sims et al. (1998) and Von Gosen and Prozzi (1998). The latter concluded that this deformation was probably caused by the accretion of the Precordillera.

2. The heavy-mineral spectra (Fig. 45) show a dominance of a low-diversity–low-quantity association in Ordovician and Silurian sedimentary rocks (Loske, 1992). Such associations are often derived from a sedimentary source that might already show a depleted spectrum. A high-diversity–high-quantity association indicates uplift and erosion of igneous and metamorphic rocks starting during the Early Devonian and being dominant during the deposition of the flysch-like Punta Negra Formation. According to Loske (1992), the increase in quantity also implies an increase in relief in the hinterland together with a less intensive transport separation. Detrital zircon populations from the Punta Negra Formation are different from those of the Talacasto Formation and include 1.1 Ga zircons (Loske, 1995). This indicates that during deposition of the Punta Negra Formation the Grenvillian-type basement of the terrane contributed significantly to the terrigenous input. Paleocurrent indicators (Gonzalez Bonorino, 1975a; Kury, 1993) and grain-size distribution prove an overall eastern source area for the detritus. Ramos et al. (1998) interpreted the evolution of the Punta Negra foreland basin as being related to the approach of Chilenia in the west. It is curious that in this model main sediment transport is toward the suture and collision zone and not away from it, as one might expect.

3. Carboniferous sedimentary rocks unconformably overlie older sedimentary rocks, not only in the Precordillera and its parental terrane, but also in many of the adjacent areas. Hence the Carboniferous sediments are the overstep succession to the lower Paleozoic deformation and to the accretion of terranes. The molasse character of the Carboniferous sediments has been well established since the time of Bodenbender (1911).

4. In the eastern Sierras Pampeanas, there was an important peak in magmatic activity during the Late Devonian and Early Carboniferous (Rapela et al., 1992). The accompanying granites are postdeformational or posttectonic granites. In contrast, in the western Sierras Pampeanas the basement was reactivated between ca. 430 and 390 Ma (Ramos et al., 1996). This reactivation is coeval to the metamorphism observed in the western basin (Buggisch et al., 1994b).

The collisional history along a transect at 32°S was outlined by Ramos et al. (1986) and Ramos (1988a). Taking these outlines and the evidence, the accretion of the Precordillera to Gondwana and to the eastern Sierras Pampeanas took place in a succession of events. The earliest signs of deformation are reflected in the metamorphism of the siliciclastic strata in the western basin. Concomitantly, the basement of the western Sierras Pampeanas was reactivated (Ramos et al., 1996). Compression in higher crustal levels is recognized in the Devonian, when deformation affected not only the basinal siliciclastic rocks, but also the siliciclastic Talacasto Formation of the eastern basin in the Jáchal area (Von Gosen, 1997, personal commun.). Timing of the deformation along the eastern margin of the Precordillera is much more difficult to elaborate because of the absence of Devonian sedimentary rocks there. As argued by Gonzalez Bonorino (1975a), a proximal facies of the Punta Negra Formation must have been deposited in the San Juan subbasin. The flysch-like sediments of both the Villavicencio Formation and the Punta Negra Formation received detrital zircons having a Grenvillian age (Loske, 1995). This indicates that erosion of the terrane basement had begun by the Early Devonian, although the age of the Villavicencio Formation is not well constrained. The composition and the heavy mineral spectra in the Punta Negra sediments mirror the uplift of an igneous and metamorphic source area to the east (Gonzalez Bonorino, 1975a; Loske, 1992; Kury, 1993). The absence of the proximal facies of the Punta Negra Formation in the San Juan subbasin and the strong deformation of Cambrian-Ordovician carbonate platform strata point to deformation and uplift subsequent to the deposition of the Punta Negra Formation. During this uplift, the proximal facies of the Punta Negra Formation was cannibalized. The posttectonic granites of the Sierras Pampeanas indicate that by Early Carboniferous time tectonic activity had essentially ceased.

At present, it seems difficult to distinguish between discrete events that led to the collision of Chilenia with the Precordillera in the west and the accretion of the Precordillera to the Sierras Pampeanas in the east. However, overall evidence (see also Rapela et al., 1998; Pankhurst and Rapela, 1998) suggests that accretion of the Precordillera terrane and the Chilenia terrane to the South American margin of Gondwana took place in the interval between the Late Silurian and earliest Carboniferous and that accretion was completed prior to the Late Carboniferous, when a laterally extensive molasse was deposited.

PLATE TECTONIC SCENARIO FOR THE EVOLUTION OF THE ARGENTINE PRECORDILLERA

Plate Tectonic Evolution of the Argentine Precordillera

Precordillera as part of Laurentia. Several lines of evidence suggest that the basement of the Precordillera and its parental terrane (Cuyania) was involved in the Grenvillian orogeny about 1.1 Ga (e.g., Abbruzzi et al., 1993; McDonough et al., 1993; Kay et al., 1996; Vujovich and Kay, 1998). If this orogeny led to the formation of the supercontinent of Rodinia, for almost 500 m.y. the Precordillera terrane shared a combined history with eastern Laurentia, first within Rodinia, later in its hypothetical successor Pannotia. Following Vujovich and Kay (1998), the basement of the Cuyania terrane was located close to the Llano uplift of the Ouachita embayment (Fig. 6). During the breakup along the Appalachian-Ouachita margin of Laurentia, a passive margin must have formed on the eastern side of the Precordillera (Fig. 59A). This margin is here interpreted to represent the original limit of the North American craton during earliest Paleozoic time. In many places of the newly formed margin of Laurentia, passive-margin sediments cover rifted-margin prisms. Rifted-margin prisms are rare in the Ouachita trough (Viele and Thomas, 1989) and seem to be absent in the Precordillera (Keller, 1995).

Following the alternative paleogeographic models, in a late stage of the breakup of Pannotia, the Precordillera started to rift and raft away from Laurentia, and the embryonic Ouachita embayment was formed (Fig. 59A–C; see also Keller and Dickerson, 1996). Crustal thinning was responsible for the formation of a marginal plateau to the southeast of the present Ouachita embayment. Both the marginal plateau and the rifted Ouachita margin are part of the same upper plate, which is in agreement with the interpretation of Thomas (1993) that the Ouachita margin is an upper plate margin. The deep basin separating the Precordillera from mainland Laurentia at that time was probably underlain by attenuated continental crust (Fig. 57A), an assumption made by Dalziel (1997) for his Texas plateau model. However, oceanic crust may have formed locally (Davis, 1997).

Several intracratonic grabens were formed on Laurentian crust during the Cambrian and remnants of a similar intracratonic graben are present in the northern Precordillera. The Ouachita off-shelf basin can be regarded as one of these grabens. The basin received calcareous detritus from the Laurentian margin, but siliciclastic input, although not exclusively, was from the Precordillera margin. The two-sided nature of the Ouachita off-shelf basin has long been recognized (King, 1937; Young, 1970) and led to the concept of "Llanoria" (Miser, 1921; King, 1937), an enigmatic landmass to the southeast of the Ouachita embayment. Keller and Dickerson (1996) pointed out that many of the enigmas are no longer enigmatic if the Precordillera is considered to be Llanoria.

As argued by Thomas (1991), the continental block formerly occupying a position within the Ouachita embayment had been removed from the embayment by Late Cambrian time. This is indicated by regionally developed and laterally continuous carbonate platform facies from the Texas promontory to the Alabama promontory. If the Precordillera was indeed this crustal block, it had moved beyond the Alabama promontory by that time. This movement apparently opened the Ouachita basin, as the oldest sediments preserved there are also of Late Cambrian age (Dagger Flat sandstone; see Dickerson and Keller, 1998). There are, however, no ophiolitic rocks of Cambrian age that might indicate the existence of ocean floor as required by the model of Thomas (1991).

During Middle Cambrian through Early Ordovician time, the Precordillera marginal plateau is almost indistinguishable from Laurentia in its sediments and faunas. Although biostratigraphic control is often poor, the major events around Laurentia are recognized. However, sections like that at Cerro Pelado, where deep-water limestones cover carbonate platform rocks, testify to renewed crustal fragmentation. Continuous or renewed crustal stretching also influenced the Ouachita off-shelf area and the sedimentation patterns there (discussion in McBride, 1989; Keller and Dickerson, 1996). One of the major clues to the juxtaposition of both areas well into the Middle Ordovician are K-bentonites found in the Precordillera (Huff et al., 1995), which have a close match in subaqueous ash-flow deposits in the Marathon area of the Ouachita embayment. A controversial discussion of both findings was given by Dickerson and Keller (1998) and Huff et al. (1998).

Argentine Precordillera as an independent microplate. Crustal extension resumed during the Llandeilo and Caradoc. The Precordillera was finally separated from Laurentia (Fig. 57B) and an Atlantic-type continental margin was formed that connected the eastern basin of the Precordillera with its western basin (Fig. 58). There, basaltic pillow lavas extruded during the Caradoc (Haller and Ramos, 1984). Faunal data indicate that during the Caradoc the Precordillera had reached maximum isolation (Fig. 43). However, conodonts, ostracods, and graptolites document the predominance of warm and temperate waters (Lehnert, 1995a; Schallreuter, 1996; Brussa, 1995; Maletz and Ortega, 1995). These faunas still show strong affinities to North American faunas, although there is still a notable difference in the graptolite faunas of the Precordillera and the remainder of Gondwana.

During the Caradoc, the Precordillera for the first time was close enough to Gondwana to permit a few organisms to cross the intervening ocean. Following Benedetto et al. (1995) and

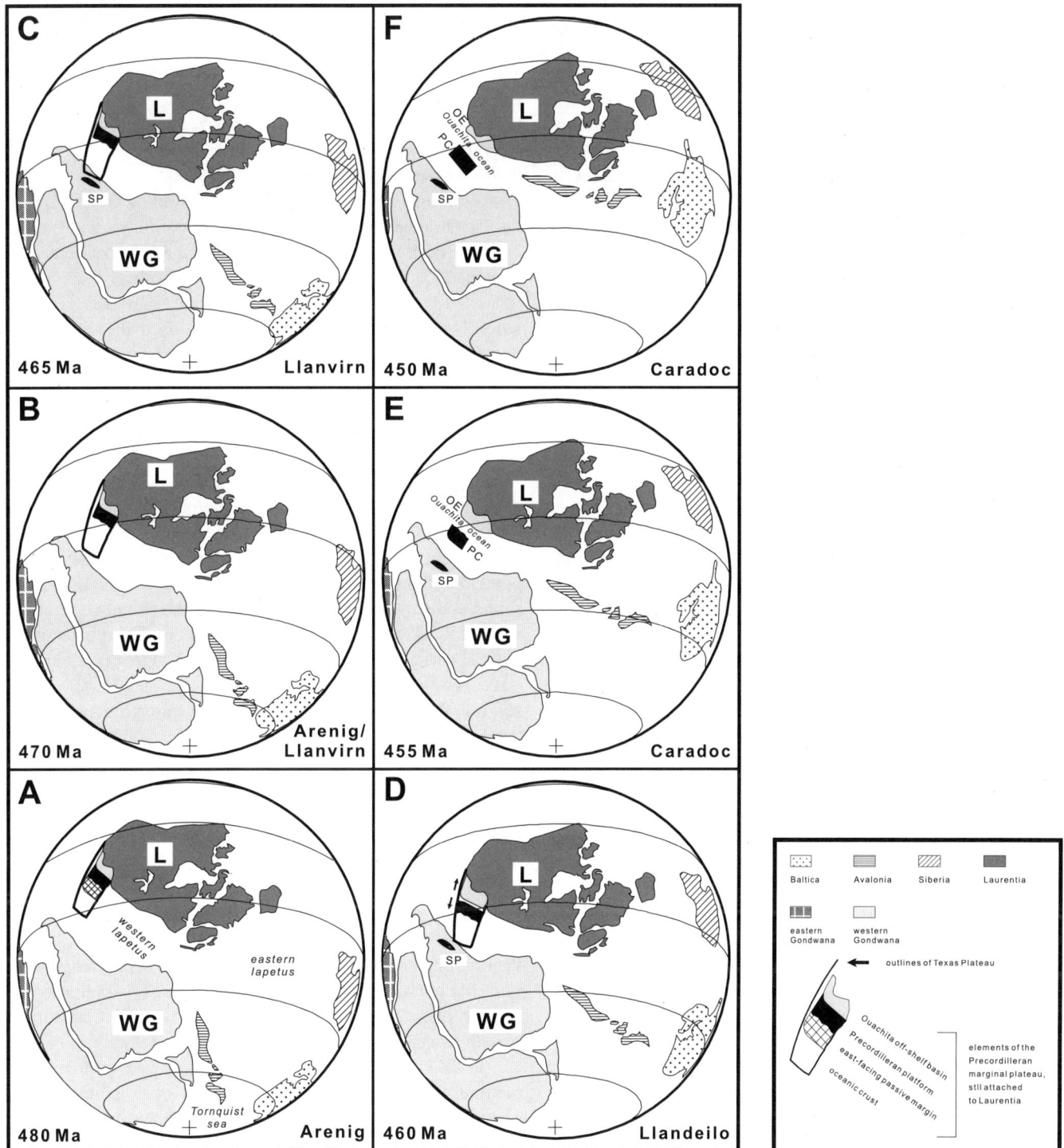

Figure 59. Plate tectonic evolution of Precordillera based on alternative paleogeographic scenario of Dalziel (1991, 1997) and Dalziel et al. (1994). In this figure outlines of Texas plateau (Dalziel 1997) are superimposed onto the elements of terrane and their size as discussed in this paper. A: Precordillera is still attached to Laurentian (L) craton and widely separated from western Gondwana (WG). No faunal exchange is possible. B and C: Texas plateau approaches Gondwana, oceanic crust is subsequently subducted, and eastern Sierras Pampeanas (SP) undergo a peak magmatic activity. However, there are no effects of this collision seen in Precordillera and there is still no faunal exchange. D: The Ouachita ocean starts to open as Precordillera finally rifts away from Laurentia. Between 465 and 460 Ma Laurentia and Gondwana are as close as paleomagnetic data permit. Still no faunal exchange is observed; however, Avalonian and Baltic faunal elements made their way into Precordillera, which is one of strongest arguments against plate tectonic reconstructions shown here. E and F: Precordillera is separated from Laurentia, the Ouachita basin (OE) fully opened. The position of Precordillera permits faunal exchange with all terranes in neighborhood. Low-latitude position of Precordillera explains the presence of abundant warm- and temperate-water faunas in Caradocian.

Benedetto (1998), the composition of the Caradocian fauna of the Precordillera is difficult to reconcile with a preceding collision of the Precordillera with Gondwana. This is true for the Ashgillian faunas. The presence of a *Hirnantia* fauna and its composition are no indications that the Precordillera was connected in any way to Gondwana (Keller and Lehnert, 1998). Instead, the elements of the Kosov province are widely distributed species of subtropical to temperate waters (Owen et al., 1991).

Concomitant to rifting and continental breakup along the western margin of the carbonate platform, strong magmatism is observed in the eastern Sierras Pampeanas. Geochemistry and especially Sr isotopes (Ramos, 1988b; Rapela et al., 1992, 1998) are typical of subduction-related magmatism. Two plate tectonic scenarios are possible to explain this magmatism. Magmatism was related to a plate of oceanic composition, independent from the Precordillera terrane; or magmatism was related to subduction of oceanic crust, which was part of the Precordillera terrane. These two scenarios require fundamentally different plate tectonic configurations for the Ordovician.

Argentine Precordillera in the conservative models. This scenario (Scotese and McKerrow, 1990; Torsvik et al., 1995, among others) assumes that Laurentia and western Gondwana were widely separated. This would have allowed the Precordillera to separate from Laurentia without the need of concomitant subduction and collision along the Precordillera's eastern margin. The Precordillera was (within the limitations of global plate tectonic processes) free to move to higher latitudes and to receive Upper Ordovician glaciomarine diamictites. These diamictites only indicate that by the Ashgillian, the Precordillera was within the reach of icebergs, and hence ice-rafted sediments. Among the pecularities of the Late Ordovician glaciation is the fact that icebergs may have drifted as far as lat 30° (Brenchley and Newall, 1984). Consequently, ice-rafted sediments alone (the glacio-marine diamictites of the Don Braulio Formation) cannot be taken as evidence of a drift toward high latitudes or for an accretion of the Precordillera to Gondwana (Keller and Lehnert, 1998). The first evidence that the Precordillera was truly close to Gondwana and in high latitudes comes from the Middle to Late Silurian *Clarkeia* fauna (Benedetto, 1998; Keller and Lehnert, 1998). According to Cocks and Fortey (1990), this fauna is restricted to high latitudes and was only present in a few areas of high-latitude Gondwana. In this scenario, the accretion of the Precordillera to Gondwana must have taken place during the Silurian or Devonian.

Argentine Precordillera as part of Gondwana: Precordillera in the alternative models. Paleomagnetically permissible reconstructions presented by Dalziel (1991, 1997) and Dalziel et al. (1994) place the Appalachian margin of Laurentia close to the proto-Andean margin of South America during the Ordovician and permit the existence of an ocean between Laurentia and northern South America (see also Torsvik et al., 1995).

The model discussed here implies that the ocean to the east of the Precordillera was the "Gondwanan Iapetus" or the "western Iapetus" (Dalziel, 1997). The ocean that opened to the west of the Precordillera was the Ouachita ocean separating the Precordillera from Laurentia. Within the western Iapetus, several terranes were trapped, among which is the Sierra de Famatina. Whatever the relative position between the Precordillera and the Sierra de Famatina, the volcanic strata of the Sierra de Famatina in this scenario may have been the source for the K-bentonites in the Precordillera without collision being recognized in the Precordillera (see also Huff et al., 1998).

Assuming that Laurentia and Gondwana were juxtaposed during Ordovician time and that the Precordillera was trapped between both plates, the separation of the Precordillera from mainland Laurentia must have been accompanied by subduction of the oceanic crust bordering the eastern passive margin. It seems logical that this Ordovician subduction, in the scenario discussed here, was subduction beneath the eastern Sierras Pampeanas causing the corresponding magmatism there. The subducted crust was part of the Precordillera terrane and represented the former eastern margin of the terrane (Fig. 59A). The intervening ocean was not yet fully closed and maintained a width greater than the maximum drift of larval stages (Fig. 59,B–E). As Laurentia continued its clockwise motion around Gondwana, the Precordillera was detached from the Ouachita embayment (Fig. 59,D and E). Dickerson and Keller (1998) pointed out that this severance might have involved a transtensional component. During the Caradoc, the approach to Gondwana slowed down and the Precordillera developed an endemic fauna. During the Late Ordovician, the Precordillera occupied a position where subtropical to temperate water faunas were able to survive. This position was also a position within the reach of icebergs during the Ashgillian glaciation.

Igneous activity and the evolution of magmas in the Sierras Pampeanas indicate continuous subduction into the Silurian and a gradual change toward collision during the Late Silurian and the Devonian (Rapela et al., 1992, 1998). This is corroborated by oblique compressional deformation in the southern Sierras Pampeanas (Sims et al., 1998; Von Gosen and Prozzi, 1998). During the Early Carboniferous posttectonic granites were emplaced. Hence the incorporation of the Precordillera (and of Chilenia in the final stage) into Gondwana seems to have been a lengthy and probably not a continuous process. With incipient collision during the latest Silurian the Precordillera became faunistically indistinguishable from the remainder of Gondwana.

In the Precordillera, the character of the heavy-mineral populations of Ordovician through Devonian strata indicates that a hinterland was composed of granitoids and metamorphic rocks was not present before the (Middle) Devonian. Prior to the Devonian, mainly a sedimentary hinterland was recycled (Loske, 1992; Kury, 1993; Astini and Maretto, 1996). The Devonian uplift of this hinterland is here interpreted as one effect of the final accretion of the Precordillera to Gondwana.

The position of the subduction zones, the one between the Precordillera (and its parental terrane) and the Sierras Pampeanas in the east as well as the one that might have existed between Chilenia and the Precordillera in the west, are debatable (Ramos et al., 1986; Ramos, 1988b; Astini et al., 1995). In addition, there is no agreement on the direction of subduction between Chilenia and the Precordillera. No attempt is made here to solve these problems because this would be far beyond the scope of this paper and the studies on which it is based.

In the light of the "alternative paleogeographic reconstruction" (Torsvik et al., 1995) which places Laurentia close to Gondwana during the Ordovician (Dalziel, 1991; Dalla Salda et al., 1992a, 1992b, 1993; Dalziel et al., 1994), the microplate stage was coeval to the early phase of the Gondwana stage. Separation from Laurentia was accompanied by the onset of accretion to Gondwana. A similar succession of events was described from the Alps by Frisch (1979). There, during the Mesozoic, the middle Penninic microcontinent was separated from the Eurasian plate (forming the North Penninic ocean) at the same time that the South Penninic ocean between the microplate and the Adriatic plate was closed.

Discussion. On the basis of sedimentologic and faunal evidence from within the Precordillera, a collision of continental crust of the Precordillera terrane (Astini et al., 1995), of the Cuyania terrane (Ramos, 1995), or of the Occidentalia terrane (Dalla Salda et al., 1992a, 1992b) with the eastern Sierras Pampeanas during the Middle Ordovician is difficult to reconcile. Consequently, the accretion of the Precordillera to Gondwana must have taken place during a later stage. The original configuration of the terrane and its approach toward Gondwana is still debatable.

If the assumption is correct that the eastern margin of Laurentia and the western margin of Gondwana were juxtaposed during the Ordovician, then it seems plausible that the continental crust of the Precordillera was detached from Laurentia; meanwhile the oceanic crust forming the eastern part of the terrane was subducted beneath the Sierras Pampeanas (Fig. 59, A–D). However, this model is only valid under the assumption of maximum separation (2500 km according to paleomagnetic data) and maximum travel distances for larvae of about 1000 km.

With all the restrictions discussed here, and explicitly excluding a Middle Ordovician Precordillera-Gondwana collision, the data from within the Precordillera *permit the alternative plate tectonic models* as described by Dalziel (1991, 1997) and Dalziel et al. (1994). If these alternative reconstructions are correct, the Precordillera must have played a key role in the interactions between Laurentia and Gondwana, the Precordillera is a tectonic tracer (Dalziel, 1993) which, for a short period during the Middle Ordovician, marked the relative positions of Laurentia and Gondwana.

Laurentia might have been relatively close to Gondwana. However, the intervening ocean was an obstacle big enough to prevent faunal exchange and the effects of subduction along the Sierras Pampeanas being recognized in the Precordillera. In contrast, faunal exchange with far more distant terranes was possible. The explosive volcanic record of eastern Laurentia and Baltica strongly suggests that these two continents were juxtaposed during Middle and Late Ordovician time (Huff et al., 1992), directly contradicting a Laurentia-Gondwana juxtaposition. Together, this evidence does not favor the alternative paleogeographic scenario, but rather *supports the conservative model.*

In its early Paleozoic history, the Precordillera displays three distinct stages: a Laurentian stage until the Caradoc; an open-ocean phase from the Caradoc at least into the Late Silurian; and a Gondwanan stage starting not earlier than the Late Silurian. Whatever global plate tectonic reconstruction for the Ordovician is the most likely, the history of the Precordillera as discussed in this paper is compatible with both but favors a conservative approach to Ordovician paleogeography.

EPILOGUE

Remaining Questions

Despite all the results presented in this paper, this study can only be regarded as a basic step to a better understanding of the early Paleozoic of the Argentine Precordillera. Many questions remain open, among which are the following.

Is the Hawke Bay event really present in the rocks of the Precordillera or is biostratigraphy in the Cambrian rocks not yet perfect enough to distinguish between tectonic and sedimentologic effects and hence, to pretend an erosional event in the early Middle Cambrian?

What is the nature of the Grand Cycles in the La Laja Formation? If the Precordillera was part of Laurentia, why are the cycles so different from their North American counterparts? Is the interpretation given here a misinterpretation, or is the difference caused by the fact that the Precordillera is part of a marginal plateau to Laurentia with a presumably different crust?

What are the age and the nature of the sequence boundaries in the Upper Cambrian rocks? Some ideas and correlations have been presented here. However, it may well be that these ideas are a mismatch of data between the Guandacol subbasin and the San Juan subbasin. A critical factor for future discussions is the correlation of the Marjuman sequence boundary in the Guandacol area to strata farther south. In addition, will it be possible to locate the Sauk II-III boundary?

In this context, the small-scale cycles present in the Upper Cambrian, and, presumably, younger rocks are interesting. Will these rocks reveal additional information via Fisher plots and other cycle analysis? Will the results of such analysis allow a better chronostratigraphic assignment and, in addition, a better correlation to Laurentian rocks?

What are the relations among the siliciclastic strata of the western basin? A variety of formations has been defined there, all of which show similar rocks and facies. Are these truly independent formations or are they all representatives of the same depositional system? Is the overall fining to the west in the western basin an artefact caused by the relatively intense deformation and the absence of reliable biostratigraphic data and stratigraphic contacts? Or is this arrangement of facies the expression of an Ordovician Atlantic-type margin in the Precordillera as interpreted here?

An important question is the nature of the hiatuses observed in the Silurian and Devonian rocks which increase in magnitude and abundance from the Jáchal area toward the Río San Juan. What is the continuation of the Silurian and Devonian toward Mendoza, where Silurian strata are virtually absent? What does the contact between the Talacasto Formation and the Punta Negra Formation represent? Is it a major erosional unconformity or a strongly diachronous evolution of the contact? Is it possible that the surface at the base of the Punta Negra Formation, which gets older toward the south, is represented by the surface at the base of the Villavicencio Formation, which is even older? If so, what is the significance of the strongly diachronous onset of flysch deposition? If not, are there two independant flysch sequences and basins? What is the significance of this overall north-south trend recognized in the Silurian-Devonian history of the Precordillera? There is an increasing number of hiatuses with an increasing magnitude from Jáchal to the south. Concomitantly, distribution of Silurian strata becomes more and more restricted, in space as well as in time, until they are absent in Mendoza. In addition, the onset of flysch deposition becomes successively older toward Mendoza. Is it possible that these patterns indicate oblique collision, starting in the south and propagating toward the north?

What was the size of the terrane? It has been documented that the Ordovician of San Rafael is part of the same terrane as the Precordillera. Are there other outcrops on the block of San Rafael–La Pampa that might be assigned to the terrane? What is the origin of the, in places mylonitic, limestones of Limay Mahuida? Do they represent part of the Cambrian-Ordovician carbonate platform or do they correspond to the limestones and marbles as exposed in the Sierra de Pie de Palo and several other places?

What is the relation between the Sierra de Famatina and the Precordillera? Is it possible that the K-bentonites in the San Juan Formation are derived from volcanic sources in the Sierra de Famatina? If so, does this suggest a proximity of both areas during the Ordovician?

What are the mechanisms of terrane transport? Nothing is known about transform faults, which might have served as pathways for the terrane. Was terrane accretion oblique or perpendicular to the margin of Gondwana?

What is the relation between Chilenia and the Precordillera? Are they totally different and independent terranes, or were they related in some way, as indicated by Grenvillian-age zircons that recently have been found in Chilenia (Ramos and Basei, 1997)?

The discussion of the evolution of the Precordillera in time and space was hampered by the two fundamentally different chronostratigraphic subdivisions of the Cambrian and Ordovician Systems based on differing philosophies of North American and European workers. One example is the upper reef-mound interval, which is Early Ordovician in age according to the Europeans. However, it is Middle Ordovician following the North Americans. In this paper, mainly the European subdivision was applied because it is the one which was traditionally applied to the rocks of the Precordillera. The Precordillera is a crustal fragment detached from Laurentia (Laurentian time scale) and subsequently accreted to Gondwana (European time scale). The investigations have demonstrated the urgent need for a globally accepted chronostratigraphic subdivision, especially of the Ordovician.

ACKNOWLEDGMENTS

This study is part of a larger project of the Erlangen Department of Geology on the evolution of the Argentine Precordillera. When this project was initiated almost 10 years ago by Werner Buggisch, my primary task was to work on the carbonate platform rocks together with him. I am grateful to him that he decided to withdraw from this part of the project and that he left the carbonates to me, and I hope I meet the confidence he placed in me.

Much of the field work in the later part of the project was carried out with Oliver Lehnert. I appreciate his patience in our discussions about any aspect of the paleontology and biostratigraphy of the Ordovician rocks.

I thank Werner Von Gosen for our discussions about the structural geology and about the tectono-sedimentary environment of the rocks in the Precordillera and for many helpful comments concerning the regional geology of western Argentina during the late stage of this project. I also acknowledge the help of all the other members of the Erlangen Institute who contributed to this study.

From the beginning, the project benefited from discussions with colleagues from other German Universities. Vicariously for all these colleagues, I thank Heinrich Bahlburg and Werner Loske for many stimulating discussions.

The carbonate platform rocks in the Precordillera are more than 2000 m thick. It would have been impossible to get to the localities and to measure all these sections without the help of Argentinian colleagues. I especially thank Osvaldo Bordonaro (Mendoza) for his hospitality and for having shared his regional knowledge of the Precordillera, his outstanding knowledge of the Cambrian rocks in the Precordillera, and many long discussions in the field. Field work with him was always a pleasure.

I have also benefited from the company and friendship of Max LaMotte (San Juan), who accompanied me during field work, carried samples, and taught me to how to get around in the Andes. I thank him and his wife Maria Silvia for their great hospitality during my stays in the city of San Juan. I had many fruitful discussions about the Upper Cambrian and Ordovician rocks of the Precordillera with Fernando Cañas, not only in the field but also during my stays in Córdoba and his stay in Erlangen. I appreciate these discussions and especially my stays with Fernando and his family.

Many discussions with other members of the Córdoba working group helped me to a better understanding of the geology of the Precordillera and I especially thank Emilio Vaccari for his part in these discussions, at the "cátedra" and in the field.

Scientific, logistic, and human support in San Juan was provided by Wilko Simon, Rubén Pelichotti, and Felisa Bercowski; I thank all of them for their help. In addition, I thank all those students from San Juan, who accompanied me during field work, for their help.

Over the years, I have had intensive discussions with colleagues from the La Plata University (Luis Spalletti, Luis Dalla Salda, Sergio Mateos, Carlos Rapela) which I enjoyed; I especially thank Carlos Cingolani for many open-minded discussions and his support.

When it became evident that the Precordillera might have been part of Laurentia during Cambrian-Ordovician time, I started field work in various parts of the southern United States. It is not possible to mention all the geologists who introduced me to individual sections or regional geology or who discussed with me aspects of lower Paleozoic sedimentation on the Laurentian craton.

I thank John D. Cooper and Tony Prave for sharing their knowledge of Great Basin geology and sedimentology with me and for introducing me to sequence stratigraphy, but especially, for their friendship and the moments that made field work in the Great Basin an outstanding episode in my life. Special thanks are due to the people of Shoshone and of China Ranch for their hospitality and their interest in my work. I was deeply impressed by and grateful to Benny Troxel and Lauren Wright for their advice and hospitality.

A two-week field trip with Allison "Pete" Palmer to the classical localities of the Cambrian in the southwestern Great Basin was unforgettable; I thank him for sharing with me his knowledge of the Cambrian of the world and for many fruitful discussions, especially about the origin of the Precordillera. His review of this book was also very benefical.

I also thank Pat Dickerson and Bill Muehlberger who, during a field trip to west Texas, sharpened my eye for the Marathon-Solitario connection; Dave LeMone, who introduced me to the geology of the El Paso area, west Texas, and New Mexico, and who gave logistic support during field work in west Texas; Jim Miller for sharing his knowledge of the Upper Cambrian and Lower Ordovician rocks during a field trip to the Llano uplift in Texas; Ray Ethington for his support during field work in Oklahoma; and Ian Dalziel, who supported these studies in various ways. His review of this paper helped improve the manuscript, especially those parts that deal with global plate tectonic reconstructions. I am especially grateful that he provided the base maps of his plate tectonic reconstructions. I thank Lisa Gahagan for preparing the different sets of maps; Isabel Montañez for her review of the manuscript and many helpful comments; and the other colleagues with whom I discussed various aspects of geology, paleontology, and volcanology of the Precordillera.

I am grateful to the Volkswagenstiftung and the German Science Foundation which gave financial support for this project through projects Bu 312-17/1 and Bu 312-17/3 (Werner Buggisch) and projects Ke 470-1/2 through Ke 470/2-4 to me.

APPENDIX 1

Alum shale
Dark gray to black, sometimes yellowish shale with abundant sulfur. Stainings are the result of weathering of alum minerals: a group of double salts $Me^I Me^{III} (SO_4)_2 \times 12 H_2O$. Most common mineral in sediments is $K Al (SO_4)_2 \times 12 H_2O$.

Andean crustal shortening, Andean orogeny
In the Precordillera, mainly Miocene deformation affects strata as young as early Tertiary. Formation of east-directed thrusts and imbrications (Baldis and Chebli, 1969; Ortiz and Zambrano, 1981; Baldis et al., 1982; Allmendinger et al., 1990; Von Gosen, 1992, 1995) and west-directed back thrusting along the eastern margin of the Precordillera occur (Cominguez and Ramos, 1990; Von Gosen, 1992; Von Gosen et al., 1995). No thermal overprint is recognizable.

Carbonate platform
This is the most general term for carbonate depositional systems (Read, 1982, 1985; Tucker and Wright, 1990). Classification is based on distinctive morphologic profile, facies, and evolutionary sequences (Read, 1985).

Carbonate ramp
This is characterized by gently inclined depositional surface, and there is no major break at the platform-slope break (Ahr, 1973; Burchette and Wright, 1992). Homoclinal ramps show a gradual deepening; however, if there is a steepening on the outer ramp due to faulting or flexuring of the lithosphere, a distally-steepened ramp is developed (Read, 1985). Storm deposition and reworking is an important factor for sedimentation on ramps (Aigner, 1985; Burchette and Wright, 1992).

Chilenia
The Chilenia terrane, accreted to the Precordillera during Silurian-Devonian time (Ramos et al., 1986; Ramos and Basei, 1997; Bahlburg and Hervé, 1997) is still poorly understood because its basement outcrops in the Cordillera Frontal were subjected to intensive magmatism and metamorphism during late Paleozoic and Mesozoic time. The presence of basic and ultrabasic magmatic rocks in the Ordovician of the Precordillera was taken as evidence of the suture zone (Haller and Ramos, 1984; Ramos et al., 1986) between the Precordillera and Chilenia (Ramos et al., 1984; Ramos, 1988b). Chilenia was interpreted as an independent terrane (Ramos and Basei, 1997) that extended from approximately the southern end of the Precordillera well into northern Chile and northwestern Argentina (Ramos et al., 1986). However, Bahlburg and Hervé (1997) claimed that Chilenia did not extend farther north than ~27°30′, hence it has a geographic extension similar to that of Precordillera. Rapela et al. (1998) pointed out that the basement of Chilenia might be a western extension of the Precordillera.

Diamictite
Throughout this paper, diamictite is used in a purely descriptive sense.

Drowning, drowned platform
This is mainly the result of a rapid rise in relative sea level. Shallow-water carbonates upsection give way rapidly to deep-water limestones or siliciclastic deposits (Kendall and Schlager, 1981; Schlager, 1981, 1989). Within the drowning succession, hardgrounds and other indicators of reduced sediment supply are locally present. In seismic lines, the lithologic change from shallow- to deep-water rocks is often recognized as an unconformity and, in consequence, is called the drowning unconformity (Schlager and Cumber, 1986).

Facies association
This combines various facies frequently occurring together (Tucker and Wright, 1990). The various facies associations identified are used to describe the architecture of a carbonate depositional system.

Famatinian orogeny, Famatinian mobile belt
These are related to east-directed Ordovician subduction between 490 and 450 Ma (Famatinian orogeny; Pankhurst and Rapela, 1998; Rapela et al., 1998) and include both the Puna and Famatina rocks and

the North Patagonian massif. Magmatic activity and formation of shear zones continued into the Devonian.

Grenvillian age
This is here used for "... rocks with ages between ~1300 to ~950 Ma that are contemporaneous with the North American Grenville province basement" (Kay et al., 1996).

Grenvillian-type basement
Grenvillian-aged basement is petrologically and isotopically similar to identical to the North American Grenvillian basement (see also Kay et al., 1996).

Isolated carbonate platform
This is carbonate accumulation on topographic highs surrounded by deep water. Adjacent slopes were generally steep and dominated by detritus from the platform. Distinctive facies patterns formed in response to differing exposure to wind and waves over relatively short distances.

Microfacies concept
Microfacies include the entirety of sedimentologic and paleontologic information as revealed by thin-section analysis (Flügel, 1982). This information is grouped into 24 standard microfacies (SMF; Wilson 1975).

Ocloyic orogeny
This is part of the Famatinian cycle. There is an angular unconformity between Lower Ordovician and Ashgillian strata in the Puna and the Eastern Cordillera. Regionally developed mylonites are in the Eastern Sierras Pampeanas (Pankhurst and Rapela, 1998).

Paleozoic deformation
This deformation affected strata as young as the Early Devonian (Talacasto Formation; Von Gosen 1997). There are no indications of Ordovician compression. Formation of west-vergent, large open folds (Cominguez and Ramos, 1990; Von Gosen, 1992, 1995, 1997; Von Gosen et al., 1995) was accompanied by a metamorphic event between 440 and 410 Ma (Buggisch et al., 1994b). The metamorphic overprint increases from east (field of diagenesis) to west (greenschist facies) and is documented by illite crystallinity and conodont CAI (Buggisch et al., 1994b; Keller et al., 1993c).

Pelagic carbonate platform
This is a special type of isolated platform, developed above structural highs (Santantonio, 1993, 1994). The sedimentary succession is pelagic deposits with abundant stratigraphic breaks.

Rimmed shelf
This is a relatively flat depositional surface, pronounced break at shelf-slope transition, often marked by continuous reef trend or sand shoals (oolitic-skeletal). The permeability of the barrier determines the nature of the landward shelf lagoon.

REFERENCES CITED

Abbruzzi, J. M., Kay, S. M., and Bickford, M. E., 1993, Implications for the nature of the Precordilleran basement from the geochemistry and age of Precambrian xenoliths in Miocene volcanic rocks, San Juan province, in Actas, XII Congreso Geológico Argentino y II Congreso de Exploración de Hidrocarburos, v. 3: Mendoza, Argentina, Asociación Geológica Argentina e Instituto Argentino de Petróleo y Gas, p. 331–339.

Aceñolaza, F. G., 1969, Características geológicas y estratigráficas del sector septentrional de la Precordillera Riojana, in Actas, 4as Jornadas Geológicas Argentinas; v. 1 Mendoza, Argentina, Asociación Geológica Argentina p. 1–13.

Aceñolaza, F. G., 1992, El sistema Ordovícico de Latinoamérica, in Gutiérrez-Marco, J. C., Saavedra, J., and Rábano, I., eds., Paleozoico Inferior de Ibero-América: Merinda Publicaciones de Universidad de Extremadura, p. 85–118.

Aceñolaza, F. G., and Peralta, S. H., 1985a, Análisis secuencial de los caracteres icnológicos del Silúrico de la Precordillera Argentina (Sa. de Talacasto, San Juan, Argentina), in Actas, 6° Congreso Geológico Chileno, v. 4, Sociedad Geológica de Chile p. 591–598.

Aceñolaza, F. G., and Peralta, S. H., 1985b, Icnofacies de Cruziana en el Silúrico de la Precordillera de San Juan, Argentina, in Anais, 11° Congresso Brasileiro do Paleontología: Fortaleza, Brasil. Sociedade Brasileira de Geologia, p 151–156.

Aceñolaza, F. G., and Peralta, S. H., 1986, Interpretación del comportamiento de la cuenca silúrica en función de sus icnofacies en el área de Talacasto, Precordillera de San Juan, in Actas, 1as Jornadas sobre Geología de Precordillera: San Juan, Argentina, 1985, Universidad Nacional de San Juan, p. 151–156.

Aceñolaza, F. G., and Toselli, A. J., 1988, El Sistema de Famatina: Su interpretación como orógeno de margen continental activo, in Actas, 5° Congreso Geológico Chileno, v. 1, Universidad de Chile, Santiago p. 55–67.

Aceñolaza, F. G., and Toselli, A. J., 1998, Precordillera Argentina: alóctono o autóctono gondwanico?: Terra Nostra, v. 5, p. 2–3.

Aceñolaza, F. G., Durand, F., and Diaz Taddei, R., 1976, Nautiloideos ordovícicos de la Precordillera Argentina. Fauna de Huaco, Provincia de San Juan: Acta Geológica Lilloana, v. 13, p. 269–284.

Ahr, W. M., 1973, The carbonate ramp: An alternative to the shelf model: Gulf Coast Association of Geological Societies Transactions, v. 23, p. 221–225.

Aigner, T., 1982, Calcareous tempestites: Storm-dominated stratification in Upper Muschelkalk limestones (Middle Triassic, SW-Germany), in Einsele, G., and Seilacher, A., eds., Cyclic and event stratification: Berlin, Heidelberg, New York, Springer, p. 180–198.

Aigner, T., 1985, Storm depositional systems: Lecture Notes in Earth Sciences, Berlin, Heidelberg, New York, Springer v. 3: 174 p.

Aitken, J. D., 1966, Middle Cambrian to Middle Ordovician cyclic sedimentation, southern Rocky Mountains of Alberta: Canadian Petroleum Geology Bulletin, v. 14, p. 405–441.

Aitken, J. D., 1967, Classification and environmental significance of cryptalgal limestones and dolomites: Journal of Sedimentary Petrology, v. 37, p. 1163–1178.

Aitken, J. D., 1978, Revised models for depositional Grand Cycles, Cambrian of the southern Rocky Mountains, Canada: Canadian Petroleum Geology Bulletin, v. 26, p. 515–542.

Albanesi, G. L., 1991, La conodontofauna y graptolitos asociados de las Formaciones San Juan y Gualcamayo en el Cerro Potrerillo, Precordillera de San Juan, Argentina [tesis de licenciatura]: Córdoba, Argentina, Universidad Nacional de Córdoba, Academia Nacional de Ciencias, 162 p.

Albanesi, G. L., and Benedetto, J. L., 1992, Late Llanvirn pebbles from the La Cantera Fm., Sierra de Villicum, San Juan Precordillera, Argentina, in Hünicken, M. A., ed., Latin-American Conodont Symposium, 2nd: Córdoba, Argentina, Abstracts, p. 92.

Albanesi, G. L., Hünicken, M. A., and Ortega, G. A., 1995, Review of Ordovician conodont-graptolite biostratigraphy of the Argentine Precordillera, in Cooper, J. D., Droser, M. L., and Finney, S. C., eds., Ordovician odyssey: Short papers for the Seventh International Symposium on the Ordovician System: Pacific Section, SEPM (Society for Sedimentary Geology) Publication 77, p. 31–35.

Alberstadt, L., and Repetski, J. E., 1989, A Lower Ordovician sponge/algal facies in the southern United States and its counterparts elsewhere in North America: Palaios, v. 4, p. 225–242.

Alfaro, M., 1988, Graptolitos del Ordovícico superior (Caradociano) de la quebrada Agua de la Cruz, Precordillera de Mendoza: Ameghiniana, v. 25, p. 299–303.

Alfaro, M., and Fernandez, R., 1985, Una graptofauna del Ordovícico superior (Caradociano) de Estancia Canota, provincia de Mendoza: Ameghiniana, v. 22, p. 63–67.

Allen, P. A., and Allen, J. R., 1990, Basin analysis: Principles and applications: Oxford, Blackwell Scientific Publications, 451 p.

Allmendinger, R. W., Figueroa, D., Snyder, D., Beer, J. A., Mpodozis, C., and Isacks, B. L., 1990, Foreland shortening and crustal balancing in the Andes at 30°S latitude: Tectonics, v. 9, p. 789–809.

Amos, A. J., 1954, Estructura de las formaciones Paleozoicas de la Rinconada, pie oriental de la Sierra Chica de Zonda: Revista de la Asociación Geológica Argentina, v. 9, p. 1–37.

Armella, C., 1989a, Microfacies trombolíticas de un biociclo ideal de la Formación La Flecha, Precordillera Oriental, Argentina: Universidad Nacional de Tucumán, Serie Correlación Geológica, v. 5, p. 45–52.

Armella, C., 1989b, Grupos morfológicos y fábrica de trombolitos del límite Cámbrico-Ordovícico de Argentina: Universidad Nacional de Tucumán, Serie Correlación Geológica, v. 5, p. 53–66.

Astini, R. A., 1991, Paleoambientes sedimentarios y secuencias depositacionales del Ordovícico clástico de la Precordillera Argentina [Ph.D. thesis]: Córdoba, Argentina, Universidad Nacional de Córdoba, 847 p.

Astini, R. A., 1992, Tectofacies ordovicicas y evolución de la cuenca eopaleozoica de la Precordillera Argentina: Estudios Geológicos, v. 48, p. 315–327.

Astini, R. A., 1993, Facies glacigénicas del Ordovícico tardío (Hirnantiense) de la Precordillera Argentina: Boletín de la Real Sociedad Española de Historía natural (Sección Geología), v. 88, p. 137–149.

Astini, R. A., and Buggisch, W., 1993, Aspectos sedimentológicos y paleoambientales de los depósitos glacigénicos de la Formación Don Braulio, Ordovícico tardío de la Precordillera argentina: Revista de la Asociación Geológica Argentina, v. 48, p. 217–232.

Astini, R. A., and Cañas, F. L., 1995, La Formación Sassito, una nueva unidad calcárea en la Precordillera de San Juan: Sedimentología y significado estratigráfico: Revista de la Asociación Sedimentológica Argentina, v. 2, p. 19–37.

Astini, R. A., and Maretto, H. M., 1996, Análisis estratigráfico del Silúrico de la Precordillera Central de San Juan y consideraciones sobre la evolución de la cuenca; *in* Actas, XIII Congreso Geológico Argentino y III Congreso de Exploración de Hidrocarburos, v. 1: Buenos Aires, Argentina, Asociación Geológica Argentina e Instituto Argentino de Petróleo y Gas, p. 351–368.

Astini, R. A., and Vaccari, N. E., 1996, Sucesión evaporítica del Cámbrico inferior de la Precordillera: significado geológico: Revista de la Asociación Geológica Argentina, v. 51, p. 97–106.

Astini, R. A., Benedetto, J. L., and Vaccari, N. E., 1995, The early Paleozoic evolution of the Argentine Precordillera as a Laurentian rifted, drifted, and collided terrane: A geodynamic model: Geological Society of America Bulletin, v. 107, p. 253–273.

Astini, R. A., Benedetto, J. L., and Vaccari, N. E., 1996a, The early Paleozoic evolution of the Argentine Precordillera as a Laurentian rifted, drifted, and collided terrane: A geodynamic model: Reply: Geological Society of America Bulletin, v. 108, p. 373–375.

Astini, R. A., Ramos, V. A., Benedetto, J. L., Vaccari, N. E., and Cañas, F. L., 1996b, La Precordillera: Un terreno exótico a Gondwana: *in* Actas, XIII Congreso Geológico Argentino y III Congreso de Exploración de Hidrocarburos, v. 5: Buenos Aires, Argentina, Asociación Geológica Argentina e Instituto Argentino de Petróleo y Gas, p. 293–324.

Bahlburg, H., and Hervé, F., 1997, Geodynamic evolution and tectonostratigraphic terranes of northwestern Argentina and northern Chile: Geological Society of America Bulletin, v. 109, p. 869–884.

Baldis, B. A., 1964, El Silúrico fosilífero de Gualilán (Provincia de San Juan): Revista de la Asociación Geológica Argentina, v. 23, p. 189–193.

Baldis, B. A., 1967, Some Devonian trilobites of the Argentine Precordillera, *in* Proceedings, International Symposium on the Devonian System: Calgary, Canada, v. 2, Canadian Society of Petroleum Geologists, p. 789–796.

Baldis, B. A., 1975, El Devónico Inferior en la Precordillera Central, Parte I: Estratigrafía: Revista de la Asociación Geológica Argentina, v. 30, p. 53–83.

Baldis, B. A., and Beresi, M. S., 1981, Biofacies de culminación del ciclo deposicional calcáreo del Arenigiano en el Oeste de Argentina, *in* Anais, 2° Congresso Latino-Americano de Paleontología v. 1: Sociedade Brasileira de Geologia, Porto Alegre, Brasil, p. 11–17.

Baldis, B., and Blasco, G., 1973, Trilobites Ordovícicos de Ponon Trehue, Sierra Pintada de San Rafael, Provincia de Mendoza: Ameghiniana, v. 10, p. 72–88.

Baldis, B., and Blasco, G., 1975, Primeros trilobites ashgillianos del Ordovícico sudamericano, *in* Actas, 1er Congreso Argentino de Paleontología y Bioestratigrafía, v. 1, Asociación Paleontológica Argentina, p. 33–48.

Baldis, B. A., and Bordonaro, O. L., 1981a, Evolución de facies carbonáticas en la cuenca Cámbrica de la Precordillera de San Juan, *in* Actas, VIII Congreso Geológico Argentino, v. 2: San Luis, Argentina, Asociación Geologica Argentina, p. 385–397.

Baldis, B. A., and Bordonaro, O. L., 1981b, Vinculación entre el Cámbrico del Noroeste de México y la Precordillera Argentina, *in* Anais, 2° Congreso Latino-Americano de Paleontología, v. 1: Porto Alegre, Brasil, Sociedade Brasileira de Geologia, p. 1–10.

Baldis, B. A., and Bordonaro, O. L., 1982, Comparación entre el Cámbrico del Great Basin Norteamericano y la Precordillera de Argentina. Su implicancia intercontinental, *in* Actas, 5° Congreso Latinoamericano de Geología, v. 1: Buenos Aires, Argentina, Servicio Geologico Nacional, p. 97–108.

Baldis, B. A., and Bordonaro, O. L., 1985, Variaciones de facies en la cuenca Cámbrica de la Precordillera Argentina, y su relación con la génesis del borde continental, *in* Actas, 6° Congreso Latinoamericano de Geología: Bogotá, Colombia, p. 31–42.

Baldis, B. A., and Chebli, G. A., 1969, Estructura profunda del área central de la Precordillera Sanjuanina, *in* Actas, 4as Jornadas Geológicas Argentinas, v. 1: Mendoza, Argentina, Asociación Geológica Argentina, p. 45–65.

Baldis, B. A., and Longobuco, M., 1977, Trilobites devónicos de la Precordillera Noroccidental (Argentina): Ameghiniana, v. 14, p. 145–161.

Baldis, B. A., and Pöthe de Baldis, E. D., 1995, Trilobites Ordovícicos de la Formación Las Aguaditas (San Juan, Argentina) y consideraciones estratigráficas: Boletín de la Academía Nacional de Ciencias de Córdoba, v. 60, p. 409–448.

Baldis, B. A., Beresi, M. S., Bordonaro, O. L., and Uliarte, E., 1981a, Estromatolitos, trombolitos y formas afines en el límite Cámbrico-Ordovícico del Oeste Argentino, *in* Anais, 2° Congreso Latino-Americano de Paleontología, v. 1: Sociedade Brasileira de Geologia, Porto Alegre, Brasil, p. 19–30.

Baldis, B. A., Bordonaro, O. L., Beresi, M. S., and Uliarte, E., 1981b, Zona de dispersión estromatolítica en la secuencia calcáreo dolomítica del Paleozoico inferior de San Juan, *in* Actas, VIII Congreso Geológico Argentino, v. 2: San Luis, Argentina, Asociación Geológica Argentina, p. 419–434.

Baldis, B. A., Beresi, M. S., Bordonaro, O. L., and Vaca, A., 1982, Síntesis evolutiva de la Precordillera Argentina, *in* Actas, 5° Congreso Latinoamericano de Geología, v. 4: Servicio Geológico Nacional, Buenos Aires, Argentina, p. 399–445.

Baldis, B. A., Beresi, M. S., Bordonaro, O. L., and Vaca, A., 1984a, The Argentine Precordillera as a key to Andean structures: Episodes, v. 7, p. 14–19.

Baldis, B. A., Peralta, S. H., and Uliarte, E. R., 1984b, Geología de la Quebrada Ancha y sus alrededores en el área de Talacasto, Precordillera Sanjuanina, *in* Actas, IX Congreso Geológico Argentino, v. 4: San Carlos de Bariloche, Río Negro, Argentina, Asociación Geológica Argentina, p. 233–254.

Baldis, B. A., Armella, C., and Cabaleri, N., 1985, Desarollo de la plataforma carbonática Ordovícica Argentina, *in* Actas, 6° Congreso Latinoamericano de Geología: Bogotá, Colombia, p. 165–174.

Baldis, B. A., Peralta, S. H., and Villegas, R., 1989, Esquematizaciones de una posible transcurrencia del terrane de Precordillera como fragmento continental procedente de areas pampeano-bonaerenses: Universidad Nacional de Tucumán, Serie Correlación Geológica, v. 5, p. 81–100.

Ball, M. M., 1967, Carbonate sand bodies of Florida and the Bahamas: Journal of Sedimentary Petrology, v. 37, p. 556–591.

Ball, M. M., Shinn, E. A., and Stockman, K. W., 1967, The geologic effects of Hurricane Donna in South Florida: Journal of Geology, v. 75, p. 556–591.

Bally, A. W., and Snelson, S., 1980, Realms of subsidence, *in* Miall, A. D., ed., Facts and principles of world petroleum occurrence: Canadian Society of Petroleum Geologists Memoir 6, p. 9–75.

Banchig, A. L., and Bordonaro, O. L., 1990, Nuevos afloramientos del talud continental Cámbrico en la Sierra del Tontal, San Juan, Argentina, *in* Actas, XI Congreso Geológico Argentino, v. 2: San Juan, Argentina, Asociación Geólogica Argentina, p. 49–52.

Banchig, A. L., and Bordonaro, O. L., 1994, Reinterpretación de la Formación Los Sombreros: Secuencia olistostrómica de talud, Precordillera Argentina, *in*

Actas, 5ª Reunión de Sedimentología, v. 2: La Plata, Argentina, Asociación Sedimentológica Argentina, p. 283–288.

Banchig, L. A., Milana, J. P., and Bordonaro, O. L., 1990a, Litofacies clásticas de la Formación Los Sombreros (Cámbrico medio) en la Quebrada Ojos de Agua, Sierra del Tontal, San Juan, *in* Actas, 3ª Reunión de Sedimentología: San Juan, Argentina, Asociación Sedimentológica Argentina, p. 25–30.

Banchig, L. A., Keller, M., and Milana, J. P., 1990b, Brechas calcáreas de la Formación Los Sombreros, Quebrada Ojos de Agua, Sierra del Tontal, San Juan, *in* Actas, XI Congreso Geológico Argentino, v. 2: San Juan, Argentina, Asociación Sedimentológica Argentina, p. 149–152.

Baraldo, J. A., Monetta, A. M., and Soechting, W., 1990, Triásico de San Juan: XI Congreso Geológico Argentino: San Juan, Argentina, Relatorio, Asociación Geológica Argentina, p. 124–139.

Bathurst, R. G. C., 1971, Carbonate sediments and their diagenesis: Developments in Sedimentology 12: Elsevier, Amsterdam, 658 p.

Beer, J. A., 1990, Steady sedimentation and lithologic completeness, Bermejo Basin, Argentina: Journal of Geology, v. 98, p. 501–517.

Beer, J. A., and Jordan, T. E., 1989, The effects of Neogene thrusting on deposition in the Bermejo Basin, Argentina: Journal of Sedimentary Petrology, v. 59, p. 330–345.

Benedetto, J. L., 1985, El hallazgo de la típica fauna de *Hirnantia* en el Ashgiliano tardío de la S. de Villicum, San Juan, *in* Actas, Reunión de Comunicaciones paleontológicas: Universidad Nacional de San Juan, p. 60–61.

Benedetto, J. L., 1986, The first typical *Hirnantia* fauna from South America (San Juan Province, Argentine Precordillera), *in* Racheboeuf, P. R., and Emig, C. C., eds., Les Brachiopodes fossiles et actuels, Biostratigraphie du Paléozoique: Brest, France, v. 4, p. 439–447.

Benedetto, J. L., 1987, Braquiópodos Clitambonitáceos de la Formación San Juan (Ordovícico temprano), Precordillera de San Juan, Argentina: Ameghiniana, v. 24, p. 95–108.

Benedetto, J. L., 1990, Los géneros *Cliftonia* y *Paromalomena* (Brachiopoda) en el Ashgilliano tardío de la Sierra de Villicum, Precordillera de San Juan: Ameghiniana, v. 27, p. 151–159.

Benedetto, J. L., 1993, La hipótesis de la aloctonía de la Precordillera Argentina: Un test estratigráfico y biogeográfico, *in* Actas, XII Congreso Geológico Argentino y II Congreso de Exploración de Hidrocarburos, v. 3: Mendoza, Argentina, Asociación Geológica Argentina e Instituto Argentino de Petróleo y Gas, p. 375–384.

Benedetto, J. L., 1995, Braquiópodos del Silúrico temprano (Llandoveriano) Malvinocáfrico, Formación La Chilca, Precordillera Argentina: Geobios, v. 28, p. 425–457.

Benedetto, J. L., 1998, Early Palaeozoic brachiopods and associated shelly faunas from western Gondwana: Their bearing on the geodynamic history of the pre-Andean margin, *in* Pankhurst, R. J., and Rapela, C. W., eds., The proto-Andean margin of Gondwana: Geological Society [London] Special Publication 142, p. 57–83.

Benedetto, J. L., and Herrera, Z., 1986, Braquiópodos del Suborden Strophomenidina de la Formación San Juan (Ordovícico temprano), Precordillera de San Juan, Argentina, *in* Actas, 4° Congreso Argentino de Paleontología y Bioestratigrafía, v. 4: Mendoza, Argentina, Asociación Paleontológica Argentina, p. 103–111.

Benedetto, J. L., and Vaccari, N. E., 1992, Significado estratigráfico y tectónico de los complejos de bloques resedimentados Cambro-Ordovícicos de la Precordillera Occidental, Argentina: Estudios Geológicos, v. 48, p. 303–313.

Benedetto, J. L., Cañas, F. L., and Astini, R. A., 1986, Braquiópodos y trilobites de la zona de transición entre las formaciones San Juan y Gualcamayo en el área de Guandacol (La Rioja, Argentina), *in* Actas, 4° Congreso Argentino de Paleontología y Bioestratigrafía, v. 1: Mendoza, Argentina, Asociación Paleontológica Argentina, p. 103–111.

Benedetto, J. L., Rachboeuf, P. R., Herrera, Z., Brussa, E. D., and Toro, B. A., 1992, Brachiopodes et biostratigraphie de la Formation Los Espejos, Siluro-Dévonien de la Précordillère (NW Argentine): Geobios, v. 25, p. 599–637.

Benedetto, J. L., Vaccari, N. E., Carrera, M. G., and Sánchez, T. M., 1995, The evolution of faunal provincialism in the Argentine Precordillera during the Ordovician: New evidence and paleogeographic implications, *in* Cooper, J. D., Droser, M. L., and Finney, S. C., eds., Ordovician odyssey: Short papers for the Seventh International Symposium on the Ordovician System: Pacific Section, SEPM (Society for Sedimentary Geology) Publication 77, p. 181–184.

Bercowski, F., and Fernandez, A. E., 1988, Hallazgo de *Nuia* sp. en la Formación Los Sombreros, Cámbrico Medio, Sierra del Tontal, Precordillera Sanjuanina: Ameghiniana, v. 25, p. 187–188.

Bercowski, F., and Figueroa, G., 1987, Flujos piroclásticos en la Formación Albarracín, Terciario, Precordillera, San Juan, Argentina, *in* Actas, X Congreso Geológico Argentino, v. 4: San Miguel de Tucumán, Argentina, Asociacion Geológica Argentina, p. 225–227.

Bercowski, F., Keller, M., and Bordonaro, O. L., 1990, Litofacies de la Formación La Laja (Cámbrico) en la Sierra Chica de Zonda, Precordillera Sanjuanina, Argentina, *in* Actas, 3ª Reunión de Sedimentología: San Juan, Argentina, Asociación Sedimentológica Argentina, p. 31–36.

Beresi, M. S., 1986a, Paleoecología y Biofacies de la Formación San Juan al sur del paralelo 30° sur, Precordillera de San Juan [Ph.D. thesis]: San Juan, Argentina, Universidad Nacional de San Juan, 400 p.

Beresi, M. S., 1986b, Capas con *Archaeoscyphia* (porífera) en los sedimentos carbonáticos ordovícicos de la Precordillera de San Juan, *in* Actas, 1as Jornadas sobre Geología de Precordillera: San Juan, Argentina, 1985, Universidad Nacional de San Juan, p. 99–102.

Beresi, M. S., 1988, Sincronismo, diacronismo y cronología de la depositación calcárea ordovícica de la Precordillera argentina en base a biozonas de conodontes, *in* Actas, 5° Congreso Geológico Chileno, v. 2, Universidad de Chile, Santiago, p. 27–36.

Beresi, M. S., 1990, El Ordovícico de la Precordillera de San Juan: XI Congreso Geológico Argentino: San Juan, Argentina, Relatorio, Asociación Geológica Argentina, p. 32–46.

Beresi, M. S., and Bordonaro, O. L., 1984, La Formación San Juan en la Quebrada de Las Lajas, Sierra Chica de Zonda, Provincia de San Juan, *in* Actas, IX Congreso Geológico Argentino: v. 1, San Carlos de Bariloche, Río Negro, Argentina, Asociación Geológica Argentina, p. 95–107.

Beresi, M. S., and Rigby, K. J., 1993, The Lower Ordovician sponges of San Juan, Argentina: Brigham Young University Geology Studies, v. 39, p. 1–64.

Beresi, M. S., and Rigby, K. J., 1994, Sponges and chancelloriids from the Cambrian of western Argentina: Journal of Paleontology, v. 68, p. 208–217.

Beresi, M. S., Bordonaro, O. L., Toro, E., and Heredia, S., 1987, Paleoecología y paleoambientes de la Formación San Juan (Ordovícico inferior) de la quebrada de Las Lajas, Sierra Chica de Zonda, Precordillera de San Juan, Argentina, *in* Actas, 4° Congreso Latinoamericano de Paleontología, v. 1: Santa Cruz de la Sierra, Bolivia, p. 17–25.

Bergström, S. M., Huff, W. D., Kolata, D. R., and Cingolani, C. A., 1994, Ordovician K-bentonites discovered in the Precordillera of Argentina. A preliminary comparison with K-bentonite complexes in Laurentia and Baltoscandia: Joint Meeting of IGCP Projects 376 and 319 and the 2nd Circum-Atlantic and Circum-Pacific Terrane Conference, Program with Abstracts, p. 6.

Bird, J. M., and Dewey, J. F., 1970, Lithosphere plate-continental margin tectonics and the evolution of the Appalachian orogen: Geological Society of America Bulletin, v. 81, p. 1031–1060.

Blasco, G., and Ramos, V. A., 1976, Graptolitos caradocianos de la Formación Yerba Loca y del Cerro La Chilca. Dep. Jáchal, San Juan: Ameghiniana, v. 13, p. 312–329.

Bodenbender, G., 1902, Contribución al conocimiento de la Precordillera y de las sierras centrales de la República Argentina: Boletín de la Academía Nacional de Ciencias de Córdoba, v. 18, p. 203–264.

Bodenbender, G., 1911, Constitución geológica de la parte meridional de la Rioja y regiones limítrofes: Boletín de la Academía Nacional de Ciencias de Córdoba, v. 19, p. 5–220.

Bond, G. C., Nickeson, P. A., and Kominz, M. A., 1984, Breakup of a supercontinent between 625 Ma and 555 Ma: New evidence and implications for continental histories: Earth and Planetary Science Letters, v. 70, p. 325–345.

Bond, G. C., Kominz, M. A., and Grotzinger, J. P., 1988, Cambro-Ordovician eustasy: evidence from geophysical modelling of subsidence in Cordilleran and Appalachian passive margins, *in* Kleinspehn, K. L., and Paola, C., eds., New perspectives in basin analysis: New York, Springer, p. 129–160.

Bond, G. C., Kominz, M. A., and Steckler, M. S., 1989, Role of thermal subsidence, flexure, and eustasy in the evolution of early Paleozoic passive-margin carbonate platforms, *in* Crevello, P. D., Wilson, J. L., Sarg, J. F., and Read, J. F., eds., Controls on carbonate platform and basin development: Society of Economic Paleontologists and Mineralogists Special Publication 44, p. 39–61.

Bordonaro, O. L., 1980, El Cámbrico en la Quebrada de Zonda, San Juan: Revista de la Asociación Geológica Argentina, v. 35, p. 26–40.

Bordonaro, O. L., 1985, El Cámbrico de San Isidro, Mendoza, en facies de borde externo y talud de la plataforma calcárea de Precordillera, *in* Actas, 1as Jornadas sobre Geología de Precordillera: San Juan, Argentina, 1985, Universidad Nacional de San Juan, p. 12–17.

Bordonaro, O. L., 1986, Bioestratigrafía del Cámbrico inferior de San Juan, *in* Actas, 4° Congreso Argentino de Paleontología y Bioestratigrafía, v. 1: Mendoza, Argentina, Asociación Paleontológica Argentina, p. 19–27.

Bordonaro, O. L., 1989, Biogeografía y evolución gondwánica durante el Paleozóico inferior en América Latina: Universidad Nacional de Tucumán, Serie Correlación Geológica, v. 5, p. 25–30.

Bordonaro, O. L., 1990a, El sistema Cámbrico de la Provincia de San Juan: XI Congreso Geológico Argentino: San Juan, Argentina, Relatorio, Asociación Geológica Argentina, p. 18–30.

Bordonaro, O. L., 1990b, Biogeografía de trilobites Cámbricos en la Precordillera Argentina: Universidad Nacional de Tucumán, Serie Correlación Geológica, v. 7, p. 25–30.

Bordonaro, O. L., 1992, El Cámbrico de Sudamerica, *in* Gutiérrez-Marco, J. C., Saavedra, J., and Rábano, I., eds., Paleozoico Inferior de Ibero-América: Publicaciones de la Universidad de Extremadura, p. 69–84.

Bordonaro, O. L., and Baldis, B. A., 1987, *Tonkinella stephensis* (Trilobita) en el Cámbrico medio de la Sierra del Tontal, Precordillera de San Juan, Argentina, y su importancia, *in* Actas, 4° Congreso Latinoamericano de Paleontología: Santa Cruz de la Sierra, Bolivia, p. 5–15.

Bordonaro, O. L., and Banchig, A. L., 1990, Nuevos trilobites del Cámbrico Medio en la Quebrada Ojos de Agua, Sierra de Tontal, San Juan, Argentina: Universidad Nacional de Tucumán, Serie Correlación Geológica, v. 7, p. 31–37.

Bordonaro, O. L., and Banchig, A. L., 1995, Trilobites laurénticos en el Cámbrico de la Precordillera argentina, *in* Actas, 6° Congreso Argentino de Paleontología y Bioestratigrafía: Trelew, Chubut, Argentina, Asociación Paleontológica Argentina, p. 59–65.

Bordonaro, O. L., and Banchig, A. L., 1996, Estratigrafía de los olistolitos Cámbricos de la Precordillera Argentina, *in* Actas, XIII Congreso Geológico Argentino y III Congreso de Exploración de Hidrocarburos, v. 5: Buenos Aires, Asociación Geológica Argentina e Instituto Argentino de Petróleo y Gas, p. 471–479.

Bordonaro, O. L., and Liñán, E., 1994, Some Middle Cambrian Agnostoids from the Precordillera Argentina: Revista Española de Paleontología, v. 9, p. 105–114.

Bordonaro, O. L., and Peralta, S. H., 1987, El Arenigiano inferior de la Formación Empozada en la localidad de San Isidro, Mendoza, Argentina, *in* Actas, X Congreso Geológico Argentino, v. 3: San Miguel de Tucumán, Argentina, Asociación Geológica Argentina, p. 81–94.

Bordonaro, O. L., Aceñolaza, G. F., Pereyra, M. E., 1992, Primeras trazas fósiles de la Sierra Pie de Palo, San Juan, Argentina: Universidad Nacional de San Juan, Argentina, Revista Ciencias, v. 1, p. 7–14.

Bordonaro, O. L., Beresi, M., and Keller, M., 1993, Reinterpretación estratigráfica del Cámbrico del area de San Isidro, Precordillera de Mendoza, *in* Actas, XII Congreso Geológico Argentino y II Congreso de Exploración de Hidrocarburos, v. 2: Mendoza, Argentina, Asociación Geológica Argentina e Instituto Argentino de Petróleo y Gas, p. 12–19.

Bordonaro, O. L., Keller, M., and Lehnert, O., 1996, El Ordovícico de Ponon Trehue en la Provincia de Mendoza (Argentina): redefiniciones estratigráficas, *in* Actas, XII Congreso Geológico Argentino y II Congreso de Exploración de Hidrocarburos, v. 1: Mendoza, Argentina, Asociación Geológica Argentina e Instituto Argentino de Petróleo y Gas, p. 541–550.

Borrello, A., 1962, Caliza La Laja (Cámbrico Medio de San Juan): Buenos Aires, Argentina, Notas y Comunicaciones de Investigaciones Científicas v. 2, p. 3–8.

Borrello, A., 1969, Los geosinclinales de la Argentina: Buenos Aires, Argentina, Dirección Nacional de Geología y Minería, Anales: v. 14, 136 p.

Borrello, A., 1971, The Cambrian of South America, *in* Holland, C., ed., Cambrian of the New World, New York, Wiley, p. 385–438.

Bova, J. A., and Read, J. F., 1987, Incipiently drowned facies within a cyclic peritidal ramp sequence, Early Ordovician Chepultepec interval, Virginia Appalachians: Geological Society of America Bulletin, v. 98, p. 714–727.

Bowman, M. B. J., 1983, The genesis of algal nodule limestones from the Upper Carboniferous (San Emiliano Formation) of N.W. Spain, *in* Peryt, T., ed., Coated grains: Berlin, Heidelberg, New York, Springer, p. 409–423.

Bowring, S. A., and Erwin, D. H., 1998, A new look at evolutionary rates in deep time: Uniting paleontology and high-precision geochronology: GSA Today, v. 8, no. 9, p. 1–8.

Brenchley, P. J., and Newall, G., 1984, Late Ordovician environmental changes and their effects on faunas, *in* Bruton, D. L., ed., Aspects of the Ordovician System: Oslo Universitetsforlaget, v. 295, p. 65–79.

Brenchley, P. J., Romano, M., Young, T. P., and Storch, P., 1991, Hirnantian glaciomarine diamictites—Evidence for the spread of glaciation and its effect on Upper Ordovician faunas, *in* Barnes, C. R., and Williams, S. H., eds., Advances in Ordovician geology: Geological Survey of Canada Paper 90-9, p. 325–336.

Brussa, E. D., 1994, Las graptofaunas Ordovícicas del sector central de la Precordillera Occidental sanjuanina, Argentina [Ph.D. thesis]: Córdoba, Argentina, Universidad Nacional de Córdoba, 323 p.

Brussa, E. D., 1995, Preliminary analysis of the distribution of early Ordovician graptolites from the Argentine Precordillera, *in* Cooper, J. D., Droser, M. L., and Finney, S. C., eds., Ordovician odyssey: Short papers for the Seventh International Symposium on the Ordovician System: Pacific Section, SEPM (Society for Sedimentary Geology) Publication 77, p. 185–188.

Buggisch, W., and Astini, R. A., 1993, The Late Ordovician Ice age—New evidence from the Argentine Precordillera, *in* Findlay, R. H., ed., Gondwana eight: Rotterdam, Balkema, p. 439–447.

Buggisch, W., Bachtadse, V., and Von Gosen, W., 1994a, The middle Carboniferous glaciation of the Argentine Precordillera (San Juan, Mendoza)—New data to basin evolution and palaeomagnetism: Zentralblatt für Geologie und Paläontologie, Teil I, p. 287–307.

Buggisch, W., Von Gosen, W., Henjes-Kunst, F., and Krumm, S., 1994b, The age of early Paleozoic deformation and metamorphism in the Argentine Precordillera—Evidence from K-Ar data: Zentralblatt für Geologie und Paläontologie, Teil I, p. 275–286.

Burchette, T. P., and Wright, V. P., 1992, Carbonate ramp depositional systems: Sedimentary Geology, v. 79, p. 3–57.

Burne, R. V., and James, N. P., 1986, Subtidal origin of club-shaped stromatolites, Hamelin Pool: International Sedimentological Congress, 12th, Abstracts, p. 49.

Burne, R. V., and Moore, L. S., 1987, Microbialites: Organosedimentary deposits of benthic microbial communities: Palaios, v. 2, p. 241–254.

Cabaleri, N. G., 1989, Fluctuaciones paleoambientales como microeventos dentro del Cámbrico de la Precordillera central de Jáchal (San Juan): Universidad Nacional de Tucumán, Serie Correlación Geológica, v. 5, p. 145–162.

Cañas, F. L., 1988, Facies perimareales del Cámbrico inferior en el área de Guandacol, *in* Actas, 2ª Reunión de Sedimentología Argentina: Buenos Aires, Argentina, Asociación Geológica Argentina, p. 46–50.

Cañas, F. L., 1995a, Estratigrafía y evolución paleoambiental de las sucesiones carbonáticas del Cámbrico tardío y Ordovícico temprano de la Precordillera Septentrional, República Argentina [Ph.D. thesis]: Córdoba, Argentina, Universidad Nacional de Córdoba, 215 p.

Cañas, F. L., 1995b, Early Ordovician carbonate platform facies of the Argentine Precordillera: Restricted shelf to open platform evolution, *in* Cooper, J. D., Droser, M. L., and Finney, S. C., eds., Ordovician odyssey: Short papers for the Seventh International Symposium on the Ordovician System: Pacific Section, SEPM (Society for Sedimentary Geology) Publication 77, p. 221–224.

Cañas, F. L., and Carrera, M. G., 1993, Early Ordovician microbial-sponge-receptaculitid bioherms of the Precordillera, western Argentina: Facies, v. 29, p. 169–178.

Cañas, F. L., and Keller, M., 1993, "Reefs" y "Reef Mounds" en la Formación San Juan (Precordillera Sanjuanina, Argentina): Los arrecifes más antiguos de Sudamérica: Boletín de la Real Sociedad de Historía Natural (Geología), v. 88, p. 127–136.

Carrera, M. G., 1994, Taxonomía, bioestratigrafía y significado paleoambiental de los poríferos y briozoos del Ordovícico de la Precordillera Argentina [Ph.D. thesis]: Córdoba, Argentina, Universidad Nacional de Córdoba, 192 p.

Cayeux, L., 1935, Les roches sédimentaires de France. Roches carbonatées (calcium et dolomies): Paris, Masson, 436 p.

Chow, N., and James, N. P., 1987, Cambrian grand cycles: A northern Appalachian perspective: Geological Society of America Bulletin, v. 98, p. 418–429.

Cingolani, C. A., and Cuerda, A., 1996, El Ordovícico del flanco oriental del Cerro Bola, en la Sierra Pintada de San Rafael, Provincia de Mendoza, in Actas, XIII Congreso Geológico Argentino y III Congreso de Exploración de Hidrocarburos, v. 1: Buenos Aires, Asociación Geológica Argentina e Instituto Argentino de Petróleo y Gas, p. 369.

Cingolani, C. A., Cuerda, A. J., Varela, R., and Schauer, O., 1987, Estratigrafía y estructura de la sierra del Tontal, Precordillera de San Juan, Argentina, in Actas, X Congreso Geológico Argentino, v. 3: San Miguel de Tucumán, Argentina, Asociación Geológica Argentina, p. 95–98.

Cingolani, C. A., Cuerda, A. J., Varela, R., and Schauer, O., 1989, Geología de la Precordillera Occidental en la Comarca de la Sierra del Tontal, Provincia de San Juan, República Argentina: Departamento de Geología, Universidad de Chile, Revista Comunicaciones, v. 40, p. 39–56.

Cocks, L. R. M., and Fortey, R. A., 1990, Biogeography of Ordovician and Silurian faunas, in McKerrow, W. S., and Scotese, C. R., eds., Palaeozoic palaeogeography and biogeography: Geological Society [London] Memoir 12, p. 97–104.

Cominguez, A. H., and Ramos, V. A., 1990, Sísmica de reflexión profunda entre Precordillera y Sierras Pampeanas, in Actas, XI Congreso Geológico Argentino, v. 2: San Juan, Argentina, Asociación Geológica Argentina, p. 311–313.

Cook, H. E., and Mullins, H. T., 1983, Basin margin environment, in Scholle, P. A., Bebout, D. G., and Moore, C. H., eds., Carbonate depositional environments: American Association of Petroleum Geologists Memoir 33, p. 539–617.

Cook, H. E., McDaniel, P. N., and Mountjoy, E. W., 1972, Allochthonous carbonate debris flows at Devonian bank ("reef") margins Alberta, Canada: Canadian Petroleum Geology Bulletin, v. 20, p. 439–497.

Cook, H. E., Field, M. E., and Gardner, J. V., 1982, Characteristics of sediments on modern and ancient continental slopes, in Scholle, P. A., and Spearing, D. R., eds., Sandstone depositional environments: American Association of Petroleum Geologists Memoir 31, p. 329–364.

Cooper, J. D., and Edwards, J. C., 1991, Cambro-Ordovician craton-margin carbonate section, southern Great Basin: A sequence-stratigraphic perspective, in Cooper, J. D., and Stevens, C. H., eds., Paleozoic paleogeography of the western United States—II: Pacific Section, SEPM (Society for Sedimentary Geology) Publication, p. 237–252.

Cooper, J. D., and Keller, M., 1995, Ordovician craton margin—miogeoclinal transition, Southern Great Basin, in Cooper J. D., ed., Ordovician of the Great Basin: Pacific Section, SEPM (Society for Sedimentary Geology) Publication 78, p. 107–132.

Criado Roque, P., and Ibañez, G., 1979, Provincia Geológica Sanrafaelino-Pampeana, in Leanza, A. F., ed., Geología Regional Argentina: Academía Nacional de Ciencias de Córdoba, v. 1, p. 837–869.

Cuerda, A., 1965, *Monograptus leintwardinensis* var. *incipiens* Wood en el Silúrico de la Precordillera: Ameghiniana, v. 4, p. 171–177.

Cuerda, A., 1969, Sobre las graptofaunas del Silúrico de San Juan: Ameghiniana, v. 6, p. 223–235.

Cuerda, A., 1971, Monograpten aus dem Unter-Ludlow aus der Vorkordillere von San Juan, Argentinien: Geologisches Jahrbuch, v. 85, p. 391–406.

Cuerda, A., 1981, Graptolitos del Silúrico inferior en la Formación Rinconada, Precordillera de San Juan: Ameghiniana, v. 18, p. 241–247.

Cuerda, A., 1985, Estratigrafía y bioestratigrafía del Silúrico de San Juan (Argentina) basada en sus faunas de graptolitos: Ameghiniana, v. 22, p. 233–241.

Cuerda, A., 1988, Investigaciones bioestratigráficas en el Grupo Villavicencio, Precordillera de Mendoza y San Juan, Argentina, in Actas, 5° Congreso Geológico Chileno, v. 2, Universidad de Chile, Santiago, p. 177–187.

Cuerda, A., and Furque, G., 1986, Graptolitos del techo de la Formación San Juan, Precordillera de San Juan, in Actas, 1as Jornadas sobre Geología de Precordillera: San Juan, Argentina, 1985, Universidad Nacional de San Juan, p. 113–118.

Cuerda, A., Cingolani, C. A., and Varela, R., 1983, Las graptofaunas de la Formación Los Sombreros, Ordovícico inferior, de la vertiente oriental de la Sierra de Tontal, Precordillera de San Juan: Ameghiniana, v. 20, p. 239–260.

Cuerda, A., Cingolani, C. A., Schauer, O., and Varela, R., 1985a, El Ordovícico de la Sierra de Tontal, Precordillera de San Juan, República Argentina, in Actas, 4° Congreso Geológico Chileno, Universidad de Chile, p. 190–192.

Cuerda, A., Cingolani, C. A., Varela, R., Schauer, O., Baldis, B. A., and Bordonaro, O. L., 1985b, Hallazgo de sedimentitas Cámbricas fosilíferas en la Sierra de Tontal (Precordillera de San Juan): Ameghiniana, v. 22, p. 281–282.

Cuerda, A., Cingolani, C. A., Varela, R., and Schauer, O., 1986, Cámbrico y Ordovícico en la Precordillera de San Juan: Formación Los Sombreros. Ampliación de su conocimiento bioestratigráfico, in Actas, 4° Congreso Argentino de Paleontología y Bioestratigrafía, v. 1: Mendoza, Argentina, Asociación Paleontológica Argentina, p. 5–17.

Cuerda, A., Cingolani, C. A., Varela, R., and Schauer, O., 1987, Graptolites Ordovícicos del "Grupo Villavicencio," flanco sudoriental de la Sierra del Tontal en el área de Santa Clara, Precordillera de San Juan-Mendoza, República Argentina, in Actas, 4° Congreso Latinoamericano de Paleontología, v. 1: Santa Cruz de la Sierra, Bolivia, p. 111–118.

Cuerda, A., Lavandaio, E., Arrondo, O., and Morel, E., 1988a, Investigaciones estratigráficas en el "Grupo Villavicencio," Canota, Provincia de Mendoza: Revista de la Asociación Geológica Argentina, v. 43, p. 356–365.

Cuerda, A., Rickards, R. B., and Cingolani, C. A., 1988b, A new Ordovician-Silurian boundary section in San Juan Province, Argentina, and its definitive graptolitic fauna: Geological Society of London Journal, v. 145, p. 749–757.

Cuerda, A., Rickards, R. B., and Cingolani, C. A., 1988c, The Ordovician-Silurian boundary in Bolivia and Argentina: British Museum of Natural History Bulletin (Geology), v. 43, p. 291–294.

Dalla Salda, L. H., and Varela, R., 1984, El metamorfísmo en el tercio sur de la sierra Pie de Palo, San Juan: Revista de la Asociación Geológica Argentina, v. 39, p. 68–93.

Dalla Salda, L. H., Cingolani, C. A., and Varela, R., 1992a, Early Paleozoic orogenic belt of the Andes in southwestern South America: Results of Laurentia-Gondwana collision?: Geology, v. 20, p. 617–621.

Dalla Salda, L. H., Dalziel, I. W. D., Cingolani, C. A., and Varela, R., 1992b, Did the Taconic Appalachians continue into southern South America?: Geology, v. 20, p. 1059–1062.

Dalla Salda, L. H., Cingolani, C. A., and Varela, R., 1993, A pre-Carboniferous tectonic model in the evolution of southern South America: International Congress on the Carboniferous-Permian, Comptes Rendus, v. 12, p. 371–384.

Dalziel, I. W. D., 1991, Pacific margins of Laurentia and East Antarctica–Australia as a conjugate rift pair: Evidence and implications for an Eocambrian supercontinent: Geology, v. 19, p. 598–601.

Dalziel, I. W. D., 1992, On the organization of American plates in the Neoproterozoic and the breakout of Laurentia: GSA Today, v. 2, p. 240–241.

Dalziel, I. W. D., 1993, Tectonic tracers and the origin of the Proto-Andean margin, *in* Actas, XII Congreso Geológico Argentino y II Congreso de Exploración de Hidrocarburos, v. 3: Mendoza, Asociación Geológica Argentina e Instituto Argentino de Petróleo y Gas, p. 367–374.

Dalziel, I. W. D., 1997, Terminal Proterozoic–early Paleozoic geography, tectonics, and environments: Geological Society of America Bulletin, v. 109, p. 16–42.

Dalziel, I. W. D., and Dalla Salda, L. H., 1996, The early Paleozoic evolution of the Argentine Precordillera as a Laurentian rifted, drifted, and collided terrane: A geodynamic model: Discussion: Geological Society of America Bulletin, v. 108, p. 372–373.

Dalziel, I. W. D., Dalla Salda, L. H., and Gahagan, L. M., 1994, Paleozoic Laurentia-Gondwana interaction and the origin of the Appalachian-Andean mountain system: Geological Society of America Bulletin, v. 106, p. 243–252.

Dalziel, I. W. D., Dalla Salda, L. H., Cingolani, C. A., and Palmer, A. R., 1996, The Argentine Precordillera: A Laurentian terrane?: GSA Today, v. 6, p. 16–18.

Davicino, R. E., and Sabalúa, J. C., 1990, El cuerpo básico de El Nihuil, Dpto. San Rafael, Pcia. de Mendoza, Rep. Argentina, *in* Actas, XI Congreso Geológico Argentino, v. 1: San Juan, Argentina, Asociación Geológica Argentina, p. 43–47.

Davis, J. S., 1997, The early to middle Paleozoic tectonic history of the SW Precordillera terrane and Laurentia-Gondwana interactions: Geological Society of America Abstracts with Programs, v. 29, no. 6, p. A380.

Demicco, R. V., 1985, Platform and off-platform carbonates of the Upper Cambrian of western Maryland: Sedimentology, v. 32, p. 1–22.

Dickerson, P. W., and Keller, M., 1998, The Argentine Precordillera: Its odyssey from the Laurentian Ouachita margin toward the Sierras Pampeanas of Gondwana, *in* Pankhurst, R. J., and Rapela, C. W., eds., The proto-Andean margin of Gondwana: Geological Society [London] Special Publication 142, p. 85–105.

Dietz, R. S., Holden, J. C., and Sproll, W. P., 1970, Geotectonic evolution and subsidence of Bahama platform: Geological Society of America Bulletin, v. 81, p. 147–151.

Dill, R. F., Shinn, E. A., Jones, A. T., Kelly, K., and Steinen, R. P., 1986, Giant subtidal stromatolites forming in normal salinity waters: Nature, v. 324, p. 55–58.

Donovan, R. N., and Ragland, D. A., 1991, Stop 10: The base of the Arbuckle Group at Bally Mountain, *in* Johnson, K. S., ed., Arbuckle Group core workshop and field trip: Oklahoma Geological Survey Special Publication, 91-3, p. 244–246.

Donovan, R. N., Ragland, D. A., Cloyd, K., Bridges, S., and Denison, R. E., 1988, Carlton Rhyolite and lower Paleozoic sedimentary rocks at Bally Mountain in the Slick Hills of southwestern Oklahoma, *in* Hayward, O. T., ed., Geological Society of America south-central section, Centennial Field Guide: v. 4, Boulder, Colorado, Geological Society of America, p. 93–96.

Dott, R. H., Jr., and Bourgeois, J., 1982, Hummocky stratification: Significance of its variable bedding sequences: Geological Society of America Bulletin, v. 93, p. 663–680.

Drake, C. L., and Burk, C. A., 1974, Geological significance of continental margins, *in* Burk, C. A., and Drake, C. L., eds., The geology of continental margins: New York, Springer, p. 3–12.

Eberlein, S., 1990, Conodontenstratigraphie und Fazies der Formation Las Aguaditas (Ordovizium/Argentinische Präkordillere) [Diploma thesis]: Germany, Universität Erlangen, 83 p.

Elliott, T., 1978, Clastic shorelines, *in* Reading, H. G., ed., Sedimentary environments and facies: Oxford, Blackwell Scientific Publications, p. 143–177.

Embry, A. F., and Dixon, J., 1990, The breakup unconformity of the Amerasia basin, Arctic ocean: evidence from Arctic Canada: Geological Society of America Bulletin, v. 102, p. 1526–1534.

Espisua, E., 1968, El Paleozoico inferior del Río de las Chacritas, Depto. de Jáchal, Prov. de San Juan, con especial referencia al Silúrico: Revista de la Asociación Geológica Argentina, v. 23, p. 297–311.

Etheridge, M. A., Symonds, P. A., and Lister, G. S., 1989, Application of the detachment model to reconstruction of conjugate passive margins, *in* Tankard, A. J., and Balkwill, H. R., eds., Extensional tectonics and stratigraphy of the North Atlantic margins: American Association of Petroleum Geologists Memoir 46, p. 23–40.

Ethington, R. L., and Clark, D. L., 1971, Lower Ordovician conodonts in North America, *in* Sweet, W. C., and Bergström, S. M., eds., Symposium on conodont biostratigraphy: Geological Society of America Memoir 127, p. 63–82.

Ethington, R. L., and Clark, D. L., 1981, Lower and Middle Ordovician conodonts from the Ibex area, western Millard County, Utah: Brigham Young University Geology Studies, v. 28, p. 1–155.

Falvey, D. A., 1974, The development of continental margins in plate tectonic theory: Australian Petroleum Exploration Association Journal, v. 14, p. 95–106.

Flügel, E., 1982, Microfacies analysis of limestones: Berlin, Heidelberg, New York, Springer, 610 p.

Fordham, B. G., 1992, Chronometric calibration of Middle Ordovician to Tournaisian conodont zones: A compilation from recent graphic-correlation and isotope studies: Geological Magazine, v. 129, p. 709–721.

Fortey, R. A., 1984, Global earlier Ordovician transgressions and regressions and their biological implications, *in* Bruton, D. L., ed., Aspects of the Ordovician System: Oslo, Universitetsforlaget, v. 295, p. 37–50.

Fortey, R. A., and Cocks, L. R. M., 1992, The early Paleozoic of the North Atlantic region as a test case for the use of fossils in continental reconstruction: Tectonophysics, v. 206, p. 147–158.

Frisch, W., 1979, Tectonic progradation and plate tectonic evolution of the Alps: Tectonophysics, v. 60, p. 121–139.

Furque, G., 1972, Descripción geológica de la hoja 16b, Cerro La Bolsa: Buenos Aires, Argentina, Boletín del Servicio Geológico Nacional, v. 125, 67 p.

Furque, G., 1979, Descripción geológica de la hoja 18c, Jáchal (provincia de San Juan): Buenos Aires, Argentina, Boletín del Servicio Geológico Nacional, v. 164, 79 p.

Furque, G., 1983, Descripción geológica de la Hoja 19c, Ciénaga de Gualilán: Buenos Aires, Argentina, Boletín del Servicio Geológico Nacional, v. 193, 111 p.

Furque, G., and Caballé, M. F., 1990, Depósitos marinos del Paleozoico Medio en la Precordillera Central de San Juan, Argentina, *in* Actas, XI Congreso Geológico Argentino, v. 2: San Juan, Argentina, Asociación Geológica Argentina, p. 81–84.

Furque, G., and Cuerda, A., 1979, Precordillera de La Rioja, San Juan y Mendoza, *in* Leanza, A. F., ed., Geología Regional Argentina: Academía Nacional de Ciencias de Córdoba, v. 1, p. 455–522.

Furque, G., and Cuerda, A., 1982, Extensión y edad de los movimientos de la Fase Guandacol en la Precordillera del oeste Argentino, *in* Actas, 5° Congreso Latinoamericano de Geología, v. 1: Buenos Aires, Argentina, Servicio Geológica Argentina, p. 191–200.

Gallardo, G., and Heredia, S., 1995, Estratigrafía y sedimentología del miembro inferior de la Formación Empozada (Ordovícico Medio y Superior), Precordillera Argentina: Boletín de la Academía Nacional de Ciencias de Córdoba, v. 60, p. 449–460.

Gallardo, G., Heredia, S., and Maldonado, A., 1988, Depósitos carbonáticos alóctonos, Miembro Superior de la Formación Empozada, Ordovícico Superior de la Precordillera de Mendoza, Argentina, *in* Actas, 5° Congreso Geológico Chileno, v. 1: Santiago, Chile, Universidad de Chile, p. 37–53.

Garrett, P., 1977, Biological communities and their sedimentary record, *in* Hardie, L. A., ed., Sedimentation on modern carbonate tidal flats of Northwest Andros Island, Bahamas: John Hopkins Studies in Geology, v. 22, p. 124–158.

Garrison, R. E., and Ramirez, P. C., 1989, Conglomerates and breccias in the Monterey Formation and related units as reflections of basin margin history, *in* Colburn, I. P., Abbott, P. L., and Minch, J., eds., Conglomerates in basin analysis: A symposium dedicated to A. O. Woodford: Pacific Section, Society of Economic Paleontologists and Mineralogists Publications 62, p. 189–206.

Goldhammer, R. K., Lehmann, E. J., and Dunn, P. A., 1993, The origin of high-frequency platform carbonate cycles and third-order sequences (Lower Ordovician El Paso Group, west Texas): Constraints from outcrop data and stratigraphic modeling: Journal of Sedimentary Petrology, v. 63, p. 318–359.

Gonzalez Bonorino, G., 1975a, Sedimentología de la Formación Punta Negra y algunas consideraciones sobre la geología regional de la Precordillera de San Juan y Mendoza: Revista de la Asociación Geológica Argentina, v. 30, p. 223–246.

Gonzalez Bonorino, G., 1975b, Acerca de la existencia de la Protoprecordillera de Cuyo, in Actas, VI Congreso Geológico Argentino, v. 1: Bahia Blanca, Buenos Aires, Asociación Geológica Argentina, p. 101–107.

Gonzalez Bonorino, G., and Gonzalez Bonorino, F., 1991, Precordillera de Cuyo y Cordillera Frontal en el Paleozoico temprano: terrenos bajo sospecha de ser autóctonos: Revista Geológica de Chile, v. 18, p. 97–107.

Haller, M. J., and Ramos, V. A., 1984, Las ofiolitas famatinianas (eopaleozóico) de la provincia de San Juan: in Actas, IX Congreso Geológico Argentino, v. 2: San Carlos de Bariloche, Argentina, Asociación Geológica Argentina, p. 66–83.

Halley, R. B., Harris, P. M., and Hine, A. C., 1983, Bank margin environment, in Scholle, P. A., Bebout, D. G., and Moore, C. H., eds., Carbonate depositional environments: American Association of Petroleum Geologists Memoir 33, p. 464–506.

Handford, C. R., and Loucks, R. G., 1993, Carbonate depositional sequences and system tracts—responses of carbonate platforms to relative sea-level changes, in Loucks, R. G., and Sarg, J. F., eds., Carbonate sequence stratigraphy—Recent developments and applications: American Association of Petroleum Geologists Memoir 57, p. 3–41.

Hardie, L. A., 1977, Sedimentation on the modern carbonate tidal flats of northwest Andros Island, Bahamas: John Hopkins University Studies in Geology, v. 22, 202 p.

Harrington, H., 1957, Ordovician Formations of Argentina, in Harrington, H., and Leanza, A. F., eds., Ordovician trilobites of Argentina: University of Kansas Special Publication 1, p. 1–59.

Harrington, H., 1971, Descripción geológica de la Hoja 22c: "Ramblón": Buenos Aires, Argentina, Boletín de la Dirección Nacional de Geología y Minería, v. 114, 81 p.

Harrington, H., and Leanza, A. F., eds., 1957, Ordovician trilobites of Argentina: University of Kansas Special Publication 1, 259 p.

Harris, P. M., 1979, Facies anatomy of a Bahamian ooid shoal: University of Miami, Comparative Sedimentology Laboratory Sedimenta, v. 7, p. 1–163.

Heezen, B. C., 1974, Atlantic-type continental margins, in Burk, C. A., and Drake, C. L., eds., The geology of continental margins: New York, Springer, p. 13–24.

Heim, A., 1948, Observaciones tectónicas en la Rinconada, Precordillera de San Juan: Buenos Aires, Argentina, Boletín de la Dirección de Mineralogía y Geología, v. 64, p. 1–38.

Heim, A., 1952, Estudios tectónicos de la Precordillera de San Juan; los ríos San Juan, Jáchal y Huaco: Revista de la Asociación Geológica Argentina, v. 7, p. 11–70.

Heredia, S., 1982, *Pygodus anserinus* Lamont & Lindström (Conodonta) en el Llandeiliano de la Formación Ponon Trehue, Provincia de Mendoza, Argentina: Ameghiniana, v. 19, p. 229–233.

Heredia, S., 1987, Zona de *Proconodontus tenuiserratus* (Conodonta), Cámbrico superior, Formación La Cruz, Mendoza, Argentina: Ameghiniana, v. 24, p. 147–150.

Heredia, S., 1990, Geología de la Cuchilla del Cerro Pelado, Precordillera de Mendoza, Argentina, in Actas, XI Congreso Geológico Argentino v. 2: San Juan, Argentina, Asociación Geológica Argentina, p. 101–104.

Heredia, S., 1995, Conodontes Cámbricos y Ordovícicos en los bloques alóctonos del conglomerado basal de la Formación Empozada, Ordovícico medio-superior, San Isidro, Precordillera de Mendoza, Argentina: Boletín de la Academía Nacional de Ciencias de Córdoba, v. 60, p. 235–243.

Heredia, S., 1996, El Cámbrico y Ordovícico de la Cuchilla del Cerro Pelado, Precordillera de Mendoza, Argentina, in Actas, XIII Congreso Geológico Argentino y III Congreso de Exploración de Hidrocarburos, v. 2: Buenos Aires, Argentina, Asociación Geológica Argentina e Instituto Argentino de Petróleo y Gas, p. 591–600.

Heredia, S., and Beresi, M., 1995, Ordovician events and sea level changes on the western margin of Gondwana: The Argentine Precordillera, in Cooper, J. D., Droser, M. L., and Finney, S. C., eds., Ordovician odyssey: Short papers for the Seventh International Symposium on the Ordovician System: Pacific Section, SEPM (Society for Sedimentary Geology) Publication 77, p. 315–318.

Heredia, S., and Bordonaro, O. L., 1986, Conodontes de la Formación la Cruz (Cámbrico Superior), San Isidro, provincia de Mendoza, R. Argentina, in Actas, 4° Congreso Argentino de Paleontología y Bioestratigrafía, v. 3: Mendoza, Argentina, Asociación Paleontologica Argentina, p. 189–202.

Heredia, S., Gallardo, G., and Maldonado, A., 1990, Conodontes Caradocianos en las calizas alóctonas del miembro superior de la Formación Empozada (Ordovícico Medio y Superior), San Isidro (Mendoza, Argentina): Ameghiniana, v. 27, p. 197–206.

Herrera, Z. A., 1985, Estratigrafía, paleontología y paleoambiente de la Formación Los Espejos (Silúrico) en el área de la Loma de los Piojos (Departamento Jáchal), Provincia de San Juan [tésis de licenciatura]: Córdoba, Argentina. Universidad Nacional Córdoba, 93 p.

Herrera, Z. A., 1993, Nuevas precisiones sobre la edad de la Formación Talacasto (Precordillera Argentina) en base a su fauna de braquiópodos, in Actas, XII Congreso Geológico Argentino y II Congreso de Exploración de Hidrocarburos, v. 2: Mendoza, Asociación Geológica Argentina e Instituto Argentino de Petróleo y Gas, p. 289–295.

Herrera, Z. A., and Benedetto, J. L., 1989, Braquiópodos del Suborden Orthidina de la Formación San Juan (Ordovícico temprano) en el área de Huaco-Cerro Viejo, Precordillera Argentina: Ameghiniana, v. 26, p. 3–22.

Herrera, Z. A., and Benedetto, J. L., 1991, Early Ordovician brachiopod faunas of the Precordillera basin, western Argentina: Biostratigraphy and paleobiogeographical affinities, in MacKinnon, D. I., Lee, D. E., and Campbell, J. D., eds., Brachiopods through time: Rotterdam, Balkema, p. 283–301.

Hine, A. C., 1977, Lily Bank, Bahamas: History of an active oolite sand shoal: Journal of Sedimentary Petrology, v. 47, p. 1554–1581.

Hiscott, R. N., 1995, Middle Ordovician clastic rocks (Humber Zone and St. Lawrence Platform), in Williams, H., ed., Geology of the Appalachian-Caledonian in Canada and Greenland: Geological Survey of Canada, Geology of Canada, v. 6, p. 87–98.

Hoffman, P. F., 1976, Stromatolite morphogenesis in Shark Bay, Western Australia, in Walter, R. M., ed., Stromatolites: Developments in Sedimentology 20: Elsevier, Amsterdam, p. 261–272.

Hoffman, P. F., 1991, Did the breakout of Laurentia turn Gondwanaland inside-out?: Science, v. 252, p. 1409–1412.

Holmer, L. E., Popov, L., and Lehnert, O., 1999, Cambrian phosphatic brachiopods from the Precordillera of western Argentina: Geologiska Föreningens i Stockholm Förhandlingar (in press).

Houseknecht, D. W., 1986, Evolution from passive margin to foreland basin: The Atoka Formation of the Arkoma basin, south-central U.S.A, in Allen, P. A., and Homewood, P., eds., Foreland basins: International Association of Sedimentologists Special Publication 8, p. 327–345.

Howell, D. G., Jones, D. L., and Schermer, E. R., 1985, Tectonostratigraphic terranes of the circum-Pacific region, in Howell, D. G., ed., Tectonostratigraphic terranes of the circum-Pacific region: Circum-Pacific Council for Energy and Mineral Resources, Earth Science Series, v. 1, p. 3–30.

Huff, W. D., Bergström, S. M., and Kolata, D. R., 1992, Gigantic Ordovician volcanic ash fall in North America and Europe: biological, tectonomagmatic, and event-stratigraphic significance: Geology, v. 20, p. 875–878.

Huff, W. D., Bergström, S. M., Kolata, D. R., Cingolani, C. A., and Davis, D. W., 1995, Middle Ordovician K-bentonites discovered in the Precordillera of Argentina: Geochemical and paleogeographical implications, in Cooper, J. D., Droser, M. L., and Finney, S. C., eds., Ordovician odyssey: Short papers for the Seventh International Symposium on the Ordovician System: Pacific Section, SEPM (Society for Sedimentary Geology) Publication 77, p. 343–349.

Huff, W. D., Bergström, S. M., Kolata, D. R., Cingolani, C. A., and Astini, R. A., 1998, Ordovician K-bentonites in the Argentine Precordillera: relations to Gondwana margin evolution, *in* Pankhurst, R. J., and Rapela, C. W., eds., The proto-Andean margin of Gondwana: Geological Society [London] Special Publication, v. 142, p. 107–126.

Hünicken, M. A., 1971, Sobre el hallazgo de conodontes en las calizas de la Formación San Juan (Ordovícico, Llanvirniano), Quebrada de Potrerillos, Sierra de Yanso, Dpto. Jáchal, (Prov. de San Juan): Ameghiniana, v. 8, p. 37–51.

Hünicken, M. A., 1975, Sobre el hallazgo de conodontes en el Silúrico de la Loma de los Piojos, Departamento Jáchal, Provincia de San Juan, *in* Actas, 1er Congreso Argentino de Paleontología y Bioestratigrafía, v. 1: San Miguel de Tucumán, Asociación Paleontológica Argentina, p. 282–291.

Hünicken, M. A., 1985, Lower Ordovician conodont biostratigraphy in Argentina: Boletín de la Academia Nacional de Ciencias de Córdoba, v. 56, p. 309–321.

Hünicken, M. A., and Ortega, G. C., 1987, Lower Llanvirn–lower Caradoc (Ordovician) conodonts and graptolites from the Argentine Central Precordillera, *in* Austin, R. L., ed., Conodonts: Investigative techniques and applications: Chichester, Ellis Horwood, p. 136–145.

Hünicken, M. A., and Pensa, M. V., 1985, Secuencia carbonática estromatolítica cambro-ordovícica (Formación Los Sapitos) en la Quebrada del Río Guandacol, La Rioja: Córdoba, Argentina, Academia Nacional de Ciencias de Córdoba, p. 1–11.

Hünicken, M. A., and Sarmiento, G., 1982, La zona baltoescandinava de *Oepikodus evae* (conodonto, Arenigiano Inferior) en el perfil del Río Guandacol, La Rioja, Argentina, *in* Actas, 5° Congreso Latinoamericano de Geología, v. 1: Buenos Aires, Argentina, Servicio Geológico Nacional, p. 791–796.

Hünicken, M. A., and Sarmiento, G., 1985, *Oepikodus evae* (lower Arenigian conodont) from Guandacol, La Rioja Province, Argentina: Boletín de la Academía Nacional de Ciencias de Córdoba, v. 56, p. 323–331.

Hünicken, M. A., and Sarmiento, G., 1988, Conodontes ludlovianos de la Formación Los Espejos, Provincia de San Juan, República Argentina, *in* Actas, 4° Congreso Argentino de Paleontología y Bioestratigrafía, v. 3: Mendoza, Argentina, Asociación Paleontológica Argentina, p. 225–233.

Hünicken, M. A., Albanesi, G. L., and Ortega, G. C., 1990, Conodonts and graptolites from the Yerba Loca Formation (Arenig-Caradoc), Ancaucha Creek, Cerro Alto de Mayo, Jáchal department, San Juan province, Argentina, *in* Hünicken, M. A., ed.: Latin-American Conodont Symposium, 1st: Córdoba, Argentina, Abstracts, Academia Nacional de Ciencias de Córdoba, p. 106–108.

Jacobi, R. D., 1981, Peripheral bulge—A causal mechanism for the Lower/Middle Ordovician unconformity along the western margin of the Northern Appalachians: Earth and Planetary Science Letters, v. 56, p. 245–251.

James, N. P., and Mountjoy, E. W., 1983, Shelf-slope break in fossil carbonate platforms: an overview, *in* Stanley, D. J., and Moore, G. T., eds., The shelfbreak: Critical interface on continental margins: Society of Economic Paleontologists and Mineralogists Special Publication 33, p. 189–206.

James, N. P., Tevens, R. K., Barnes, C. R., and Knight, I., 1989, Evolution of a lower Paleozoic continental-margin carbonate platform, Northern Canadian Appalachians, *in* Crevello, P. D., Wilson, J. L., Sarg, J. F., and Read, J. F., eds., Controls on carbonate platform and basin development: Society of Economic Paleontologists and Mineralogists Special Publication 44, p. 123–146.

Johnson, H. D., 1978, Shallow siliciclastic seas, *in* Reading, H. G., ed., Sedimentary environments and facies: Oxford, Blackwell Scientific Publications, p. 207–258.

Jordan, T. E., Isacks, B. L., Allmendinger, R. W., Brewer, J. A., Ramos, V. A., and Ando, C. J., 1983, Andean tectonics related to geometry of subducted Nazca plate: Geological Society of America Bulletin, v. 94, p. 341–361.

Kay, S. M., Ramos, V. A., and Kay, R., 1984, Elementos mayoritarios y trazas de las vulcanitas ordovícicas de la Precordillera occidental: basaltos de rift oceánico temprano(?) proximos al margen continental, *in* Actas, IX Congreso Geológico Argentino, v. 2: San Carlos de Bariloche, Argentina, Asociación Geológica Argentina, p. 48–65.

Kay, S. M., Orrell, S., and Abbruzzi, J. M., 1996, Zircon and whole rock Nd-Pb isotopic evidence for a Grenville age and Laurentian origin for the basement of the Precordilleran terrane in Argentina: Geological Society of America Abstracts with Programs, v. 28, no. 1, p. 21–22.

Kayser, E., 1876, Ueber Primordiale und untersilurische Fossilien aus der Argentinischen Republik: Beiträge zur Geologie und Paläontologie der Argentinischen Republik II, Palaeontologie Theil 1, Abteilung Palaeontologie Supplement, v. 3, p. 1–33.

Keidel, J., 1921, Observaciones geológicas en la Precordillera de San Juan y Mendoza: Buenos Aires, Argentina, Ministerio de Agricultura sección Geología y Mineralogía, Anales: v. 15, p. 7–102.

Keller, G. R., Braile, L. W., McMechan, G. A., Thomas, W. A., Harder, S. H., Chang, W. F., and Jardine, W. G., 1989, Paleozoic continent-ocean transition in the Ouachita Mountains imaged from PASSCAL wide-angle seismic reflection-refraction data: Geology, v. 17, p. 119–122.

Keller, M., 1995, Continental slope deposits in the Argentine Precordillera: Sediments and geotectonic significance, *in* Cooper, J. D., Droser, M. L., and Finney, S. C., eds., Ordovician odyssey: Short papers for the Seventh International Symposium on the Ordovician System: Pacific Section, SEPM (Society for Sedimentary Geology) Publication 77, p. 211–215.

Keller, M., 1996, Anatomy of the Precordillera (Argentina) during Cambro-Ordovician times: Implications for the Laurentia-Gondwana transfer of the Cuyania Terrane: Orstom, Paris, International Symposium on Andean Geodynamics, 3rd, Saint Malo, France, extended abstracts, p. 775–778.

Keller, M., and Bordonaro, O. L., 1993, Arrecifes de estromatopóridos en el Ordovícico inferior del Oeste Argentino y sus implicaciones paleogeográficas: Revista Española de Paleontología, v. 8, p. 165–169.

Keller, M., and Cooper, J. D., 1995, Paleokarst in the Lower and Middle Ordovician of southeastern California and adjacent Nevada and its bearing on the Sauk-Tippecanoe Boundary problem, *in* Cooper J. D., Droser, M. L., and Finney, S. C., eds., Ordovician odyssey: Short papers for the Seventh International Symposium on the Ordovician System: Pacific Section, SEPM (Society for Sedimentary Geology) Publication 77, p. 323–327.

Keller, M., and Dickerson, P. W., 1996, The missing continent of Llanoria—Was it the Argentine Precordillera?, *in* Actas, XIII Congreso Geológico Argentino y III Congreso de Exploración de Hidrocarburos, v. 5: Buenos Aires, Asociación Geológica Argentina e Instituto Argentino de Petróleo y Gas, p. 355–367.

Keller, M., and Flügel, E., 1996, Early Ordovician reefs from Argentina: Stromatoporoid vs. stromatolite origin: Facies, v. 34, p. 177–192.

Keller, M., and Lehnert, O., 1998, The Rio Sassito pelagic carbonate platform: A pinpoint in the geodynamic evolution of the Argentine Precordillera: Geologische Rundschau, v. 87, p. 326–344.

Keller, M., Buggisch, W., and Bercowski, F., 1989, Facies and sedimentology of Upper Cambrian shallowing-upward cycles in the La Flecha Formation (Argentine Precordillera): Zentralblatt für Geologie und Paläontologie, Teil 1, p. 999–1011.

Keller, M., Bordonaro, O. L., and Beresi, M., 1993a, The Cambrian of San Isidro, Mendoza, Argentina: Facies and sedimentology at the platform slope transition: Neues Jahrbuch für Geologie und Paläontologie Monatshefte, v. 1993/6, p. 373–383.

Keller, M., Eberlein, S., and Lehnert, O., 1993b, Sedimentology of Middle Ordovician carbonates in the Argentine Precordillera: Evidence for regional relative sea-level changes: Geologische Rundschau, v. 82, p. 362–377.

Keller, M., Lehnert, O., and Buggisch, W., 1993c, The transition from diagenesis to low grade metamorphism in the Argentine Precordillera: An application of the conodont colour alteration index, *in* Actas, XII Congreso Geológico Argentino y II Congreso de Exploración de Hidrocarburos, v. 1: Mendoza, Asociación Geológica Argentina e Instituto Argentino de Petróleo y Gas, p. 294–299.

Keller, M., Cañas, F., Lehnert, O., and Vaccari, N. E., 1994, The Upper Cambrian and Lower Ordovician of the Precordillera (Western Argentina): Some stratigraphic reconsiderations: Newsletters in Stratigraphy, v. 31, p. 115–132.

Keller, M., Buggisch, W., and Lehnert, O., 1998, The stratigraphic record of the Argentine Precordillera and its plate tectonic background, *in* Pankhurst, R. J., and Rapela, C. W., eds., The proto-Andean margin of Gondwana: Geological Society [London] Special Publication, v. 142, p. 35–56.

Kendall, G. C. S. C., and Schlager, W., 1981, Carbonates and relative changes in sea level: Marine Geology, v. 44, p. 181–212.

Keppie, J. D., 1991, Avalon, an exotic Appalachian-Caledonide terrane of western South American provenance: Departamento de Geología, Universidad de Chile, Revista Comunicaciones, v. 42, p. 109–111.

Kerlleñevich, S. C., 1967, Hallazgo del Devónico marino en la zona de Calingasta, Provincia de San Juan: Revista de la Asociación Geológica Argentina, v. 22, p. 291–294.

King, P. B., 1937, Geology of the Marathon region, Texas: U.S. Geological Survey Professional Paper 187, 148 p.

Knight, I., James, N. P., and Williams, H., 1995, Cambrian-Ordovician carbonate sequence, *in* Williams, H., ed., Geology of the Appalachian-Caledonian orogen in Canada and Greenland: Geological Survey of Canada, Geology of Canada, v. 6, p. 67–87.

Kobayashi, T., 1937, The Cambro-Ordovician shelly faunas of South America: Faculty of Science Journal, Imperial University of Tokyo, section II, Geology, Mineralogy, Geography, Seismology, v. 4, p. 369–522.

Kolata, D. R., Huff, W. D., Bergström, S. M., and Cingolani, C. A., 1994, Ordovician K-bentonite beds discovered in the Precordillera of Argentina: Geological Society of America Abstracts with Programs, v. 26, p. 503.

Kominz, M. A., and Bond, G. C., 1986, Geophysical modelling of the thermal history of foreland basins: Nature, v. 320, p. 252–256.

Krekeler, M. P. S., Huff, W. D., Kolata, D. R., Bergström, S. M., and Cingolani, C. A., 1995, Mineralogy and grain characteristics of Middle Ordovician K-bentonites from the Precordillera of Argentina, *in* Cooper, J. D., Droser, M. L., and Finney, S. C., eds., Ordovician odyssey: Short papers for the Seventh International Symposium on the Ordovician System: Pacific Section, SEPM (Society for Sedimentary Geology) Publication 77, p. 355–356.

Kukal, Z., and Saadallah, A., 1973, Aeolian admixtures in the sediments of the Northern Persian Gulf, *in* Purser, B. H., ed., The Persian Gulf: Berlin, Heidelberg, New York, Springer, p. 117–121.

Kury, W., 1993, Características composicionales de la Formación Villavicencio, Devónico, Precordillera de Mendoza, *in* Actas, XII Congreso Geológico Argentino y II Congreso de Exploración de Hidrocarburos, v. 1: Mendoza, Argentina, Asociación Geológica Argentina e Instituto Argentino de Petróleo y Gas, p. 321–328.

Lehnert, O., 1990, Conodontenstratigraphie und Fazies der Formation San Juan bei Niquivil (Unteres Ordoviz, Argentinien, Präkordillere) [Diploma thesis]: Germany, Universität Erlangen, 171 p.

Lehnert, O., 1993, Bioestratigrafía de los conodontes Arenigianos de la Formación San Juan en la localidad de Niquivil (Precordillera Sanjuanina, Argentina) y su correlación intercontinental: Revista Española de Paleontología, v. 8, p. 153–164.

Lehnert, O., 1994, A *Cordylodus proavus* fauna from West-Central Argentina (Los Sombreros Fm., Sierra del Tontal, San Juan Province): Zentralblatt für Geologie und Paläontologie, Teil 1, p. 245–262.

Lehnert, O., 1995a, Ordovizische Conodonten aus der Präkordillere Westargentiniens: Ihre Bedeutung für Stratigraphie und Paläogeographie: Erlanger Geologische Abhandlungen, v. 125, 193 p.

Lehnert, O., 1995b, Geodynamic processes in the Ordovician of the Argentine Precordillera: New biostratigraphic constraints, *in* Cooper, J. D., Droser, M. L., and Finney, S. C., eds., Ordovician odyssey: Short papers for the Seventh International Symposium on the Ordovician System: Pacific Section, SEPM (Society for Sedimentary Geology) Publication 77, p. 75–80.

Lehnert, O., 1995c, The Tremadoc/Arenig transition in the Argentine Precordillera, *in* Cooper, J. D., Droser, M. L., and Finney, S. C., eds., Ordovician odyssey: Short papers for the Seventh International Symposium on the Ordovician System: Pacific Section, SEPM (Society for Sedimentary Geology) Publication 77, p. 145–148.

Lehnert, O., and Keller, M., 1993a, The conodont record of the Argentine Precordillera: Problems and possibilities: Zentralblatt für Geologie und Paläontologie, Teil 1, p. 231–244.

Lehnert, O., and Keller, M., 1993b, Posición estratigráfica de los arrecifes Arenigianos en la Precordillera Argentina: Documents du Laboratoire de Géologie de l'Université de Lyon, v. 125, p. 263–275.

Lehnert, O., Miller, J. F., and Repetski, J. E., 1997, Paleogeographic significance of *Clavohamulus hintzei* Miller (Conodonta) and other Ibexian conodonts in an early Paleozoic carbonate platform facies of the Argentine Precordillera: Geological Society of America Bulletin, v. 109, p. 429–443.

Lehnert, O., Keller, M., and Bordonaro, O. L., 1998, Early Ordovician conodonts from the southern Cuyania terrane (Mendoza province, Argentina): Palaeontologia Polonica, v. 58, p. 47–65.

LeMone, D. V., 1988a, Precambrian and Paleozoic stratigraphy; Franklin Mountains, west Texas, *in* Hayward, O. T., ed., Geological Society of America south-central section, Centennial Field Guide v. 4, Boulder, Colorado, Geological Society of America, p. 387–394.

LeMone, D. V., 1988b, *Pulchrilamina*, the Early Ordovician labechiid stromatoporoid and its mound, *in* LeMone, D. V., ed., Franklin Mountains, Tobosa Mountains, and related basins: El Paso Geological Society and American Association of Petroleum Geologists southwest section, Field Trip Guide Book, p. 142–156.

Leveratto, M. A., 1968, Geología de las zonas al oeste de Ullum-Zonda, borde oriental de la Precordillera de San Juan, eruptividad subvolcánica y estructura: Revista de la Asociación Geológica Argentina, v. 23, p. 129–157.

Leveratto, M. A., 1976, Edad de intrusivos cenozoicos en la Precordillera de San Juan y su implicancia estratigráfica: Revista de la Asociación Geológica Argentina, v. 31, p. 53–58.

Levy, R., and Nullo, F., 1974, La fauna Ordovícica (Ashgilliano) de Villicum, San Juan, Argentina: Ameghiniana, v. 9, p. 173–194.

Levy, R., and Nullo, F., 1975, Braquiópodos ordovícicos de Ponon Trehue, Bloque de San Rafael (Provincia de Mendoza), *in* Actas, 1er Congreso Argentino de Paleontología y Bioestratigrafía: San Miguel de Tucumán, v. 1, Asociación Paleontológica Argentina, p. 23–32.

Lindström, M., 1971, Lower Ordovician conodonts of Europe, *in* Sweet, W. C., and Bergström, S. M., eds., Symposium on Conodont Biostratigraphy: Geological Society of America Memoir 127, p. 21–61.

Lister, G. S., Etheridge, M. A., and Symonds, P. A., 1991, Detachment models for the formation of passive continental margins: Tectonics, v. 10, p. 1038–1064.

Löfgren, A., 1978, Arenigian and Llanvirnian conodonts from Jämtland, northern Sweden: Fossils and Strata, v. 13, p. 1–129.

Löfgren, A., 1996, Lower Ordovician conodonts; reworking and biostratigraphy of Orreholmen quarry, Västergötland, south-central Sweden: Geologiska Föreningens i Stockholm Förhandlingar, v. 118, p. 169–183.

Logan, B. W., Rezak, R., and Ginsburg, R. N., 1964, Classification and environmental significance of algal stromatolites: Journal of Geology, v. 72, p. 68–83.

Loske, W. P., 1992, Sedimentologie, Herkunft und geotektonische Entwicklung paläozoischer Gesteine der Präkordillere West-Argentiniens: Münchener Geologische Hefte, v. 7, 155 p.

Loske, W. P., 1995, 1.1Ga old zircons in W Argentina: implications for sedimentary provenance in the Palaeozoic of Western Gondwana: Neues Jahrbuch für Geologie und Paläontologie Monatshefte, v. 1995/1, p. 51–64.

Loutit, T. S., Hardenbol, J., Vail, P. R., and Baum, G. R., 1988, Condensed sections: The key to age dating and correlation of continental margin sections, *in* Wilgus, C. K., Hastings, B. S., Kendall, C. G. S. C., Posamentier, H. W., Ross, C. A., and Van Wagoner, J. C., eds., Sea-level changes: An integrated approach: Society of Economic Paleontologists and Mineralogists Special Publication 42, p. 183–213.

Lowe, D. R., 1985, Ouachita trough; part of a Cambrian failed rift system: Geology, v. 13, p. 790–793.

Lowe, D. R., 1989, Stratigraphy, sedimentology, and depositional setting of pre-orogenic rocks of the Ouachita Mountains, Arkansas and Oklahoma, *in* Hatcher, R. D., Jr., Thomas, W. A., and Viele, G. W., eds., The Appalachian-Ouachita orogen in the United States: Boulder, Colorado, Geological Society of America, Geology of North America, v. F-2, p. 575–590.

MacNiocaill, C., Van der Pluijm, B. A., and Van der Voo, R., 1997, Ordovician paleogeography and the evolution of the Iapetus ocean: Geology, v. 25, p. 159–162.

Maletz, J., and Ortega G., 1995, Ordovician graptolites of South America: Paleogeographic implications, *in* Cooper, J. D., Droser, M. L., and Finney, S. C., eds., Ordovician odyssey: Short papers for the Seventh International Symposium on the Ordovician System: Pacific Section, SEPM (Society for Sedimentary Geology) Publication 77, p. 189–192.

Markello, J. R., and Read, J. F., 1982, Upper Cambrian intrashelf basin, Nolichuky Formation, southwest Virginia Appalachians: American Association of Petroleum Geologists Bulletin, v. 66, p. 860–878.

McBride, E. F., 1989, Stratigraphy and sedimentary history of pre-Permian Paleozoic rocks of the Marathon uplift, *in* Hatcher, R. D., Jr., Thomas, W. A., and Viele, G. W., eds., The Appalachian-Ouachita orogen in the United States: Boulder, Colorado, Geological Society of America, Geology of North America, v. F-2, p. 603–620.

McDonough, M. R., Ramos, V. A., Isachsen, C. E., Bowring, S. A., and Vujovich, G. I., 1993, Nuevas edades de circones del basamento de la sierra de Pie de Palo, Sierras Pampeanas Occidentales de San Juan: sus implicancias para los modelos del supercontinente proterozoico de Rodinia, *in* Actas, XII Congreso Geológico Argentino y II Congreso de Exploración de Hidrocarburos, v. 3: Mendoza, Asociación Geológica Argentina e Instituto Argentino de Petróleo y Gas, p. 340–342.

McIlreath, I. A., and James, N. P., 1984, Carbonate slopes, *in* Walker, R. G., ed., Facies models: Geoscience Canada Reprint series, v. 1, p. 245–257.

McKerrow, W. S., and Cocks, L. R. M., 1986, Oceans, islands arcs and olistostromes: The use of fossils in distinguishing sutures, terranes and environments around the Iapetus: Geological Society of London Journal, v. 143, p. 185–191.

McKerrow, W. S., Dewey, J. F., and Scotese, C. R., 1991, The Ordovician and Silurian development of the Iapetus ocean: Special Papers in Palaeontology, v. 44, p. 165–178.

McKerrow, W. S., Scotese, C. R., and Brasier, M. D., 1992, Early Cambrian continental reconstructions: Geological Society of London Journal, v. 149, p. 599–606.

McMenamin, M. A. S., and McMenamin, D. L. S., 1990, The emergence of animals: The Cambrian break-trough: New York, Columbia University Press, 217 p.

Meert, J. G., Van der Voo, R., Powell, C. M., Li, Z. X., McElhinny, M. W., Chen, Z., and Symons, D. T. A., 1993, A plate tectonic speed limit?: Nature, v. 363, p. 216–217.

Miller, J. F., 1984, Cambrian and earliest Ordovician conodont evolution, biofacies, and provincialism, *in* Clark, D. L., ed., Conodont biofacies and provincialism: Geological Society of America Special Paper 196, p. 43–68.

Miller, J. F., 1988, Conodonts as biostratigraphic tools for the redefinition and correlation of the Cambro-Ordovician boundary: Geological Magazine, v. 125, p. 349–362.

Miller, J. F., 1992, The Lange Ranch eustatic event: A regressive-transgressive couplet near the base of the Ordovician System, *in* Webby, B. D., and Laurie, J. R., eds., Global perspectives on Ordovician geology: Rotterdam, Balkema, p. 395–407.

Miser, H. D., 1921, Llanoria, the Paleozoic land areas in Louisiana and eastern Texas: American Journal of Science, v. 2, p. 61–89.

Mitchell, C. E., 1997, Biogeography of Middle and Upper Ordovician graptolites, Precordilleran terrane, Argentina: plate tectonic implications: Geological Society of America Abstracts with Programs, v. 29, no. 6, p. A379.

Monty, C. L., 1967, Distribution and structure of recent stromatolitic algal mats, eastern Andros Island, Bahamas: Annales de la Sociètè Geologique Belge, v. 90, p. 55–100.

Moores, E. M., 1991, Southwest U.S.–East Antarctica (SWEAT) connection: A hypothesis: Geology, v. 19, p. 425–428.

Mount, J. F., Hunt, D. L., Greene, L. R., and Dienger, J., 1991, Depositional systems, biostratigraphy and sequence stratigraphy of Lower Cambrian grand cycles, southwestern Great Basin, *in* Cooper, J. D., and Stevens, C. H., eds., Paleozoic paleogeography of the Western United States—II: Pacific Section, SEPM (Society for Sedimentary Geology) Publication 67, p. 209–229.

Mountjoy, E. W., Cook, H. E., Pray, L. C., and McDaniel, P. N., 1972, Allochthonous carbonate debris flows—worldwide indicators of reef complexes, banks or shelf margins: International Geological Congress, 24th, Proceedings, section 6, p. 172–189.

Mullins, H. T., and Lyntz, G. W., 1977, Origin of the northwestern Bahama Platform: Review and reinterpretation: Geological Society of America Bulletin, v. 88, p. 1447–1461.

Mullins, H. T., and Newman, A. C., 1979, Deep carbonate bank margin structure and sedimentation in the Northern Bahamas, *in* Doyle, L. J., and Pilkey, O. H., eds., Geology of continental slopes: Society of Economic Paleontologists and Mineralogists Special Publication 27, p. 165–192.

Mussman, W. J., and Read, J. F., 1986, Sedimentology and development of a passive- to convergent-margin unconformity: Middle Ordovician Knox unconformity, Virginia Appalachians: Geological Society of America Bulletin, v. 97, p. 282–295.

Mutti, E., and Ricci Lucchi, F., 1972, Le torbiditi dell' Appennino settentrionale: introduzione all' analisi di facies: Memorie della Societá Geologica Italiana, v. 11, p. 161–199.

Nardin, T. R., Heinar, F. J., Gorsline, D. S., and Edwards, B. D., 1979, A review of mass movement processes, sediment and acoustic characteristics, and contrasts in slope and base-of-slope systems versus canyon-fan-basin floor systems, *in* Doyle, L. J., and Pilkey, O. H., eds., Geology of continental slopes: Society of Economic Paleontologists and Mineralogists Special Publication 27, p. 61–73.

Newell, N. D., Imbrie, J., Purdy, E. G., and Thurber, D. L., 1959, Organism communities and bottom facies, Great Bahama Bank: American Museum of Natural History Bulletin, v. 117, p. 177–228.

Nielsen, A. T., 1992, Intercontinental correlation of the Arenigian (Early Ordovician) based on sequence and ecostratigraphy, *in* Webby, B. D., and Laurie, J. R., eds., Global perspectives on Ordovician geology: Rotterdam, Balkema, p. 367–379.

Nuñez, E., 1979, Descripción geológica de la hoja 28d, Estación Soitué, Provincia de Mendoza: Buenos Aires, Argentina, Boletín del Servicio Geológico Nacional, v. 166, 67 p.

Ortega, G. C., 1987, Las graptofaunas y los conodontes de la Formación Los Azules, Cerro Viejo, Zona de Huaco, Departamento Jáchal, San Juan [Ph.D. thesis]: Córdoba, Argentina, Universidad Nacional de Córdoba, 209 p.

Ortega, G., Cañas, F. L., and Hünicken, M. A., 1985, Sobre la presencia de *Isograptus victoriae* Harris en la Formación Gualcamayo, La Rioja, Argentina: Revista Técnica Yacimientos Petrolíferos Fiscales Bolivia, v. 9, p. 215–221.

Ortega, G. C., Brussa, E., and Astini, R. A., 1991, Nuevos hallazgos de graptolitos en la Formación Yerba Loca y su implicancia estratigráfica (Precordillera de San Juan, Argentina): Ameghiniana, v. 28, p. 163–178.

Ortiz, A., and Zambrano, J. J., 1981, La provincia geológica Precordillera oriental, *in* Actas, VIII Congreso Geológico Argentino v. 3: San Luis, Argentina, Asociación Geológica Argentina, p. 59–74.

Osleger, D. A., and Montañez, I. P., 1996, Cross-platform architecture of a sequence boundary in mixed siliciclastic-carbonate lithofacies, Middle Cambrian, southern Great Basin, USA: Sedimentology, v. 43, p. 197–217.

Osleger, D. A., and Read, J. F., 1993, Comparative analysis of methods used to define eustatic variations in outcrop: Late Cambrian interbasinal sequence development: American Journal of Science, v. 293, p. 157–216.

Owen, A. L., Harper, D. A. T., and Rong, J., 1991, Hirnantian trilobites and brachiopods in space and time, *in* Barnes, C. R., and Williams, S. H., eds., Advances in Ordovician geology: Geological Survey of Canada Professional Paper 90-9, p. 179–190.

Padula, E., Rolleri, E., Mingramm, A. R., Criado Roque, P., Flores, M., and Baldis, B. A., 1967, Devonian of Argentina: Proceedings, International

Symposium on the Devonian System v. 2: Calgary, Canada, Canadian Society of Petroleum Geologists, p. 165–199.
Palmer, A. R., 1971a, The Cambrian of the Great Basin and adjacent areas, western United States, *in* Holland, C. H., ed., Lower Paleozoic rocks of the New World: New York, Wiley, p. 1–78.
Palmer, A. R., 1971b, The Cambrian of the Appalachian and eastern New England regions, eastern United States, *in* Holland, C. H., ed., Lower Paleozoic rocks of the New World: New York, Wiley, p. 169–218.
Palmer, A. R., 1981, On the correlatibility of Grand Cycle tops: U.S. Geological Survey Open-File Report, v. 81-743, p. 156–159.
Palmer, A. R., 1998, A proposed nomenclature for stages and series for the Cambrian of Laurentia: Canadian Journal of Earth Sciences, v. 35, p. 323–328.
Palmer, A. R., and Halley, R., 1979, Physical stratigraphy and trilobite biostratigraphy of the Carrara Formation (Lower and Middle Cambrian) in the southern Great Basin: U.S. Geological Survey Professional Paper 1047, 130 p.
Palmer, A. R., and James, N. P., 1980, The Hawke Bay event: A circum-Iapetus regression event near the Lower to Middle Cambrian boundary, *in* Wones, D. R., ed., Proceedings—Caledonides in the U.S.A.: Blacksburg, Virginia, Polytechnic Institute and State University Memoir 2, p. 15–18.
Palmer, A. R., DeMis, W. D., Muehlberger, W. R., and Robinson, R. A., 1984, Geological implications of Middle Cambrian boulders from the Haymond Formation (Pennsylvanian) in the Marathon basin, west Texas: Geology, v. 12, p. 91–94.
Pankhurst, R. J., and Rapela, C. W., 1998, The proto-Andean margin of Gondwana: an introduction, *in* Pankhurst, R. J., and Rapela, C. W., eds., The proto-Andean margin of Gondwana: Geological Society [London] Special Publication 142, p. 1–9.
Pankhurst, R. J., Rapela, C. W., Saavedra, J., Baldo, E., Dahlquist, J., Pascua, I., and Fanning, C. M., 1998, The Famatinian magmatic arc in the central Sierras Pampeanas: An Early to Middle Ordovician continental arc on the Gondwana margin, *in* Pankhurst, R. J., and Rapela, C. W., eds., The proto-Andean margin of Gondwana: Geological Society [London] Special Publication 142, p. 343–367.
Peralta, S. H., 1984, Ludloviano en la Precordillera Oriental sanjuanina, *in* Actas, 9° Congreso Geológico Argentino, v. 4: San Carlos de Bariloche, Argentina, Asociación Geológica Argentina, p. 296–304.
Peralta, S. H., 1986, Graptolitos del Llandoveriano inferior en el Paleozoico inferior clástico del pie oriental de la Sa. del Villicum, Precordillera Oriental de San Juan, *in* Actas, 1as Jornadas sobre Geología de Precordillera: San Juan, Argentina, 1985, Universidad Nacional de San Juan, p. 134–138.
Peralta, S. H., 1990, Silúrico de la Precordillera de San Juan—Argentina: XI Congreso Geológico Argentino: San Juan, Argentina, Relatorio, Asociación Geológica Argentina, p. 48–64.
Peralta, S. H., 1993, Estratigrafía y consideraciones paleoambientales de los depósitos marino-clásticos eopaleozoicos de la Precordillera Oriental de San Juan, *in* Actas, XI Congreso Geológico Argentino v. 1: San Juan, Argentina, Asociación Geológica Argentina, p. 128–137.
Peralta, S. H., 1995, La Formación Gualcamayo en la Sierra de Villicum: Sus graptolitos y faunas asociadas: Boletín de la Academia Nacional de Ciencias de Córdoba, v. 60, p. 401–408.
Peralta, S. H., and Aceñolaza, F. G., 1989, *Palaeohelminthoida* Ruchholz (traza fosil) en los estratos basales de la Formación Punta Negra (Devónico Medio—Superior?), en el perfíl del Río San Juan, Precordillera Central del Oeste Argentino: Universidad Nacional de Tucumán, Serie Correlación Geológica, v. 5, p. 195–198.
Peralta, S. H., and Baldis, B. A., 1990, Devónico de la provincia de San Juan: XI Congreso Geológico Argentino: San Juan, Argentina, Relatorio, Asociación Geológica Argentina, p. 66–76.
Peralta, S. H., and Carter, C. H., 1990a, La glaciación Gondwánica del Ordovícico tardío: Evidencias en fangolitas guijarrosas de la Precordillera de San Juan, *in* Actas, XI Congreso Geológico Argentino: San Juan, Argentina, v. 2: Asociación Geológica Argentina, p. 181–185.
Peralta, S. H., and Carter, C. H., 1990b, Facies de plataforma e icnofacies de la Formación Tambolar (Silúrico) en su localidad tipo, Precordillera Central sanjuanina, Argentina, *in* Actas, 3ª Reunión de Sedimentología Argentina: San Juan, Argentina, Asociación Sedimentológica Argentina, p. 339–344.
Peralta, S. H., and Medina, E., 1986, Estratigrafía de la Formación Rinconada en el borde oriental del Cerro Pedernal, Precordillera Oriental de San Juan, *in* Actas, 1as Jornadas sobre Geología de Precordillera: San Juan, Argentina, 1985, Universidad Nacional de San Juan, p. 157–162.
Peralta, S. H., and Uliarte, E. R., 1986, Estructura de la Formación Rinconada (Eopaleozoico) en su localidad tipo, Precordillera de San Juan, *in* Actas, 1as Jornadas sobre Geología de la Precordillera San Juan, Argentina, 1985, Universidad Nacional de San Juan, p. 237–242.
Peralta, S. H., Gamboa, L. A., and Baldis, B. A., 1989, Icnofacies de *Zoophycos*: su posición estratigráfica e interpretación de su recurrencia en el Silúrico-Devónico de la Precordillera Central sanjuanina: Universidad Nacional de Tucumán, Serie Correlación Geológica, v. 5, p. 199–209.
Pereyra, B. R., 1986, Análisis estratigráfico y genético de los niveles Cámbricos del Cerro de Zonda, Provincia de San Juan, *in* Actas, 1as Jornadas sobre Geología de Precordillera: San Juan, Argentina, 1985, Universidad Nacional de San Juan, p. 24–29.
Perkins, R. D., and Enos, P., 1967, Hurricane Betsy in the Florida-Bahamas area; geologic effects and comparison with Hurricane Donna: Journal of Geology, v. 76, p. 710–717.
Peryt, T. M., 1981, Phanerozoic oncoids—an overview: Facies, v. 4, p. 197–214.
Pina, L., Baldis, B. A., and Bordonaro, O. L., 1986, Formación Estancia San Martín (nov. nom.) del Cámbrico inferior-Cámbrico medio en la Comarca de San Isidro, Mendoza, *in* Actas, 1as Jornadas sobre Geología de Precordillera: San Juan, Argentina, 1985, Universidad Nacional de San Juan, p. 7–11.
Pohler, S. M. L., and James, N. P., 1989, Reconstruction of a Lower/Middle Ordovician carbonate shelf margin: Cow Head Group, Western Newfoundland: Facies, v. 21, p. 189–262.
Posamentier, H. W., and Vail, P. R., 1988, Eustatic controls on clastic deposition II—sequence and systems tract models, *in* Wilgus, C. K., Hastings, B. S., Kendall, C. G. S. C., Posamentier, H. W., Ross, C. A., and Van Wagoner, J. C., eds., Sea-level changes: An integrated approach: Society of Economic Paleontologists and Mineralogists Special Publication 42, p. 125–142.
Posamentier, H. W., Jervey, M. T., and Vail, P. R., 1988, Eustatic controls on clastic deposition II—Sequence and systems tract model, *in* Wilgus, C. K., Hastings, B. S., Kendall, C. G. S. C., Posamentier, H. W., Ross, C. A., and Van Wagoner, J. C., eds., Sea-level changes: An integrated approach: Society of Economic Paleontologists and Mineralogists Special Publication 42, p. 109–124.
Poulsen, C., 1960, Fossils from the late Middle Cambrian *Bolaspidella* zone of Mendoza, Argentina: Matematisk-Fysiske Meddedelser, Kongelike Danske Videnskabernes Selskab, v. 32, p. 1–42.
Powell, C. M., 1995, Are Neoproterozoic glacial deposits preserved on the margins of Laurentia related to the fragmentation of two supercontinents: Comment: Geology, v. 23, p. 1053–1054.
Powell, C. M., Li, Z. X., McElhinny, M. W., Meert, J. G., and Park, J. K., 1993, Paleomagnetic constraints on timing of the Neoproterozoic breakup of Rodinia and the Cambrian formation of Gondwana: Geology, v. 21, p. 889–892.
Powell, C. M., Dalziel, I. W. D., Li, Z. X., and McElhinny, M. W., 1995, Did Pannotia, the latest Neoproterozoic southern supercontinent, really exist?: Eos (Transactions, American Geophysical Union), v. 76, p. 172.
Pratt, B. R., and James, N. P., 1982, Cryptalgal-metazoan bioherms of Early Ordovician age in the St. George Group, western Newfoundland: Sedimentology, v. 29, p. 543–569.
Pratt, B. R., and James, N. P., 1986, The St. George Group (Lower Ordovician) of western Newfoundland: Tidal flat island model for carbonate sedimentation in shallow epeiric seas: Sedimentology, v. 33, p. 313–343.
Quartino, B. J., Zardini, R. A., and Amos, A. J., 1971, Estudio y exploración geológica de la región Barreal-Calingasta, Provincia de San Juan—República Argentina: Buenos Aires, Argentina, Asociación Geológica Argentina Monografías: v. 1, 184 p.

Ramos, V. A., 1988a, The tectonics of the Central Andes; 30° to 33° S latitude, in Clark, S., and Burchfiel, D., eds., Processes in continental lithospheric deformation: Geological Society of America Special Paper 218, p. 31–54.

Ramos, V. A., 1988b, Late Proterozoic–early Paleozoic of South America: A collisional history: Episodes, v. 11, p. 168–173.

Ramos, V. A., 1995, Sudamérica: un mosaico de continentes y océanos: Ciencias Hoy, v. 6, p. 24–29.

Ramos, V. A., and Basei, M., 1997, The basement of Chilenia: An exotic continental terrane to Gondwana during the early Paleozoic, in Bradshaw, J. D., and Weaver, S. D., eds., Terrane Dynamics 97, International Conference on Terrane Geology: Christchurch, New Zealand, Abstracts, p. 140–143.

Ramos, V. A., Jordan, T. E., Allmendinger, R. W., Kay, S. M., Cortes, J. M., and Palma, M., 1984, Chilenia: un terreno alóctono en la evolución paleozoica de los Andes centrales, in Actas, IX Congreso Geológico Argentino, v. 2: San Carlos de Bariloche, Argentina, Asociación Geológica Argentina, p. 84–106.

Ramos, V. A., Jordan, T. E., Allmendinger, R. W., Mpodozis, C., Kay, S. M., Cortes, J. M., and Palma, M., 1986, Paleozoic terranes of the Central Argentine-Chilean Andes: Tectonics, v. 5, p. 855–880.

Ramos, V. A., Vujovich, G., Kay, S. M., and McDonough, M., 1993, La orogénesis de Grenville en las Sierras Pampeanas occidentales: La Sierra Pie de Palo y su integración al supercontinente Proterozoico, in Actas, XII Congreso Geológico Argentino y II Congreso de Exploración de Hidrocarburos, v. 3: Mendoza, Asociación Geológica Argentina e Instituto Argentino de Petróleo y Gas, p. 343–357.

Ramos, V. A., Vujovich G. I., and Dallmeyer, R. D., 1996, Los klippes y ventanas tectónicas preándicas de la Sierra de Pie de Palo (San Juan): Edad e implicaciones tectónicas, in Actas, XIII Congreso Geológico Argentino y III Congreso de Exploración de Hidrocarburos, v. 5: Buenos Aires, Asociación Geológica Argentina e Instituto Argentino de Petróleo y Gas, p. 377–391.

Ramos, V. A., Dallmeyer, R. D., and Vujovich G. I., 1998, Time constraints on the early Palaeozoic docking of the Precordillera, central Argentina, in Pankhurst, R. J., and Rapela, C. W., eds., The proto-Andean margin of Gondwana: Geological Society [London] Special Publication 142, p. 143–158.

Rankin, D. W., Drake, A. A., Jr., Glover III, L., Goldsmith, R., Hall, M. L., Murray, D. P., Ratcliffe, N. M., Read, J. F., Secor, D. T., and Stanley, R. S., 1989, Pre-orogenic terranes, in Hatcher, R. D., Jr., Thomas, W. A., and Viele, G. W., eds., The Appalachian-Ouachita orogen in the United States: Boulder, Colorado, Geological Society of America, Geology of North America, v. F-2, p. 7–100.

Rapela, C. W., Pankhurst, R. J., and Harrison, S. M., 1989, Gondwanan plutonism of northern Patagonia: International Geological Congress, 28th, Abstracts, v. 2, p. 675.

Rapela, C. W., Coira, B., Toselli, A., and Saavedra, J., 1992, El magmatismo del Paleozoico inferior en el sudoeste de Gondwana, in Gutiérrez-Marco, J. C., Saavedra, J., and Rábano, I., eds., Paleozoico Inferior de Ibero-América: Publicaciones de la Universidad de Extremadura, p. 21–68.

Rapela, C. W., Pankhurst, R. J., Casquet, C., Baldo, E., Saavedra, J., and Galindo, C., 1998, Early evolution of the Proto-Andean margin of South America: Geology, v. 26, p.707–710.

Read, J. F., 1982, Carbonate platforms of passive (extensional) continental margins: Types, characteristics and evolution: Tectonophysics, v. 81, p. 195–212.

Read, J. F., 1985, Carbonate platform facies models: American Association of Petroleum Geologists Bulletin, v. 66, p. 860–878.

Read, J. F., 1989, Controls on evolution of Cambrian-Ordovician passive margin, U.S. Appalachians, in Crevello, P. D., Wilson, J. L., Sarg, J. F., and Read, J. F., eds., Controls on carbonate platform and basin development: Society of Economic Paleontologists and Mineralogists Special Publication 44, p. 167–186.

Reineck, H.-E., and Singh, I. B., 1980, Depositional sedimentary environments: Berlin, Heidelberg, New York, Springer, 549 p.

Ross, C. A., and Ross, J. R. P., 1995, North American Ordovician depositional sequences and correlations, in Cooper, J. D., Droser, M. L., and Finney, S. C., eds., Ordovician odyssey: Short papers for the Seventh International Symposium on the Ordovician System: Pacific Section, SEPM (Society for Sedimentary Geology) Publication 77, p. 309–313.

Ross, R. J., Jr., Hintze, L. F., Ethington, R. L., Miller, J. F., Taylor, M. E., and Repetski, J. E., 1993, The Ibexian Series (Lower Ordovician), a replacement for "Canadian Series" in North American chronostratigraphy: U.S. Geological Survey Open-File Report 93-598, p. 1–75.

Sánchez, T. M., Benedetto, J. L., and Brussa, E., 1991, Late Ordovician stratigraphy, paleoecology, and sea level changes in the Argentine Precordillera, in Barnes, C. R., and Williams, S. H., eds., Advances in Ordovician geology: Geological Survey of Canada Professional Paper 90-9, p. 245–258.

Santantonio, M., 1993, Facies associations and evolution of pelagic carbonate platform/basin systems: Examples from the Italian Jurassic: Sedimentology, v. 40, p. 1039–1067.

Santantonio, M., 1994, Pelagic carbonate platforms in the geologic record: Their classification and sedimentary and paleotectonic evolution: American Association of Petroleum Geologists Bulletin, v. 78, p. 122–141.

Sarg, J. F., 1988, Carbonate sequence stratigraphy, in Wilgus, C. K., Hastings, B. S., Kendall, C. G. S. C., Posamentier, H. W., Ross, C. A., and Van Wagoner, J. C., eds., Sea-level changes: An integrated approach: Society of Economic Paleontologists and Mineralogists Special Publication 42, p. 155–181.

Sarmiento, G. N., 1986, La biozona de *Amorphognathus variabilis–Eoplacognathus pseudoplanus* (Conodonta), Llanvirniano inferior, en el flanco oriental de la Sierra de Villicum, in Actas, 1as Jornadas sobre Geología de Precordillera: San Juan, Argentina, 1985, Universidad Nacional de San Juan, p. 119–123.

Sarmiento, G. N., 1990, Conodontes Ordovícicos de Argentina: Treballs Museu Geologia Barcelona, v. 1, p. 135–161.

Sarmiento, G. N., and Garcia Lopez, S., 1993, Síntesis sobre las faunas de conodontos del Paleozoico Inferior de Ibero-América y de la Península Ibérica (1958–1992): Revista Española de Paleontología, v. 8, p. 191–205.

Sarmiento, G. N., and Rábano, I., 1992, Nuevas precisiones bioestratigráficas sobre la Formación Gualcamayo (Ordovícico Inferior) en la Sierra de Villicum, San Juan, Argentina: Zentralblatt für Geologie und Paläontologie, Teil 1, p. 1785–1797.

Sarmiento, G. N., Vaccari, N. E., and Peralta, S. H., 1986, Conodontes Ordovícicos de la Rinconada, Precordillera de San Juan, Argentina, in Actas, 4° Congreso Argentino de Paleontología y Bioestratigrafía, v. 3: Mendoza, Argentina, Asociación Paleontológica Argentina, p. 219–224.

Schallreuter, R., 1996, Ordovizische Ostrakoden Argentiniens II: Mitteilungen Geologisch-Paläontologisches Institut der Universität Hamburg, v. 79, p. 139–169.

Schlager, W., 1981, The paradox of drowned reefs and carbonate platforms: Geological Society of America Bulletin, v. 92, p. 197–211.

Schlager, W., 1989, Drowning unconformities on carbonate platforms, in Crevello, P. D., Wilson, J. L., Sarg, J. F., and Read, J. F., eds., Controls on carbonate platform and basin development: Society of Economic Paleontologists and Mineralogists Special Publication 44, p. 15–25.

Schlager, W., and Chermak, A., 1979, Sediment facies of platform-basin transition, Tongue of the Ocean, Bahamas, in Doyle, L. J., and Pilkey, O. H., eds., Geology of continental slopes: Society of Economic Paleontologists and Mineralogists Special Publication, v. 27, p. 193–208.

Schlager, W., and Cumber, O., 1986, Submarine slope angles, drowning unconformities, and self-erosion of limestone escarpments: Geology, v. 14, p. 762–765.

Schlager, W., and Enos, P., 1996, Holocene systems tracts on carbonate platforms: American Association of Petroleum Geologists Abstracts with Program, p. 125.

Schlager, W., and James, N. P., 1978, Low-magnesian calcite limestones forming at the deep-sea floor, Tongue of the Ocean, Bahamas: Sedimentology, v. 25, p. 675–702.

Scholle, P. A., Arthur, M. M., and Ekdale, A. A., 1983, Pelagic, in Scholle, P. A.,

Bebout, D. G., and Moore, C. H., eds., Carbonate depositional environments: American Association of Petroleum Geologists Memoir 33, p. 619–691.

Schutter, S. R., 1992, Ordovician hydrocarbon distribution in North America and its relationship to eustatic cycles, *in* Webby, B. D., and Laurie, J. R., eds., Global perspectives on Ordovician geology: Rotterdam, Balkema, p. 421–432.

Scotese, C. R., and McKerrow, W. S., 1990, Revised world maps and introduction, *in* McKerrow, W. S., and Scotese, C. R., eds., Palaeozoic palaeogeography and biogeography: Geological Society [London] Memoir 12, p. 1–21.

Seilacher, A., 1969, Fault-graded beds interpreted as seismites: Sedimentology, v. 13, p. 155–159.

Selles Martinez, J., 1986, Caracterización litoestructural de la Formación El Planchón, Devónico de la Precordillera Occidental de San Juan, *in* Actas, 1as Jornadas sobre Geología de Precordillera: San Juan, Argentina, 1985, Universidad Nacional de San Juan, p. 53–58.

Serpagli, E., 1974, Lower Ordovician conodonts from Precordilleran Argentina (Province of San Juan): Bolletin della Societá di Paleontologia Italiana, v. 13, p. 17–98.

Sessarego, H. L. F., 1983, La posición estratigráfica y edad del conglomerado atribuido a la Formación del Salto, Río San Juan, Provincia de San Juan: Revista de la Asociación Geológica Argentina, v. 38, p. 494–497.

Sheehan, P. M., and Coorough, P. J., 1990, Brachiopod zoogeography across the Ordovician-Silurian extinction event, *in* McKerrow, W. S., and Scotese, C. R., eds., Palaeozoic palaeogeography and biogeography: Geological Society [London] Memoir 12, p. 181–190.

Shergold, J. H., Bordonaro, O. L., and Liñan, E., 1995, Late Cambrian agnostid trilobites from Argentina: Palaeontology, v. 38, p. 241–257.

Shinn, E. A., 1983a, Tidal flat environment, *in* Scholle, P. A., Bebout, D. G., and Moore, C. H., eds., Carbonate depositional environments: American Association of Petroleum Geologists Memoir 33, p. 173–210.

Shinn, E. A., 1983b, Birdseyes, fenestrae, shrinkage pores and loferites: A re-evaluation: Journal of Sedimentary Petrology, v. 53, p. 619–629.

Shinn, E. A., Lloyd, R. M., and Ginsburg, R. N., 1969, Anatomy of a modern carbonate tidal flat, Andros Island, Bahamas: Journal of Sedimentary Petrology, v. 39, p. 1202–1228.

Sims, J. P., Ireland, T. R., Camacho, A., Lyons, P., Pieters, P. E., Skirrow, R. G., Stuart-Smith, P. G., and Miró, R., 1998, U-Pb, Th-Pb and Ar-Ar geochronology from the southern Sierras Pampeanas, Argentina: Implications for the Palaeozoic tectonic evolution of the western Gondwana margin, *in* Pankhurst, R. J., and Rapela, C. W., eds., The proto-Andean margin of Gondwana: Geological Society [London] Special Publication 142, p. 259–281.

Skehan, J. W., 1988, Evolution of the Iapetus Ocean and its border in pre-Arenig times: A synthesis, *in* Harris, A. L., and Fettes, D. J., eds., The Caledonian-Appalachian orogen: Geological Society [London] Special Publication 38, p. 185–229.

Spalletti, L. A., Cingolani, C. A., Varela, R., and Cuerda, A. J., 1989, Sediment gravity flow deposits of an Ordovician deep-sea fan system (western Precordillera, Argentina): Sedimentary Geology, v. 61, p. 287–301.

Stappenbeck, R., 1910, La Precordillera de San Juan y Mendoza, Buenos Aires, Argentina: Ministerio de Agricultura, sección Geología y Mineralogía, Anales, v. 4, p. 3–179.

Stelzner, A., 1873, Mitteilungen an Professor H. B. Gleinitz über seine Reise durch die argentinischen Provinzen San Juan und Mendoza und die Cordillere zwischen dem 35. und 32.° S: Neues Jahrbuch für Geologie und Paläontologie, v. 5, p. 726–744.

Strasser, A., 1984, Black-pebble occurrence and genesis in Holocene carbonate sediments (Florida Key, Bahamas and Tunisia): Journal of Sedimentary Petrology, v. 54, p. 1097–1109.

Sweet, W. C., 1979, Conodonts and conodont biostratigraphy of post-Tyrone Ordovician rocks of the Cincinnati region: U.S. Geological Survey Professional Paper 1066G, p. 1–26.

Taylor, J. F., Repetski, J. E., and Orndorff, R. C., 1992, The Stonehenge transgression: A rapid submergence of the central Appalachian platform in the Early Ordovician, *in* Webby, B. D., and Laurie, J. R., eds., Global perspectives on Ordovician geology: Rotterdam, Balkema, p. 409–418.

Thomas, W. A., 1991, The Appalachian-Ouachita rifted margin of southeastern North America: Geological Society of America Bulletin, v. 103, p. 415–431.

Thomas, W. A., 1993, Low-angle detachment geometry of the late Precambrian-Cambrian Appalachian-Ouachita rifted margin of southeast North America: Geology, v. 21, p. 921–924.

Thomas, W. A., 1996, Rifting of the Argentine Precordillera from the Ouachita Embayment of the Laurentian margin: Geological Society of America Abstracs with Programs, v.28, no. 1, p. 66.

Thomas, W. A., and Astini, R. A., 1996, Asymmetric conjugate rift margins of the Ouachita rift and the Argentine Precordillera: Geological Society of America Abstracts with Programs, v.28, no. 1, p. 66.

Toomey, D. F., 1970, An unhurried look at a Lower Ordovician mound horizon, southern Franklin Mountains, west Texas: Journal of Sedimentary Petrology, v. 40, p. 1318–1334.

Toomey, D. F., and Nitecki, M. H., 1979, Organic buildups in the Lower Ordovician (Canadian) of Texas and Oklahoma: Fieldiana, v. 2, p. 1–181.

Torsvik, T. H., Tait, J., Moralev, V. M., McKerrow, W. S., Sturt, B. A., and Roberts, D., 1995, Ordovician paleogeography of Siberia and adjacent continents: Geological Society of London Journal, v. 152, p. 279–287.

Torsvik, T. H., Smethurst, M. A., Meert, J. G., Van der Voo, R., McKerrow, W. S., Brasier, M. D., Sturt, B. A., and Walderhaug, H. J., 1996, Continental break-up and collision in the Neoproterozoic and Palaeozoic—a tale of Baltica and Laurentia: Earth Science Reviews, v. 40, p. 229–258.

Tortello, M. F., and Bordonaro, O. L., 1997, Cambrian agnostid trilobites from Mendoza, Argentina: A systematic revision and biostratigraphic implications: Journal of Palaeontology, v. 71, p. 74–86.

Tucker, M. E., and Wright, V. P., 1990, Carbonate sedimentology: Oxford, Blackwell Scientific Publications, 482 p.

Tucker, R. D., Bradley, D. C., Krogh, R. J., Ross, S. H., and Williams, S. H., 1990, Time-scale calibration by high-precision U-Pb zircon dating of interstratified volcanic ashes in the Ordovician and Lower Silurian stratotypes of Britain: Earth and Planetary Science Letters, v. 100, p. 51–58.

Tucker, R. D., Bradley, D. C., Ver Straeten, C. A., Harris, A. G., Evert, J. R., and McCutcheon, S. R., 1998, New U-Pb zircon ages and the duration and division of Devonian time: Earth and Planetary Science Letters, v. 158, p. 175–186.

Vaccari, N. E., 1987, Perfil geológico a lo largo del Río Jáchal al Oeste de la Quebrada Caracol (Sierra de Los Túneles), Departamento Jáchal, Provincia de San Juan [tésis de licenciatura]: Córdoba, Argentina, Universidad Nacional de Córdoba, 38 p.

Vaccari, N. E., 1994, Las faunas de trilobites de las sucesiones carbonáticas del Cámbrico y Ordovícico temprano de la Precordillera Septentrional Argentina [Ph.D. thesis]: Córdoba, Argentina, Universidad Nacional de Córdoba, 271 p.

Vaccari, N. E., 1995, Early Ordovician trilobite biogeography of Precordillera and Famatina, western Argentina: Preliminary results, *in* Cooper, J. D., Droser, M. L., and Finney, S. C., eds., Ordovician odyssey: Short papers for the Seventh International Symposium on the Ordovician System: Pacific Section, SEPM (Society for Sedimentary Geology) Publication 77, p. 193–196.

Vaccari, N. E., and Bordonaro, O. L., 1993, Trilobites en los olistolitos cámbricos de la Formación Los Sombreros (Ordovícico), Precordillera de San Juan, Argentina: Ameghiniana, v. 30, p. 383–393.

Vail, P. R., Mitchum, R. M., Todd, R. G., Widmier, J. M., Thompson, S., III, Sangree, J. B., Bubb, J. N., and Hatlelid, W. G., 1977, Seismic stratigraphy and global changes in sea level, *in* Payton, C. E., ed., Seismic stratigraphy—applications to hydrocarbon exploration: American Association of Petroleum Geologists Memoir 26, p. 49–212.

Van der Pluijm, B. A., Van der Voo, R., and Torsvik, T. H., 1995, Convergence and subduction at the Ordovician margin of Laurentia, *in* Hibbard, J. P., Van Staal, C. R., and Carwood, P. A., eds., Current perspectives in the Appalachian-Caledonian orogen: Geological Society of Canada Special Paper 41, p. 127–136.

Van Wagoner, J. C., Posamentier, H. W., Mitchum, R. M., Vail, P. R., Sarg, J. F., Loutit, T. S., and Hardenbol, J., 1988, An overview of the key definitions of sequence stratigraphy and key definitions, *in* Wilgus, C. K., Hastings, B. S., Kendall, C. G. S. C., Posamentier, H. W., Ross, C. A., and Van Wagoner, J. C., eds., Sea-level changes: An integrated approach: Society of Economic Paleontologists and Mineralogists Special Publication 42, p. 39–45.

Varela, R., 1973, Estudio geotectónico del extremo sudoeste de la Precordillera de Mendoza, República Argentina: Revista de la Asociación Geológica Argentina, v. 28, p. 241–267.

Varela, R., and Dalla Salda, L. H., 1993, Geocronología Rb-Sr de metamorfitas y granitoides del tercio sur de la Sierra de Pie de Palo, San Juan, Argentina: Revista de la Asociación Geológica Argentina, v. 47, p. 271–275.

Varela, R., Cuerda, A., and Schauer, O., 1982, Graptolitos ordovicianos en la Formación Cabeceras, vertiente occidental de la Precordillera de San Juan: Revista de la Asociación Geológica Argentina, v. 37, p. 384–387.

Varela, R., Lopez de Luchi, M. G., Cingolani, C., and Dalla Salda, L. H., 1996, Geocronología de gneises y granitoides de la Sierra de Umango, La Rioja, *in* Actas, XIII Congreso Geológico Argentino y III Congreso de Exploración de Hidrocarburos, v. 3: Buenos Aires, Argentina, Asociación Geológica Argentina e Instituto Argentino de Petróleo y Gas, p. 519–527.

Viele, G. W., 1989, The Ouachita orogenic belt, *in* Hatcher, R. D., Jr., Thomas, W. A., and Viele, G. W., eds., The Appalachian-Ouachita orogen in the United States: Boulder, Colorado, Geological Society of America, Geology of North America, v. F-2, p. 555–561.

Viele, G. W., and Thomas, W. A., 1989, Tectonic synthesis of the Ouachita orogenic belt, *in* Hatcher, R. D., Jr., Thomas, W. A., and Viele, G. W., eds., The Appalachian-Ouachita orogen in the United States: Boulder, Colorado, Geological Society of America, Geology of North America, v. F-2, p. 695–728.

Von Gosen, W., 1992, Structural evolution of the Precordillera (Argentina): The Rio San Juan section: Journal of Structural Geology, v. 14, p. 643–667.

Von Gosen, W., 1995, Polyphase structural evolution of the southwestern Argentine Precordillera: Journal of South American Earth Sciences, v. 8, p. 377–404.

Von Gosen, W., 1997, Early Paleozoic and Andean structural evolution in the Río Jáchal section of the Argentine Precordillera: Journal of South American Earth Sciences, v. 10, p. 361–388.

Von Gosen, W., and Prozzi, C., 1998, Structural evolution of the Sierras de San Luis (Eastern Sierras Pampeanas, Argentina): Implications for the proto-Andean margin of Gondwana, *in* Pankhurst, R. J., and Rapela, C. W., eds., The proto-Andean margin of Gondwana: Geological Society [London] Special Publication 142, p. 235–258.

Von Gosen, W., Buggisch, W., and Lehnert, O., 1995, Evolution of the early Paleozoic mélange at the eastern margin of the Argentine Precordillera: Journal of South American Earth Sciences, v. 8, p. 405–424.

Vujovich, G. I., and Kay, S. M., 1998, A Laurentian? Grenville-age oceanic arc/back-arc terrane in the Sierra de Pie de Palo, Western Sierras Pampeanas, Argentina, *in* Pankhurst, R. J., and Rapela, C. W., eds., The proto-Andean margin of Gondwana: Geological Society [London] Special Publication 142, p. 159–179.

Walker, R. G., 1984a, Shelf and shallow marine sands, *in* Walker, R. G., ed., Facies models: Geoscience of Canada Reprint series, v. 1, p. 140–170.

Walker, R. G., 1984b, Turbidites and associated coarse clastic deposits, *in* Walker, R. G., ed., Facies models: Geoscience of Canada Reprint series, v. 1, p. 171–188.

Webby, B. D., 1992, Global biogeography of Ordovician corals and stromatoporoids, *in* Webby, B. D., and Laurie, J. R., eds., Global perspectives on Ordovician geology: Rotterdam, Balkema, p. 261–276.

Wernicke, B., 1985, Uniform-sense normal simple shear of the continental lithosphere: Canadian Journal of Earth Sciences, v. 22, p. 108–125.

Wichmann, R., 1928, Datos geológicos sobre la región comprendida entre el Cerro Nevado y Cerro Nihuil (Provincia de Mendoza): Buenos Aires, Argentina, Dirección Nacional de Minería, Geología e Hidrología, 11 p.

Williams, K. E., 1997, Early Paleozoic paleogeography of Laurentia and western Gondwana: Evidence from tectonic subsidence analysis: Geology, v. 25, p. 747–750.

Willner, A., and Miller, H., 1982, Polyphase metamorphism in the Sierra de Ancasti (Pampean Ranges, NW Argentina) and its relation to deformation, *in* Actas, 5° Congreso Latinoamericano de Geología: Buenos Aires, Argentina, v. 3, Sernvicio Geológico Nacional, p. 441–455.

Wilson, J. L., 1969, Microfacies and sedimentary structures in "deeper water" lime mudstones, *in* Friedman, G. M., ed., Depositional environments in carbonate rocks: Society of Economic Paleontologists and Mineralogists Special Publication 14, p. 4–19.

Wilson, J. L., 1975, Carbonate facies in geologic history: Berlin, Heidelberg, New York, Springer, 471 p.

Wuellner, D. E., Lehtonen, L. R., and James, W. C., 1986, Sedimentary-tectonic development of the Marathon and Val Verde basins, west Texas, U.S.A.: A Permo-Carboniferous migrating foredeep, *in* Allen, P. A., and Homewood, P., eds., Foreland basins: International Association of Sedimentologists Special Publication 8, p. 347–368.

Young, L. M., 1970, Early Ordovician sedimentary history of the Marathon geosyncline, Trans-Pecos Texas: American Association of Petroleum Geologists Bulletin, v. 54, p. 2303–2316.

MANUSCRIPT ACCEPTED BY THE SOCIETY FEBRUARY 10, 1999